AF443239

ELECTROMAGNETIC NONDESTRUCTIVE EVALUATION (VI)

Studies in
Applied Electromagnetics and Mechanics

Editors

K. Miya, A.J. Moses, Y. Uchikawa, A. Bossavit, R. Collins, T. Honma,
G.A. Maugin, F.C. Moon, G. Rubinacci, H. Troger and S.-A. Zhou

Volume 23

Previously published in this series:

Volumes 1-6 have been published by Elsevier Science under the series title "Elsevier Studies in Applied Electromagnetics in Materials".

ISSN: 1383-7281

Electromagnetic Nondestructive Evaluation (VI)

Edited by

F. Kojima
Kobe University, Japan

T. Takagi
Tohoku University, Japan

S.S. Udpa
Michigan State University, USA

and

J. Pávó
Budapest University of Technology and Economics, Hungary

IOS Press

Ohmsha

Amsterdam • Berlin • Oxford • Tokyo • Washington, DC

ISBN 1 58603 245 3 (IOS Press)
ISBN 4 274 90505 5 C3042 (Ohmsha)
Library of Congress Control Number: 2002104883

Publisher
IOS Press
Nieuwe Hemweg 6B
1013 BG Amsterdam
The Netherlands
fax: +31 20 620 3419
e-mail: order@iospress.nl

Distributor in the UK and Ireland
IOS Press/Lavis Marketing
73 Lime Walk
Headington
Oxford OX3 7AD
England
fax: +44 1865 75 0079

Distributor in the USA and Canada
IOS Press, Inc.
5795-G Burke Centre Parkway
Burke, VA 22015
USA
fax: +1 703 323 3668
e-mail: iosbooks@iospress.com

Distributor in Germany, Austria and Switzerland
IOS Press/LSL.de
Gerichtsweg 28
D-04103 Leipzig
Germany
fax: +49 341 995 4255

Distributor in Japan
Ohmsha, Ltd.
3-1 Kanda Nishiki-cho
Chiyoda-ku, Tokyo 101-8460
Japan
fax: +81 3 3233 2426

LEGAL NOTICE
The publisher is not responsible for the use which might be made of the following information.

PRINTED IN THE NETHERLANDS

Foreword

This volume includes a selection of papers presented at the Seventh International Workshop on Electromagnetic Nondestructive Evaluation (ENDE), held in Kobe, Japan from May 17 through 19, 2001. The workshop was organized jointly by the Japan Society of Applied Electromagnetics and Mechanics (JSAEM) and by Kobe University, in cooperation with the Japan Society of Mechanical Engineers, the Magnetic Society of Japan, the Japan Society for Simulation Technology, the Japan Biomagnetism and Biomagnetics Society, the Society of Instrument and Control Engineers, the Institute of Systems, Control and Information Engineers, the Japanese Society for Non-Destructive Inspection, the Institute of Electrical Engineers of Japan, the Japan Society of Plasma Science and Nuclear Fusion Research, the Atomic Energy Society of Japan, the Iron and Steel Institute of Japan, the Japan Society for Precision Engineering and the Cryogenic Association of Japan. The event was co-sponsored by the Inoue Foundation for Science, the Kajima Foundation and the Hyogo Science and Technology Association.

The workshop was held in the Centennial Memorial Hall, which is a brand new building of Kobe University. Sixty participants from twelve countries officially registered for the Workshop. Countries that were represented include China (1), the Czech Republic (1), Egypt (1), France (1), Germany (4), Hungary (1), Italy (2), Korea (2), New Zealand (1), Poland (1), the United States (4) and Japan (41). 49 papers were presented altogether and full papers were due at the Workshop. The papers were reviewed by at least two referees as in the past.

The aim of the ENDE Workshop was to provide a forum for discussing recent developments in the growing field of electromagnetic nondestructive evaluation methods. The workshop talks were organized into fourteen sessions including one plenary talk, two special sessions, nine oral sessions, and two poster sessions. Professor H.T. Banks presented the plenary talk entitled, "A reduced order computational methodology for eddy current based nondestructive evaluation techniques", on the first day, following the opening address. For the first Workshop of the new millennium, two special sessions were scheduled, one was the "overview of ENDE" and the other was the "near future". On the first day of the Workshop, the survey of the benchmark activities in JSAEM was presented in a special session entitled "The advanced eddy current technologies – survey". The other special session entitled "Future direction on electromagnetic nondestructive evaluations" was presented at the final session of this Workshop. The contributed papers that were presented in oral and poster sessions cover several interesting aspects of the field including direct and inverse problems to electromagnetic nondestructive testing, new developments in eddy current testing, evaluation of degradation mechanism in magnetic materials, advanced electromagnetic sensors and industrial applications of electromagnetic problems. The Workshop emphasized both basic science and early engineering developments in the field, thereby allowing a wide spectrum of experts ranging from theoreticians to research engineers in industry to participate. It is the belief of the Workshop organizers that, as in the case of previous Workshops, this new millennium Workshop allowed a better understanding of underlying issues and contributed to advancement of the subject in all of the above areas. Finally, at the closing remarks of the Workshop, Dr. G. Dobmann announced that the Fraunhofer-Institut für Zerstörungsfreie Prüfverfahren would organize the next ENDE Workshop in Saarbrücken, Germany.

The organizers express their sincere thanks to the speakers, to the chairpersons and to all participants who contributed to the success of the meeting. The financial and moral support from the sponsors is also gratefully acknowledged. Thanks are due to the members of the standing committee who provided considerable guidance and advice. The editors are indebted to the reviewers for their careful work. Special thanks are due to Dr. Futoshi Kobayashi for his invaluable help given to the editors during the preparation of this book. Many thanks are also due to the organizing committee.

F. Kojima T. Takagi, S.S. Udpa and J. Pávó
Editors

List of Referees

Shanker Balasubramaniam (Iowa State University, USA)
John Bowler (Iowa State University, USA)
Gangzhu Chen (Japan Power Engineering and Inspection Corporation, Japan)
Zhenmao Chen (Japan Society of Applied Electromagnetics and Mechanics, Japan)
Hiroyuki Fukutomi (Central Research Institute of Electric Power, Japan)
Antal Gasparics (Research Institute for Technical Physics and Materials Science, Hungary)
Attila Gilanyi (Ericsson Hungary Ltd., Hungary)
Szabolcs Gyimothy (Budapest University of Technology and Economics, Hungary)
Mitsuo Hashimoto (Polytechnic University, Japan)
Hidetoshi Hashizume (Tohoku University, Japan)
Haoyu Huang (Tohoku University, Japan)
Plamen Ivanov (Iowa State University, USA)
Marcus Johnson (Iowa State University, USA)
Gregory Kobidze (Iowa State University, USA)
Fumio Kojima (Kobe University, Japan)
Kiyoshi Koyama (Nihon University, Japan)
Chester Lo (Iowa State University, USA)
R. Ludwig (Worchester Polytechnic Institute, USA)
Shreekanth Mandayam (Rowan University, USA)
Norio Nakagawa (Iowa State University, USA)
Jozsef Pávó (Budapest University of Technology and Economics, Hungary)
Pradeep Ramuhalli (Iowa State University, USA)
Yoshifuru Saito (Hosei University, Japan)
Imre Sebestyen (Budapest University of Technology and Economics, Hungary)
Young-Kil Shin (Kunsan National University, USA)
Yushi Sun (Iowa State University, USA)
Toshiyuki Takagi (Tohoku University, Japan)
Antonello Tamburrino (Universita Degli Studi di Cassino, Italy)
Lalita Udpa (Michigan State University, USA)
Satish Udpa (Michigan State University, USA)
Gabor Vértesy (Research Institute for Technical Physics and Materials Science, Hungary)
Masahiro Yamamoto (The University of Tokyo, Japan)
Reza Zoughi (University of Missouri, USA)

ENDE – Kobe
The 7th International Workshop on Electromagnetic Nondestructive Evaluation
Kobe, Japan, May 17–19, 2001

Organized by:
- Japan Society of Applied Electromagnetics and Mechanics
- Kobe University

In Cooperation with:
- The Japan Society of Mechanical Engineers
- The Magnetic Society of Japan
- Japan Society for Simulation Technology
- Japan Biomagnetism and Biomagnetics Society
- The Society of Instrument and Control Engineers
- Institute of Systems, Control and Information Engineers
- The Japanese Society for Non-Destructive Inspection
- The Institute of Electrical Engineers of Japan
- The Japan Society of Plasma Science and Nuclear Fusion Research
- The Atomic Energy Society of Japan
- The Iron and Steel Institute of Japan
- Japan Society for Precision Engineering
- Cryogenic Association of Japan

Co-sponsors:
- Inoue Foundation for Science, Japan
- The Kajima Foundation, Japan
- Hyogo Science and Technology Association, Japan

Standing Committee:
- J.R. Bowler, Iowa State University, U.S.A., Chairman
- R. Albanese, Universita Reggio Calabria, Italy
- G. Dobmann, Fraunhofer-Institute for NDT, Germany
- R. Grimberg, National Institute of R&D for Technical Physics, Romania
- H.K. Jung, Seoul National University, South Korea
- F. Kojima, Kobe University, Japan
- D. Lesselier, DRE-LSS CNRS-SUPELEC, France
- V. Lunin, Moscow Power Engineering Institute, Russia
- K. Miya, International Institute of Universum, Japan
- G.Z. Ni, Zhejing University, China
- J. Pávó, Budapest University of Technology and Economics, Hungary
- A. Razek, LGEP CNRS-SUPELEC, France

- G. Rubinacci, Universita Cassino, Italy
- J.N. Sheng, X'ian Jiatong University, China
- T. Sollier, CEA Paris, France
- T. Takagi, Tohoku University, Japan
- H. Takamatsu, KEPCO, Japan
- S.S. Udpa, Iowa State University, U.S.A.
- V. Vengrinovich, Institute of Applied Physics, Belarus

Secretary:
Futoshi Kobayashi
Graduate School of Science and Technology, Kobe University
1-1 Rokkodai, Nada 657-8501, JAPAN
Tel: +81-78-803-6489, Fax: +81-78-803-6493
E-mail: ende2001@buna.scitec.kobe-u.ac.jp

E'NDE – Kobe, List of Participants

ALBANESE, Raffaele
Univ. Reggio Calabria
DIMET, Fac. Ingegneria, Via Graziella, Loc. Feo
di Vito, I-89100 Reggio Calabria, ITALY
+39 081 7683243
+39 081 7683171
albanese@unirc.it

ARA, Katsuyuki
Applied Superconductivity Research Laboratory
Tokyo Denki University
2-1200, Gakuendai, Busei, Inzai
Chiba, 270-1382 JAPAN
ara@hp.asrl.dendai.ac.jp

BANKS, H. T.
Center for Research in Scientific Computation,
North Carolina State University
Raleigh, NC 27695-8205 USA
htbanks@eos.ncsu.edu

BEUKER, Thomas
H. Rosen Engineering GmbH,
Research & Development Dept.
Am Seitenkanal 8, D-49811 Lingen (Ems)
GERMANY
+49-591-9136-414
+49-591-9136-121

BOWLER, John
Electrical & Computer Engineering,
Iowa State University
1915 Scholl Road, Ames Iowa, 50011-3042 USA
515 294 2093
515 294 7771
jbowler@cnde.iastate.edu

CHAMONINE, Mikhail
H. Rosen Engineering GmbH,
Research & Development Dept.
Am Seitenkanal 8, D-49811 Lingen (Ems)
GERMANY
+49-591-9136-414
+49-591-9136-121
MChamonine@RosenGermany.de

CHEN, Zhenmao
International Institute of Universum
SB Bldg. 801,
1-4-6 Nezu, Bunkyo-ku, Tokyo 113-0031, JAPAN
+81-3-5814-5330
+81-3-5814-6705
chen@jsaem.gr.jp

DOBMANN, Gerd
Fraunhofer-Institut für Zerstörungsfreie
Prüfvergahren
Universitat, Geb. 37, 66123 Saarbrücken,
GERMANY
+49 681 9302 3855
+49 681 9302 5933
dobmann@izfp.fhg.de

DOI, Tatsuya
Ashikaga Institute of Technology
Ohmae, Ashikaga, Tochigi, 326-8558 JAPAN
+81-284-62-0605
+81-284-62-4633
t-doi@ashitech.ac.jp

ENDO, Hisashi
c/o Professor Yoshifuru Saito
Graduate School of Engineering, Hosei University
3-7-2 Kajino Koganei,Tokyo 184-8584 JAPAN
+81-42-387-6200
+81-42-387-6213
endo@ysaitoh.k.hosei.ac.jp

ENOKIZONO, Masato
Faculty of Engineering, Oita University
700 Dannoharu, Oita 870-1192 JAPAN
+81-97-554-7821
+81-97-554-7822
enoki@cc.oita-u.ac.jp

GOTTVALD, Ales
Academy of Sciences of the CR,
Inst. of Scientific Instruments
Kralovopolska 147, 612 64 Brno
CZECH REPUBLIC
+420 5 415 14 224
+420 5 415 14 402
gott@isibrno.cz

HARADA, Yutaka
Nuclear Engineering Ltd., Japan
1-3-7 Tosabori Nishi-ku Osaka 550-0001 JAPAN
+81-6-6446-9363
+81-6-6446-1746

HASHIMOTO, Mitsuo
Electronic Engineering, Polytechnic University
4-1-1 Hasimotodai, Sagamihara, Kanagawa
229-1196, JAPAN
+81-42-763-9133
+81-42-763-9150
hasimoto@uitec.ac.jp

HATSUKADE, Yoshimi
c/o Professor A. Ishiyama
Graduate School of Science and Engineering,
Waseda University
3-4-1 Ohkubo, Shinjuku-ku, Tokyo
169-8555 JAPAN
+81-3-5286-3376
+81-3-3208-9337
hatsu@mn.waseda.ac.jp

HOPPE, Ronald H.W.
Inst. of Mathematics, University of Augsburg
Universitaetsstrasse 14, D-86159 Augsburg
GERMANY
+49-821-598-2194, +49-171-4517057
+49-821-598-2339
hoppe@math.uni-augsburg.de

HUANG, Haoyu
Institute of Fluid Science
Tohoku University
Katahira 2-1-1, Aoba-ku, Sendai 980-8577 JAPAN
+81-22-217-5298
+81-22-217-5298
huang@ifs.tohoku.ac.jp

IGARASHI, Hajime
Graduate School of Engineering,
Hokkaido University
Kita 13, Nishi 8, Kita-ku, Sapporo
060-8628 JAPAN

ISHIBASHI, Kazuhisa
Mechanical Engineering, Faculty of Engineering,
Tokai University
2-28-4 Tomigaya, Shibuya-ku, Tokyo
151-8677 JAPAN
isibasi@keyaki.cc.u-tokai.ac.jp

JU, Yang
Department of Mechanical Engineering,
Tohoku University
Aoba 01, Aramaki, Aoba-ku, Sendai
980-8579 JAPAN
+81-022-217-6898
+81-022-217-6893
ju@ism.mech.tohoku.ac.jp

KAMADA, Yasuhiro
c/o Professor Seiki Takahashi
Iwate University
4-3-5, Ueda, Morioka-shi, 020-8551 JAPAN
+81 19 621 6431
+81 19 621 6431
kamada@iwate-u.ac.jp

KASAI, Naoko
AIST
1-1-4 Umezono, Tsukuba, Ibaraki, 305-8568
JAPAN
+81-298-61-5565
+81-298-61-5530
kasai-naoko@aist.go.jp

KOJIMA, Fumio
Graduate School of Science and Technology,
Kobe University
1-1 Rokkodai, Nada, Kobe 657-8501 JAPAN
+81-78-803-6493
+81-78-803-6493
kojima@cs.kobe-u.ac.jp

KUROZUMI, Yasuo
Institute of Nuclear Safety Systems, Inc.
6464 Sata,Mihama-cho,Fukui 919-1205 JAPAN
+81-770-37-9114
+81-770-37-2009
kurozumi@inss.co.jp

LESSELIER, Dominique
C.N.R.S. Plateau de Moulon,
91192 Gif Sur Vverre FRANCE
33 1 69 85 17 12
01 69 85 17 69
dominique.lesselier@lss.supelec.fr

MATSUMOTO, Eiji
Department of Energy Conversion Science,
Graduate School of Energy Science,
Kyoto University, Kyoto 606-8501 JAPAN
+81-(0)75-753-5247
+81-(0)75-753-5247
matumoto@energy.kyoto-u.ac.jp

MIYA, Kenzo
International Institute of Universum
SB Bldg. 801,
1-4-6 Nezu, Bunkyo-ku, Tokyo 113-0031, JAPAN
+81-3-5814-5330
+81-3-5814-6705
miya@jsaem.gr.jp

MUKHOPADHYAY, S. C.
Institute of Information Sciences and Technology,
Massey University (Turitea)
RIDDET BUILDING ROOM No. 2.06, Massey
University (Turitea), Private Bag 11222,
Palmerston North NEW ZEALAND
+64-6-350-5799 ext. 2480
+64-6-350-5604
S.C.Mukhopadhyay@massey.ac.nz

MUMTAZ, Khalid
Faculty of Engineering, Iwate University
Ueda 4-3-5, Morioka, Iwate 020-8551 JAPAN
+81-19-621-6350
+81-19-621-6373
fahadkm@iwate-u.ac.jp

NAGATA, Shoichiro
Faculty of Engineering, Miyazaki University
1-1 Gakuenkibanadai-nishi,
Miyazaki, 889-2195 JAPAN
nagata@ee.miyazaki-u.ac.jp

NAKASONE, Yuji
Department of Mechanical Engineering
Science University of Tokyo
1-3, Kagurazaka, Shinjyuku-ku, Tokyo,
162-8601 JAPAN
+81-3-3260-4272 (EXT 3356)
+81-3-3260-4291
nakasone@rs.kagu.sut.ac.jp

NISHIKAWA, Masahiro
Graduate School of Engineering, Osaka University
2-1 Yamada-oka, Suita, Osaka, 565-0871 JAPAN
+81-6-6879-7234
+81-6-6879-7235
nisikawa@ppl.eng.osaka-u.ac.jp

OHE, Takashi
Department of Simulation Physics,
Faculty of Informatics,
Okayama University of Science
1-1 Ridai-cho, Okayama 700-0005 JAPAN
+81-86-256-9616
ohe@sp.ous.ac.jp

OKA, Mohachiro
Dept. of Computer and Control Engineering,
Oita National College of Technology
1666 Maki, Oita, 870-0152. JAPAN
+81-97-552-7464
+81-97-552-7464
oka@oita-ct.ac.jp

OKAJIMA, Nobuyuki
c/o Professor Fumio Kojima
Graduate School of Science and Technology,
Kobe University
1-1 Rokkodai, Nada, Kobe 657-8501, JAPAN
+81-78-803-2449
+81-78-803-5349
okajima@buna.scitec.kobe-u.ac.jp

OKINO, Yuu
c/o Professor Seiki Takahashi
Faculity of Engineering, Iwate University
Ueda 4-3-5, Morioka, Iwate 020-8551, JAPAN
+81-19-621-6350
+81-19-621-6373
t2200004@iwate-u.ac.jp

PAVO, Jozsef
Department of Electromagnetic Theory,
Budapest University of Technology and Economics
H-1521 Budapest, Egry J. u. 18. HUNGARY
+36 1 4632913
+36 1 4633189
pavo@evtsz1.evt.bme.hu

RUBINACCI, Guglielmo
University of Cassino
DAEIMI, Via di Biasio 43 - I03043 Cassino,
ITALY
+39 0776 299 626
+39 0776 299 628
rubinacci@unicas.it

SAKAMOTO, Shintaro
Shinryo Co.
41 Wadai Tsukuba Ibaraki JAPAN
+81-298-64-6110
+81-298-64-6127
sakamoto.sh@shinryo.com

SARUTA, Kenji
c/o Professor Ju, Yang
Department of Mechanical Engineering,
Tohoku University
Aoba 01, Aramaki, Aoba-ku, Sendai
980-8579 JAPAN

SATO, Yasumoto
National Space Development Agency of Japan
2-4-1 Hamamatsu, Minato, Tokyo 105-8060
JAPAN
+81-3-3438-6186
+81-3-5402-6515
satoh.yasumoto@nasda.go.jp

SATOH, Hidenobu
c/o Professor Seiki Takahashi
Faculty of Engineering, Iwate University
Ueda 4-3-5, Morioka, Iwate 020-8551, JAPAN
+81-19-621-6350
+81-19-621-6373
t2200012@iwate-u.ac.jp

SHIMONE, Junri
Nuclear Engineering Ltd., Japan
1-3-7 Tosabori Nishi-ku Osaka 550-0001, JAPAN
+81-6-6446-9363
+81-6-6446-1746
jshimone@sg-gr.neltd.co.jp

SHIN, Young-Kil
School of Electronic and Information Engineering,
Kunsan National University
Kunsan, Chonbuk, 573-701 KOREA
ykshin@kunsan.ac.kr

SIKORA, Ryszard
Department of Electrical Engineering
Technical University of Szczecin
ul. Sikorskiego 37, 70-310 Szczecin, POLAND
(48 91) 449 49 67
(48 91) 434 09 26
rs@main.tuniv.szczecin.pl

SONG, Sung-Jin
School of Mechanical Engineering,
Sungkyunkwan University
Suwon, KOREA
+82-31-290-7451
+82-31-290-5276
sjsong@yurim.skku.ac.kr

SUKEGAWA, Toshio
Graduate School of Engineering,
The University of Tokyo
2-22, Shirane, Shirakata, Tokai, Naka, Ibaraki,
319-1106 JAPAN
+81-29-287-8445
+81-29-287-8488
sukegawa@tokai.t.u-tokyo.ac.jp

TAKAGI, Toshiyuki
Institute of Fluid Science, Tohoku University
Katahira 2-1-1, Aoba-ku, Sendai 980-8577 JAPAN
+81-22-217-5248
+81-22-217-5248
takagi@ifs.tohoku.ac.jp

TAKAHASAHI, Seiki
Faculity of Engineering, Iwate University
Ueda 4-3-5, Morioka, Iwate 020-8551 JAPAN
+81-19-621-6348
+81-19-621-6348
seiki.t@iwate-u.ac.jp

TAKAMATSU, Hiroshi
The Kansai Electric Power, Co. Inc.
Nuclear Power Division
3-3-22 Nakanoshima, Osaka 530-8270 JAPAN
+81-6-6441-8821
+81-6-6444-6279
K529581@kepco.co.jp

TANIGUCHI, Tetsuki
Department of Electronic Engineering
University of Electro-Communications
1-5-1 Chofugaoka, Chofu, Tokyo, 182-8585
JAPAN

TSUBOI, Hajime
Department of Information Processing
Engineering,
Fukuyama University
Gakuen-cho, Fukuyama, Hiroshima
729-0292 JAPAN
+81-849-36-2112 ext. 4767
+81-849-36-0080
tsuboi@fuip.fukuyama-u.ac.jp

TSUCHIDA, Yuji
Faculty of Engineering, Oita University
700 Dannoharu, Oita 870-1192 JAPAN
tsuchida@cc.oita-u.ac.jp

UDPA, Lalita
Department of Electrical and Computer
Engineering, Michigan State University
East Lansing, MI 48824 USA
517-355-9261
udpal@egr.msu.edu

UDPA, Satish
Department of Electrical and Computer
Engineering, Michigan State University
East Lansing, MI 48824 USA
517-432-4787
udpa@egr.msu.edu

WAHSH, Said
Electronics Research Institute
12311 Dokki, Giza, EGYPT
wahsh@eri.sci.eg

YAMADA, Satoshi`
MAGCAP, Faculty of Engineering,
Kanazawa University
2-40-20 Kodatsu-no, Kanazawa 920-8667 JAPAN
+81-76-234-4942
+86-76-234-4946
yamada@magstr.ec.t.kanazawa-u.ac.jp

YAMAMOTO, Masahiro
The University of Tokyo
3-8-1 komaba, Meguro, Tokyo 153-8914, JAPAN
+81-3-5465-8328
+81-3-5465-7017
myama@ms.u-tokyo.ac.jp

YAMAZAKI, Katsumi
Department of Electrical Engineering,
Chiba Institute of Technology
2-17-1, Tsudanuma, Narashino, Chiba,
275-0016 JAPAN
+81-47-478-0373
+81-47-478-0379
yamazaki@pf.it-chiba.ac.jp

YASUNISHI, Michio
c/o Professor Masahiro Nishikawa
Graduate School of Engineering
Osaka University
2-1 Yamada-oka, Suita, Osaka 565-0871 JAPAN
+81-6-6877-5111 (ext. 3699)
+81-6-6879-7867
michio@ppl.eng.osaka-u.ac.jp

YUSA, Noritaka
International Institute of Universum
SB Bldg. 801,
1-4-6 Nezu, Bunkyo-ku, Tokyo 113-0031 JAPAN
+81-3-5814-5330
+81-3-5814-6705
yusa@jsaem.gr.jp

ZOU, Jun
Department of Mathematics,
The Chinese University of Hong Kong
Shatin, N.T., HONG KONG
(852) 2609 7985
(852) 2603 5154
zou@math.cuhk.edu.hk

Contents

Materials Characterization

Novel E'NDE Techniques

New Sensors

Signal Detection

Inverse Problems

Electromagnetic Interrogation Techniques for Damage Detection

H. T. Banks[a], Michele L. Joyner[a], Buzz Wincheski[b], and William P. Winfree[b]

[a] *Center for Research in Scientific Computation, North Carolina State University, Raleigh, NC 27695*
[b] *NASA Langley Research Center, Hampton, VA 23681*

Abstract.

This paper introduces a computational method for use with eddy current damage detection techniques. To identify the geometry of a subsurface damage, an optimization algorithm is employed which requires solving the forward problem numerous times. In order for the method to be effective in a practical setting, i.e., in real-time applications, the forward algorithm must be extremely fast and accurate. Therefore, we have chosen an approach based on reduced order Proper Orthogonal Decomposition (POD) methods. This allows one to create a set of basis elements using snapshots with either numerical simulations or experimental data. The data is organized in an optimal way allowing one to use a reduced number of basis elements, resulting in a fast algorithm while still obtaining an accurate approximation to the solution. We first derive the model associated with the chosen nondestructive evaluation (NDE) technique and prove some well-posedness results for the model. We then introduce the proposed computational methodology and test it on both numerically simulated data as well as experimental data obtained from a GMR (Giant Magnetoresistive) sensor. The results demonstrate that the method is extremely efficient and accurate.

1 Introduction and Problem Formulation

As technology continually advances, the field of nondestructive evaluation is in continual need of new techniques and instruments to improve the accuracy and efficiency of locating and characterizing subsurface damages. We attempt to develop a new methodology which, when coupled with already existing techniques, can help decrease the total computational time required to detect and explicitly characterize a damage within a material. This is necessary in practical settings where the methods must be fast and accurate, producing real-time results. Given data obtained from a measuring device such as the GMR (Giant Magnetoresistive) sensor, we seek to locate and parameterize the damage while minimizing the amount of time required to complete this task. To this end, we formulate and develop an appropriate model used in describing the variation in the data as a function of a damage within the sample and present computational methods along with numerical results to support the efficacy of our approach.

1.1 Description of the Test Problem

An advanced method of damage detection uses a device such as the GMR sensor in the context of eddy current methods [1, 2]. In a standard experimental setting, as depicted in Figure 1, a thin conducting sheet carrying a uniform current is placed above (or below) the sample. The current within the sheet induces a magnetic field perpendicular to it that in turn produces a current within the sample, called an eddy current. When a flaw is present within the sample, the flaw disrupts the eddy current flow near the flaw and this disturbance is manifested in the magnetic flux density detected by the measuring device.

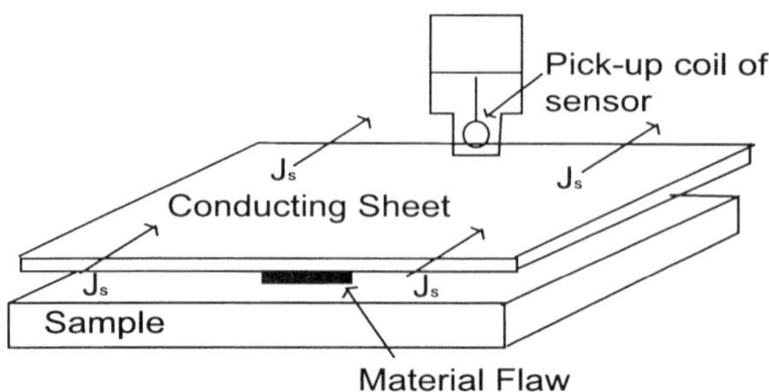

Figure 1: Inspection Process Using a GMR Sensor

For illustrative purposes, we will assume uniformity along the width (z direction) of the sample, therefore reducing the three-dimensional problem to a two-dimensional problem. To test the feasibility of reconstructing the geometry of the damage, we consider the damage (which we shall refer to as a "crack") to be rectangular in shape. In other words, we assume the crack, located at a certain depth within the sample, has a fixed length and thickness. To further simplify the test problem, we disregard the boundary effects of the materials in the x direction (sample length) by assuming an infinite sample and conducting sheet in that direction. Because we are considering materials of infinite extent, we will construct our forward problem by focusing on a small "window", called our computational domain $\Omega = \{(x, y, z) \in \mathbb{R}^3 : 0mm \leq x \leq 50mm, -35mm \leq y \leq 35mm\}$, centered such that the left boundary of the computational domain, at location $x = 0$, is positioned in the center of the crack in the x direction, i.e., the crack is symmetric through the yz plane at $x = 0$. A schematic of the resulting two-dimensional problem is depicted in Figure 2 where it is assumed that the sample (which is $20mm$ thick) is composed of aluminum, the conducting sheet (which is $0.1mm$ thick) is made up of copper and the crack is centered in the y direction around the center of the sample (i.e., around $y = -10mm$).

1.2   Resulting Equations for the Test Problem

In our computational efforts, we employ the use of the software package Ansoft Maxwell 2D Field Simulator. Therefore, our equations are formulated to correspond to those used by the software. (For a full derivation of the resulting equations, see [3].) Although the GMR sensor measures the magnetic flux density above the sample at a certain location, our equations are formulated in terms of the magnetic vector potential $\mathbf{A}$, where all the field quantities are assumed to be phasor quantities [4, 5, 6]. However, one can easily obtain the resulting magnetic flux density $\mathbf{B}$ through the relationship $\mathbf{B} = \nabla \times \mathbf{A}$.

Using Maxwell's equations in conjunction with Ohm's law and constitutive laws, we obtain an equation for the magnetic vector potential $\mathbf{A}$ given by

$$\nabla \times \left(\frac{1}{\mu(x, y)} \nabla \times \mathbf{A}(x, y) \right) = (\sigma(x, y) + i\omega\epsilon)(\mathbf{A}(x, y) - \nabla\phi) \quad \forall x, y \in \Omega, \quad (1)$$

where μ represents the magnetic permeability, σ represents the conductivity, ω is the angular frequency, and ϕ is a scalar potential.

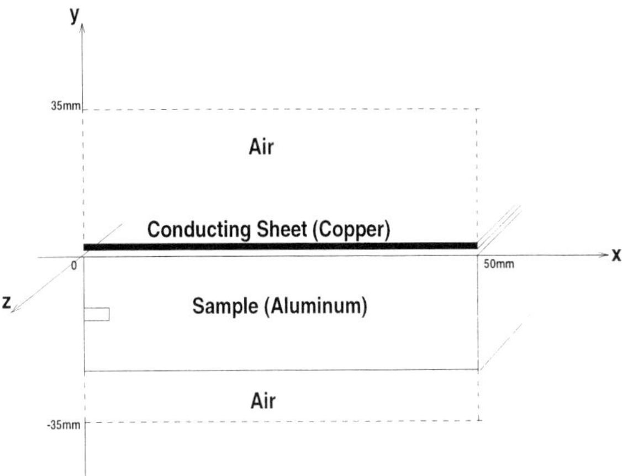

Figure 2: 2-D Schematic of Problem

Since Eq.(1) contains two unknowns, $\mathbf{A}$ and ϕ, an additional equation is needed to uniquely determine the solution for $\mathbf{A}$. For this we use an integral constraint given by

$$I_{cs} = \int_{cs} \mathbf{J}_t \cdot \mathbf{n} da = \int_{cs} (\sigma(x, y) + i\omega\epsilon(x, y))(\mathbf{A}(x, y) - \nabla\phi) \cdot \mathbf{n} da \qquad (2)$$

between the total current I_{cs} flowing in the conducting sheet (cs) and the total current density $\mathbf{J}_t$ within the conducting sheet. This is the second equation used in the software package Ansoft Maxwell 2D Field Simulator which we use in our computational efforts. However, this only gives us equations which completely describe the magnetic vector potential and electric scalar potential in the conducting sheet. Nonetheless, the conducting sheet is the only region in which a source current of the form $J_s = -\sigma\nabla\phi$ is present and hence is the remaining regions we intuitively assume there is no change in potential, in other words $\nabla\phi \equiv 0$. This gives us appropriate equations in which the magnetic vector potential $\mathbf{A}$ can be uniquely determined if appropriate boundary conditions on $\mathbf{A}$ are specified.

In Section 1.1 we assumed the sample was of infinite extent with the crack being symmetric in the x direction. In other words on the x boundaries, we assume the fields on both sides of the boundary oscillate in the same direction. To account for the even symmetry, we assign Neumann boundary conditions to these boundaries. In a similar manner, we assume the y boundaries are "sufficiently far" away from the sample and scanning area so as to not effect the overall measurements. Indeed, as one moves farther away from the sample and conducting sheet, the magnetic vector potential $\mathbf{A}$ tends to zero. Therefore, on the y boundaries we assign null Dirichlet boundary conditions. The magnetic vector potential $\mathbf{A}$ is thus determined by

$$\nabla \times \left(\frac{1}{\mu(x, y)}\nabla \times \mathbf{A}(x, y)\right) = (\sigma(x, y) + i\omega\epsilon(x, y))(-i\omega\mathbf{A}(x, y) - \nabla\phi) \quad \forall x, y \in \Omega, \quad (3)$$

$$I_{cs} = \int_{cs} \mathbf{J}_t \cdot \mathbf{n} da = \int_{cs} (\sigma(x, y) + i\omega\epsilon(x, y))(-i\omega\mathbf{A}(x, y) - \nabla\phi) \cdot \mathbf{n} da \qquad (4)$$

and

$$\nabla\phi = 0 \quad \forall x, y \in \Omega \setminus cs \qquad (5)$$

with

$$\mathbf{A}(x, -35) \;=\; 0 \;=\; \mathbf{A}(x, 35)$$
$$\nabla \mathbf{A} \cdot \mathbf{n}|_{(0,y)} \;=\; 0 \;=\; \nabla \mathbf{A} \cdot \mathbf{n}|_{(50,y)}.$$

2 Well-Posedness

In this section we consider the existence and uniqueness of a weak solution to the above boundary value problem on a general domain given by

$$\tilde{\Omega} = \{(x, y, z) \in \mathbb{R}^3 : x_{min} \le x \le x_{max}, y_{min} \le y \le y_{max}\}$$

for which our test problem is a specific example. Let $H = L_2(\tilde{\Omega})$ and $V = \{\psi \in H^1(\tilde{\Omega})|\ \psi(x, y_{min}) = 0 = \psi(x, y_{max})\}$ where we use the standard Sobolev space notation, $H^1(\tilde{\Omega}) = \{\psi \in L^2(\tilde{\Omega}) : \nabla\psi \in L^2(\tilde{\Omega})\}$ and note that we interpret pointwise evaluation of functions (along the boundary and elsewhere) in terms of a trace operator for which we suppress notation throughout this paper [7]. We denote by $\langle \phi, \psi \rangle \equiv \int_{\tilde{\Omega}} \phi\overline{\psi}da$ the standard inner product in H and $\langle \phi, \psi \rangle_V \equiv \int_{\tilde{\Omega}} \nabla\phi \cdot \overline{\nabla\psi}da$ the (H^1-equivalent) inner product in V.

We note that in this two-dimensional problem, the term $\nabla\phi$ can be proven to be piecewise constant as done in [3]. In doing this, $\nabla\phi$ can be written in terms of the magnetic vector potential in the conducting sheet by solving for $\nabla\phi$ in (4). We can then reduce the system into an integro-differential equation. Using integration by parts together with natural boundary conditions and imposed conditions on test functions $\psi \in V$, the variational form is given by

$$\langle \nabla A, \nabla\psi \rangle + \langle \beta_1 A, \psi \rangle + \beta_2 \int_{cs} A da \int_{cs} \overline{\psi}da = \int_{cs} f\overline{\psi}da. \tag{6}$$

where $\beta_1 = i\omega\mu(\sigma + i\omega\epsilon)$, $\beta_2 = -\frac{i\omega\mu_{cu}(\sigma_{cu}+i\omega\epsilon_{cu})}{\Delta_{cs}}$, and $f = \frac{\mu_{cu}I_{cs}}{\Delta_{cs}}$.

We consider both the existence and uniqueness of the solution A to (6) as well as the continuous dependence on the parameters which represent the damage in the context of a Gelfand triple setting $V \hookrightarrow H \simeq H^* \hookrightarrow V^*$ where we have that the embedding $V \hookrightarrow H$ is dense and continuous with

$$|\psi|_H \le k|\psi|_V \quad \text{for all } \psi \in V. \tag{7}$$

where the norm in V will be denoted by $|\cdot|_V$ and $|\cdot|$ will denote the norm in H for the rest of this section.

Using this notation, we have the following theorem. For full details and proof, we refer the reader to [3, 8].

Theorem 2.1. *There exists $\mathcal{F} = \mathcal{F}(\mu, \epsilon, \tilde{\Omega})$ such that for source frequencies $f_s < \mathcal{F}(\mu, \epsilon, \tilde{\Omega})$, there exists a unique weak solution A to (6).*

Furthermore, we let $\tilde{\mathbf{q}}$ represent the four corners of any quadrilateral damage, i.e., $\tilde{\mathbf{q}} \equiv [(x_1, y_1), (x_2, y_2), (x_3, y_3), (x_4, y_4)]$ is a vector in $\mathbb{R}^4 \times \mathbb{R}^2$ or equivalently $\mathbb{R}^8$. We denote by $\tilde{Q}_{ad}$ the set of admissible parameters $\tilde{\mathbf{q}}$ where it is assumed $\tilde{Q}_{ad}$ is a compact subset of $\mathbb{R}^8$. Then for any two damages given by parameters $\hat{\mathbf{q}}$ and $\tilde{\mathbf{q}}$ in $\tilde{Q}_{ad}$, let

$$d(\hat{\mathbf{q}}, \tilde{\mathbf{q}}) = \|\hat{\mathbf{q}} - \tilde{\mathbf{q}}\| = \left[(\hat{x}_1 - x_1)^2 + (\hat{y}_1 - y_1)^2 + ... + (\hat{x}_4 - x_4)^2 + (\hat{y}_4 - y_4)^2\right]^{1/2} \tag{8}$$

to be the standard Euclidean norm in $\mathbb{R}^8$. We denote by $\delta\hat{\Omega}$ (see Figure 3) the points in Ω which are either in the damage represented by $\hat{\mathbf{q}}$ or $\tilde{\mathbf{q}}$ but which are not in both. In other words, let $\Omega_{\hat{\mathbf{q}}}$ represent the points (x, y) in Ω within the damage given by $\hat{\mathbf{q}}$ and $\Omega_{\tilde{\mathbf{q}}}$ the points (x, y) in Ω within the damage given by $\tilde{\mathbf{q}}$. Then $\delta\hat{\Omega} = \Omega_{\hat{\mathbf{q}}} \cup \Omega_{\tilde{\mathbf{q}}} - \Omega_{\hat{\mathbf{q}}} \cap \Omega_{\tilde{\mathbf{q}}}$.

Using the terminology above, we have the following theorem. Again, we refer the reader to [3, 8] for details and proof.

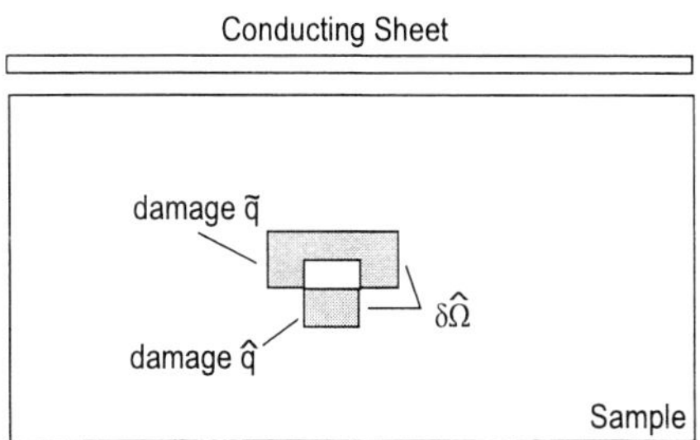

Figure 3: The Area Represented by $\delta\hat{\Omega}$

Theorem 2.2. *Assume the admissible parameter set $\tilde{Q}_{ad}$ is a compact subset of $\mathbb{R}^8$. Then there exists $\mathcal{F} = \mathcal{F}(\mu, \epsilon, \tilde{\Omega})$ such that for source frequencies $f_s < \mathcal{F}(\mu, \epsilon, \tilde{\Omega})$, $\tilde{q} \to A(\tilde{q})$ is continuous from $\tilde{Q}_{ad}$ to V.*

3 Computational Method

To enable the techniques used in nondestructive evaluation to be implemented in a practical setting, we must not only locate and characterize subsurface damages; we need to do so in a fast and efficient manner. To develop a fast and efficient forward algorithm, we use the reduced order Karhunen-Loeve or Proper Orthogonal Decomposition (POD) methodology. A unique feature of the POD technique is its ability to create an ordered basis that models experimental or simulated data, capturing most of the important aspects of the data in the first few elements. Typically the number of basis elements required in the forward algorithm is reduced, resulting in the development of a faster algorithm that still maintains the accuracy of traditional finite element algorithms.

3.1 The POD Method

In this section we discuss the concepts and method of implementation of the POD method in the context of damage detection. For details on the general POD method, we refer the reader to [9, 10, 11, 12, 13, 14, 15, 16, 17, 18, 19, 20] and the extensive list of references contained therein. In the process of detecting a subsurface damage, a device, such as the GMR sensor, is scanned above (or below) a sample and data is taken. The variation in the field due to a damage is manifested in the data taken. Therefore, in forming a reduced basis, we want to incorporate the effects of a damage in the reduced basis. This is accomplished by taking "snapshots" of the data across various damages. In other words, let $\mathbf{q}$ be a vector parameter characterizing physical properties of the damage such as length, thickness, depth, etc. of the damage. Then given an ensemble of damages $\{\mathbf{q}_j\}_{j=1}^{N_s}$, we obtain corresponding solutions, $\{\mathbf{A}(\mathbf{q}_j)\}_{j=1}^{N_s}$, of the boundary value problem, for magnetic vector potentials which we call our "snapshots". We then use these snapshots to generate a basis incorporating the properties of the various damages. (Without loss of generality, we will denote the vector $\mathbf{A}$ by its scalar nonzero component A, i.e., the A_3 component of $\mathbf{A}$.)

The formation of the POD basis can be summarized in a few steps. In order to successfully reduce the number of basis elements while maintaining accuracy, a single basis element must contain aspects of each damage $\mathbf{q}_j$. Therefore, as explained in [20, 21], we seek basis elements of the form

$$\Phi_i = \sum_{j=1}^{N_s} V_i(j) A(\mathbf{q}_j) \tag{9}$$

where the coefficients $V_i(j)$ are chosen such that each POD basis element Φ_i, $i = 1, 2, ..., N_s$, maximizes

$$\frac{1}{N_s} \sum_{j=1}^{N_s} |\langle A(\mathbf{q}_j), \Phi_i \rangle_{L^2(\Omega, \mathbb{C})}|^2 \tag{10}$$

subject to $\langle \Phi_i, \Phi_i \rangle_{L^2(\Omega, \mathbb{C})} = ||\Phi_i||^2 = 1$. Forming the basis elements in such a way assures that a single basis element will contain information from each of the snapshots. From standard arguments, it can be seen that the coefficients $V_i(j)$ can be found by solving the eigenvalue problem $CV = \lambda V$ where C is given by

$$[C]_{ij} = \frac{1}{N_s} \langle A(\mathbf{q}_i), A(\mathbf{q}_j) \rangle_{L^2(\Omega, \mathbb{C})}. \tag{11}$$

Hence, the next step in forming the POD basis is to form the covariance matrix C and find the associated eigenvalues and eigenvectors. Since C is Hermitian positive semi-definite, we know it possesses a complete set of orthogonal eigenvectors and corresponding nonnegative eigenvalues. In forming the POD basis, we want to be able to readily decide which basis elements to use in the reduced basis. To this end, we order the eigenvalues along with their corresponding eigenvectors such that the eigenvalues are in decreasing order,

$$\lambda_1 \geq \lambda_2 \geq ... \geq \lambda_{N_s} \geq 0. \tag{12}$$

We then normalize the eigenvectors corresponding to the rule

$$V_i \cdot V_j = \frac{\delta_{ij}}{N_s \lambda_j}. \tag{13}$$

Consequently, the ith POD basis element is defined by Eq. (9) where $V_i(j)$ represents the jth component of the ith eigenvector of C.

We now have the full POD basis and need a criterion to decide how many basis elements are required to accurately portray the data. In other words, we want to choose N such that $span\{\Phi_i\}_{i=1}^{N} \approx span\{A(\mathbf{q}_j)\}_{j=1}^{N_s}$. In choosing this number N, we compute

$$\sum_{j=1}^{N} \lambda_j \bigg/ \sum_{j=1}^{N_s} \lambda_j \tag{14}$$

which represents the percentage of "energy" in $span\{A(\mathbf{q}_j)\}_{j=1}^{N_s}$ that is captured in $span\{\Phi_j\}_{j=1}^{N}$. Then the reduced basis consists of the first N POD basis elements where N is chosen to capture the desired amount of energy.

We now note that $span\{\Phi_i\}_{i=1}^{N_s} = span\{A(\mathbf{q}_j)\}_{j=1}^{N_s}$. Indeed, given any $A(\mathbf{q}_j)$, we have

$$A(\mathbf{q}_j) = \sum_{k=1}^{N_s} \alpha_k(\mathbf{q}_j) \Phi_k \tag{15}$$

where

$$\alpha_k(\mathbf{q}_j) = \langle A(\mathbf{q}_j), \Phi_k \rangle_{L^2(\Omega, \mathbb{C})} \tag{16}$$

as $\{\Phi_j\}_{j=1}^{N_s}$ are orthonormal in $L^2(\Omega, \mathbb{C})$.

However, to complete the analysis, one needs to be able to calculate $A^N(\mathbf{q})$ where $\mathbf{q}$ is a given parameter *not* in the set $\{\mathbf{q}_j\}_{j=1}^{N_s}$. To this end, we extend the approximation formula to obtain

$$A^N(\mathbf{q}) = \sum_{k=1}^{N} \alpha_k(\mathbf{q})\Phi_k \tag{17}$$

where $\alpha_k(\mathbf{q})$ can be evaluated through different methods. In [3], we explored two different methods one might choose: a POD/Galerkin method or a POD/Interpolation method. However, in this paper, we only present results in which we use the POD/Interpolation method with linear interpolation for the one-parameter simulated results and cubic spline interpolation for all other results. For details on these methods, we refer the reader to [22] and [23].

3.2　Simulated Results

In [4], we performed several trials in which we assumed we had access to various types of data, such as the A field or the $\mathbf{B}$ field at various points (x_i, y_j) in Ω. We compared and contrasted the accuracy to which we could estimate the length l of the damage based on whether the A field or $\mathbf{B}$ field was used and whether we considered the field along a single line, multiple lines or within the entire region (which is not experimentally possible and was only tested for initial comparisons). From the results presented in [4], we concluded that extremely accurate results were obtained *only* when the y component B_2 of the magnetic flux density was used in the cost criterion, i.e., when we used

$$J(\mathbf{q}) = \frac{1}{2}\sum_{i=1}^{n}\sum_{j=1}^{m} |10^8 B_2^N(x_i, y_j, \mathbf{q}) - 10^8 \hat{B}_2(x_i, y_j, \mathbf{q}^*)|^2 \tag{18}$$

where 10^8 is a scaling factor accounting for the low order of magnitude of the field (B_2 is on the order of $10^{-8} Wb/m^2$), $B^N(\mathbf{q}) = \nabla \times A^N(\mathbf{q})$ is the reduced order POD approximation in which $A^N(\mathbf{q})$ is given by (17), and $\hat{B}_2$ is "data" from a sample we wish to characterize. (In this section, $\hat{B}_2$ is obtained from Ansoft finite element simulations to which noise has been added in the usual manner (see [3, 21]). Furthermore, performing multi-line scans or using full region data improved the results only marginally and hence did not warrant the extra effort and time in collecting more extensive data sets. Consequently the results presented in this section involve only the least squares difference in the B_2 field given by (18) along a single line located $1mm$ above the conducting sheet.

In the sample results presented, we focus on estimating a single parameter of the crack assuming all other parameters are fixed or estimating two parameters with one parameter fixed. The results below summarize the feasibility of determining the length and depth of a crack. In a specific trial run, ten different data sets (exact data with ten different sets of added random noise) are used where the relative noise is chosen at a 10% noise level with a confidence level of 99.7% (3 standard deviations). (Details can be found in [3] and [4].)

In determining the length of the damage alone, we first followed the steps outlined in Section 3.1 and generated an ensemble of "snapshots". Keeping the crack fixed at a depth of $9mm$ with a thickness of $2mm$, we varied the crack length l from $0mm$ to $4mm$ in increments of $0.2mm$. Using the snapshots, $\{A(l_j)\}_{j=1}^{N_s=21}$, we formed the POD basis and determined that 99.99% of the energy of the system was captured with a single basis element [4]. However, in the specific trial runs, we found that using 4 POD basis elements improved the parameter estimation with data containing no noise, while there was no significant difference between the use of 5 basis elements over the use of 4 basis elements. Therefore, the results are based on the use of 4 basis elements in the algorithm.

To test the inverse methodology, we first try to identify the length of the damage, $l^* = 1.3mm$. Using data on a single line above the conducting sheet with 10% relative noise, an average estimated length of $1.2977mm$ was obtained with a variance of $0.3237 \times 10^{-4}mm^2$ across the ten trials.

Proceeding as we did in estimating the length of a damage, to estimate the depth of the damage, we fixed the thickness of the crack at $0.5mm$ with length $2mm$ and varied the depth of the crack d from $0.25mm$ to $16.25mm$ in increments of $0.5mm$. Using 5 basis elements, we were able to accurately estimate a crack depth of $8mm$ in the presence of 10% relative noise; an average depth of $8.0631mm$ was estimated with a variance $0.1180 \times 10^{-3}mm^2$. Thus, as in the case of estimating the length of a crack, we can also recapture the depth of a crack quite accurately and efficiently.

Finally, we estimated both length and depth simultaneously. However, first, in [3, 8], we discussed in detail a need to modify the assumptions made in the original test problem to more accurately describe the behavior of experimental data used in the next section. In short, the computational domain was expanded beyond the edges of the sample and *snapshots* were taken of the magnetic flux density data *on a single line* above the conducting sheet (instead of the whole region). Recall that for the previous trials, we took snapshots of the magnetic vector potential for the entire computational domain even though in the inverse problem we only considered those data points along a single line. Furthermore, we considered data across the entire length of the sample, instead of just half the sample as done previously. (For more details , see [3, Chapter 6].) As a result, we also implement these changes in the two-parameter estimation problem.

We proceed as in the previous estimation problems by first generating an ensemble of damages. We consider damages with depths ranging from $1mm$ to $11mm$ in increments of $2mm$ in combination with lengths from $0.5cm$ to $3.5cm$ in increments of $1cm$ (we now consider longer damages similar to those used in obtaining experimental data). We keep the thickness fixed at $1mm$. A total of 24 snapshots, $\{B_2(d_i, l_j)\}$, $i = 1, ..., 6$, $j = 1, ..., 4$ were generated using Ansoft. Using 4 basis elements with 10% relative noise added, we estimated a damage with depth $d^* = 4mm$ and length $l^* = 2cm$. We obtained average estimates of $d = 4.0855mm$ with variance $0.5516 \times 10^{-4}mm^2$ and length $l = 1.980cm$ with variance $0.9632 \times 10^{-4}mm^2$. Thus, even when we estimated two parameters simultaneously, the computational method proposed produced extremely accurate results.

3.3 *Experimental Results*

Since the results using simulated data were extremely promising, we designed an experiment in which we tried to detect and parameterize a damage within an aluminum sample using a giant magnetoresistive (GMR) sensor. The sample was constructed of 17 layers of $1mm$ thick aluminum plates with a slice cut out of one of the layers to simulate a damage within the sample (see [3, 8] for graphical representations). The "damaged" piece of aluminum is moved from one layer to another to simulate damages within the sample at different depths, and the length of the damage is varied by producing "gaps" of varying size from the aluminum plate (the thickness of the damage is always fixed at $1mm$). As a means of inducing current within the sample, a thin sheet of copper carrying a uniform current of 3 Amps is placed above the sample on top of a thin sheet of paper (to avoid direct physical contact between the sample and the conducting sheet). The GMR sensor measures the amplitude and phase of the magnetic flux density across a $2in$ line (along the length of the sample) every $0.635mm$. The data is then filtered through a lock-in amplifier and saved to a file.

The experimental data did not resemble Ansoft simulations from our model; therefore, we chose to snapshot on the data itself in forming the POD basis elements. Furthermore, it was necessary to filter out the background noise (data for a sample containing no damage) in order to obtain a representative pattern in the data as a function of the damage within the sample (see [3] for full details). In filtering out the background noise, we magnified boundary effects from the edges of the sample. Therefore, in the cost criterion, we only consider data across the center of the sample. In other words, we use the cost criterion given by

$$J(\mathbf{q}) = \frac{1}{2} \sum_{j=a-2}^{b+2} \left| 10^9 B_2^N(x_j; \mathbf{q}) - 10^9 \hat{B}_2^j \right|^2, \tag{19}$$

where $\mathbf{q}$ is the vector containing the parameters we wish to estimate, $B_2^N(\mathbf{q})$ is the POD approximation formed using snapshots on the data itself, $\hat{B}_2^i$ is GMR data at grid points x_j, $j = 1, ..., n$ with n total grid points and a and b indicate the indices of the grid points we consider in our cost criterion.

In estimating a depth of $d^* = 3mm$ with corresponding length $l^* = 1.5cm$, we obtained an estimate of $d = 2.9759mm$ and $l = 1.4612cm$, resulting in a relative error for depth of $R_d = 0.80\%$ and length $R_l = 2.59\%$. Again, even with experimental data, the results were accurate.

4　Conclusion

In this paper we presented a reduced order computational algorithm which contributes to the overall field of nondestructive evaluation. In detecting subsurface damages, there is a need for a fast and efficient inverse problem methodology. The reduced order POD method is an attractive method, because it allowed us to create a reduced basis of less than 10 basis elements in the trials presented while still capturing 99% of the total energy of the system. This results in an accurate, as well as fast, forward algorithm.

We then gave an overview of the POD methodology in the context of subsurface damage detection or parameter identification. We discussed the implementation of the inverse problem for both simulated results as well as experimental results, providing examples for each case. Based upon the results presented, we showed that using either simulated data or experimental data on either a one-parameter estimation problem or a two-parameter estimation problem, we achieved accurate results.

However, the most *significant* finding is in regard to reduction in computational time which can be summarized as follows. If one were to use a software package such as Ansoft's Maxwell 2D Field Simulator to calculate the forward problem each time it is required in the inverse problem, it would take approximately 5-10 minutes for a *single* forward solve. The forward solve is typically required anywhere from as few as 10 times to as many as 500-1000 times. Assuming the forward algorithm is called approximately 100 times with an average of 7 minutes for each forward run, the inverse problem would take approximately 42,000 seconds. In the examples provided, the *entire* inverse problem ran in approximately 8-10 seconds. Consequently, on average the computational time is reduced by a factor of 10^3. This suggests that an appropriate sensing device, when coupled with reduced order modeling in the inverse problem, might prove feasible in practical damage detection applications.

Acknowledgments

This research was supported (MLJ) by the NASA Langley Graduate Researcher's Program under grant NGT-1-52196 and in part (HTB) by the Air Force Office of Scientific Research under grant AFOSR F49620-01-1-0026.

References

[1] Buzz Wincheski and Min Namkung. Development of very low frequency self-nulling probe for inspection of thick layered aluminum structures. In *1998 Review of Progress in Quantitative NDE*, Snowbird, Utah, Aug. 1998.

[2] Buzz Wincheski and Min Namkung. Deep flaw detection with giant magnetoresistive (GMR) based self-nulling probe. In *1999 Review of Progress in Quantitative NDE*, Montreal, Canada, July 1999.

[3] Michele L. Joyner. *An Application of a Reduced Order Computational Methodology for Eddy Current Based Nondestructive Evaluation Techniques.* PhD thesis, North Carolina State University, 2001.

[4] H.T. Banks, M.L. Joyner, B. Wincheski, and W.P. Winfree. Evaluation of material integrity using reduced order computational methodology. CRSC Tech. Rep. CRSC-TR99-30, North Carolina State University, 1999.

[5] David K. Cheng. *Field and Wave Electromagnetics.* Addison-Wesley, Reading, MA, second edition, 1992.

[6] Ansoft Corporation. *Maxwell 2D Field Simulator - Technical Notes*, 1995-1999.

[7] Lawrence C. Evans. *Partial Differential Equations.* American Mathematical Society, Providence, 1991.

[8] H.T. Banks, M.L. Joyner, B. Wincheski, and W.P. Winfree. Real time computational algorithms for eddy current bas based damage detection. In progress.

[9] H.T. Banks, R.C. del Rosario, and R.C. Smith. Reduced order model feedback control design: Numerical implementation in a thin shell model. *IEEE Trans. Auto. Control*, 45:1312–1324, July 2000.

[10] G. Berkooz. Observations on the proper orthogonal decomposition. In *Studies in Turbulence*, pages 229–247. Springer-Verlag, New York, 1992.

[11] G. Berkooz, P. Holmes, and J.L. Lumley. The proper orthogonal decomposition in the analysis of turbulent flows. *Annual Review of Fluid Mechanics*, 25(5):539–575, 1993.

[12] R.C. del Rosario. *Computational Methods for Feedback Control in Structural Systems.* PhD thesis, North Carolina State University, 1998.

[13] K. Karhunen. Zur spektral theorie stochasticher prozesse. *Ann. Acad. Sci. Fennicae*, 37(A1), 1946.

[14] M. Kirby and L. Sirovich. Application of the Karhunen-Loeve procedure for the characterization of human faces. *IEEE Transactions on Pattern Analysis and Machine Intelligence*, 12(1):103–108, 1990.

[15] M. Kirby, J.P. Boris, and L. Sirovich. A proper orthogonal decomposition of a simulated supersonic shear layer. *International Journal for Numerical Methods in Fluids*, 10:411–428, 1990.

[16] K. Kunisch and S. Volkwein. Control of Burgers' equation by a reduced-order approach using proper orthogonal decomposition. *J. Optimization Theory and Applic.*, 102(2):345–371, 1999.

[17] M. Loeve. *Functions aleatoire de second ordre.* Compte rend. Acad. Sci., Paris, 1945.

[18] J.L. Lumley. The structure of inhomogeneous turbulent flows. *Atmospheric Turbulence and Radio Wave Propagation*, pages 166–178, 1967.

[19] J.L. Lumley. *Stochastic Tools in Turbulence.* Academic Press, New York, 1970.

[20] H.V. Ly and H.T. Tran. Proper orthogonal decomposition for flow calculations and optimal control in a horizontal cvd reactor. CRSC Tech. Rep. CRSC-TR98-13, North Carolina State University, 1998, *Quart. Appl. Math*, to appear.

[21] H.T. Banks, M.L. Joyner, B. Wincheski, and W.P. Winfree. Nondestructive evaluation using a reduced-order computational methodology. *Inverse Problems*, 16, 2000.

[22] Alfio Quarteroni, Riccardo Sacco, and Fausto Saleri. *Numerical Mathematics.* Springer-Verlag, New York, 2000.

[23] J. Stoer and R. Burlisch. *Introduction to Numerical Analysis.* Springer-Verlag, New York, 2nd edition, 1993.

Electromagnetic Nondestructive Evaluation (VI)
F. Kojima et al. (Eds.)
IOS Press, 2002

Shape identification in conductive materials by electrical resistance tomography

Guglielmo RUBINACCI, Antonello TAMBURRINO, Salvatore VENTRE, Fabio VILLONE
*Ass. EURATOM/ENEA/CREATE, DAEIMI, Lab. of Computational
Electromagnetics, Università di Cassino - 03043 Cassino, Italy*

Abstract. Electrical Resistance Tomography (ERT) uses measurements (in steady-state conditions) of mutual resistances (conductances) between electrodes held in contact with the boundary of the conductive material under testing. This paper is focused on a new inversion procedure for shape identification in conductors made of two different conducting phases. Specifically, it is shown that if the region with high resistivity is enlarged, then the difference between the resistance matrix related to the enlarged region and the resistance matrix before the enlargement, is a positive semi-definite matrix. This property forms the basis for the proposed inversion procedure.

1. Introduction

Electrical resistance tomography (ERT) is a technique used to reconstruct the unknown resistivity of a conductive body starting from noisy boundary measurements in steady-state conditions. This technique has been considered in a variety of applications covering different fields such as non-destructive testing, process tomography, geophysical prospecting, and biomedical imaging.

In the typical configuration, a set of electrodes located at the boundary of the specimen is used to inject steady-state currents into the specimen. The resistivity of the conductive body affects both the current density distribution and the potential at the boundary. The unknown resistivity can be reconstructed once the boundary potential is known for any injected current density.

In the present work we address the problem of non invasive imaging of a conductor whose resistivity is a two-level function of the spatial position. This problem occurs, for instance, in industrial process tomography when the non-invasive imaging of a two-phases fluid is required [1].

2. Solution of the forward problem

For the sake of simplicity we assume that the resistivity at spatial coordinate $\mathbf{r}$ assumes the value η_i or η_b, being η_i and η_b fixed reference values and $\eta_i > \eta_b$. The forward problem consists of computing the resistance matrix $\mathbf{R}$ when the resistivity $\eta(\mathbf{r})$ is η_i in V and assumes the value η_b in $D \backslash V$ for a given $V \subset D$, D being the region containing the conductor. The forward problem realizes the mapping $V \subset D \rightarrow \mathbf{R}$. The electrodes involved in the definition of the conductance matrix are located on the boundary of D. Each electrode is assumed to be to be equi-potential; we also neglect the contact resistance (shunt model). Let S_k be the part of $S = \partial D$ in contact with the k-th electrode; we assume also that the current density has no normal component on S_J, the part of S that is not in contact with any electrodes.

Here we numerically compute the resistance matrix $\mathbf{R}$ by means of the complementary formulations as in [2]:

$$\mathbf{R} = \frac{\mathbf{R}^A + \mathbf{R}^\Phi}{2} \tag{1}$$

where $\mathbf{R}^\Phi$ arises from the scalar potential ($\mathbf{E} = -\nabla\varphi$) based formulation and $\mathbf{R}^A$ from the electric vector potential ($\mathbf{J} = \nabla\times\mathbf{T}$) based formulation. Specifically, $\mathbf{R}^A$ is defined by $P_A = \mathbf{I}^T\mathbf{R}^A\mathbf{I}$ where $\mathbf{I}$ is the column vector containing the net currents flowing into the electrodes and P_A is the value of the ohmic power losses computed using the electric vector potential formulation. Similarly, $\mathbf{R}^\Phi$ is defined as $\mathbf{R}^\Phi = (\mathbf{G}^\Phi)^{-1}$ where $P_\Phi = \mathbf{v}^T\mathbf{G}^\Phi\mathbf{v}$ represents the ohmic power losses and $\mathbf{v}$ is the column vector of the electrode voltages. In particular, φ and $\mathbf{T}$ are the numerical solutions of the variational formulations of the field problem here briefly summarized ([2-3]):

$$\text{find } \varphi \in \Phi^V \text{ such that } \int_D \sigma \|\nabla\varphi\|^2 \, dV \text{ is minimum} \tag{2}$$

$$\text{find } \mathbf{T} \in A^I \text{ such that } \int_D \eta \|\nabla\times\mathbf{T}\|^2 \, dV \text{ is minimum} \tag{3}$$

where

$$\Phi^V \triangleq \left\{ \varphi' \in \Phi \mid \varphi'|_{S_{M+1}} = 0, \, \varphi'|_{S_k} = v_k, \, k = 1,\ldots,M \right\} \tag{4}$$

$$A^I \triangleq \left\{ \mathbf{T}' \in A \mid \int_{S_k} \nabla\times\mathbf{T}'\cdot\hat{\mathbf{n}}\,dS = i_k, \, k = 1,\ldots,M \right\} \tag{5}$$

$$\Phi \triangleq \left\{ \varphi' \in L^2_{grad}(D) \mid \hat{\mathbf{n}}\times\nabla\varphi' = \mathbf{0} \text{ on } S_1 \cup\ldots\cup S_{M+1} \right\} \tag{6}$$

$$A \triangleq \left\{ \mathbf{T}' \in \mathbf{L}^2_{rot}(D) \mid \hat{\mathbf{n}}\cdot\nabla\times\mathbf{T}' = 0 \text{ on } S_\perp \right\} \tag{7}$$

The numerical solution of problem (2) and (3) is then obtained by restricting the minimization to Φ^V_N and A^I_N, finite dimensional subset of Φ^V and A^I, respectively, and by using standard isoparametric shape functions for the scalar potential and edge-elements based shape functions [2] for the electric vector potential.

3. Solution of the inverse problem

The inverse problem consists of identifying the subset V where the resistivity $\eta(\mathbf{r})$ assumes the value η_i, from the knowledge of the measured (and noisy) resistance matrix $\tilde{\mathbf{R}}$. We recall that the resistivity in $D\backslash V$ is equal to a known value η_b.

The key issue for the solution of the inverse problem is the following properties of the resistance matrices.

<u>Proposition 1</u>. $\eta_1 \geq \eta_2 \Rightarrow \mathbf{R}_1 \geq \mathbf{R}_2$ where $\mathbf{R}_1$ is the resistance matrix related to η_1, $\mathbf{R}_2$ is the resistance matrix related to η_2 and $\mathbf{R}_1 \geq \mathbf{R}_2$ means that $\mathbf{R}_1 - \mathbf{R}_2$ is a positive semi-definite matrix, i.e. $\mathbf{x}^T(\mathbf{R}_1 - \mathbf{R}_2)\mathbf{x} \geq 0, \forall\mathbf{x}$.

Let $\mathbf{T}_1$ and $\mathbf{T}_2$ be defined as the solution of (3) when the column vector $\mathbf{I}$ of the electrode voltages is given and the resistivities are η_1 and η_2, respectively. Then it follows that

$$\mathbf{I}^T\mathbf{R}_1\mathbf{I} = \int_D \eta_1\|\nabla\times\mathbf{T}_1\|^2 \, dV \geq \int_D \eta_2\|\nabla\times\mathbf{T}_1\|^2 \, dV \geq \int_D \eta_2\|\nabla\times\mathbf{T}_2\|^2 \, dV = \mathbf{I}^T\mathbf{R}_2\mathbf{I} \tag{8}$$

where the first and last equality follow from the definition of resistance matrix, the first inequality follows from the hypothesis and the second inequality follows from (3) when particularized to η_2. The thesis follows from the arbitrariness of **I**. $\blacklozenge$

In a similar way it is possible to show :

Proposition 2.
$$\sigma_1 \geq \sigma_2 \Rightarrow \mathbf{G}_1 \geq \mathbf{G}_2 \tag{9}$$
$$\sigma_1 \geq \sigma_2 \Rightarrow \mathbf{G}_1^\Phi \geq \mathbf{G}_2^\Phi \tag{10}$$
$$\eta_1 \geq \eta_2 \Rightarrow \mathbf{R}_1^A \geq \mathbf{R}_2^A \tag{11}$$
where $\mathbf{G}_1$ (resp. $\mathbf{G}_2$) is the conductance matrix due to σ_1 (resp. σ_2), $\mathbf{G}_1^\varphi$ (resp. $\mathbf{G}_2^\varphi$) is the conductance matrix due to σ_1 (resp. σ_2) but numerically computed by using the scalar potential formulation, and $\mathbf{R}_1^A$ (resp. $\mathbf{R}_2^A$) is the resistance matrix due to η_1 (resp. η_2) but numerically computed by using the electric vector potential. $\blacklozenge$

In addition, we notice that (see Appendix) if **A** and **B** are two real, symmetric and invertible matrices, then
$$\mathbf{A} \geq \mathbf{B} \quad \Leftrightarrow \quad \mathbf{A}^{-1} \leq \mathbf{B}^{-1} \tag{12}$$

Therefore, from (10) and (11) it follows:

Corollary 1.
$$\sigma_1 \geq \sigma_2 \Rightarrow \mathbf{R}_1^\Phi \leq \mathbf{R}_2^\Phi \tag{13}$$
$$\eta_1 \geq \eta_2 \Rightarrow \mathbf{G}_1^A \leq \mathbf{G}_2^A \tag{14}$$
$\blacklozenge$

Finally, by combining (10), (11), (13) and (14) it results:

Corollary 2.
$$\eta_1 \geq \eta_2 \Rightarrow \frac{\mathbf{R}_1^\Phi + \mathbf{R}_1^A}{2} \geq \frac{\mathbf{R}_2^\Phi + \mathbf{R}_2^A}{2} \tag{15}$$
$$\sigma_1 \geq \sigma_2 \Rightarrow \frac{\mathbf{G}_1^\Phi + \mathbf{G}_1^A}{2} \geq \frac{\mathbf{G}_2^\Phi + \mathbf{G}_2^A}{2} \tag{16}$$
$\blacklozenge$

These results leads, in particular, to the following inequality:
$$D_0 \subseteq D_1 \subseteq D \Rightarrow \mathbf{R}_0 \leq \mathbf{R}_1 \tag{17}$$
where $\mathbf{R}_0$ ($\mathbf{R}_1$) is the resistance matrix when $\eta = \eta_i$ in D_0 (D_1) and $\eta = \eta_b$ in $D\backslash D_0$ ($D\backslash D_1$). Property (17) is the key for the inversion procedure. In fact from (17) it follows that:
$$\mathbf{R}_1 - \mathbf{R}_0 \text{ is positive semi-definite} \Rightarrow D_1 \not\subset D_0 \tag{18}$$
$$\mathbf{R}_1 - \mathbf{R}_0 \text{ is negative semi-definite} \Rightarrow D_0 \not\subset D_1 \tag{19}$$
$$\mathbf{R}_1 - \mathbf{R}_0 \text{ is not positive nor negative semi-definite} \Rightarrow D_1 \not\subset D_0 \text{ and } D_0 \not\subset D_1 \tag{20}$$
Therefore, we can establish if $D_0 \not\subset D_1$, or $D_1 \subseteq D_0$ or ($D_1 \not\subset D_0$ and $D_0 \not\subset D_1$) by simply analyzing the eigenvalues of the matrix $\mathbf{R}_1 - \mathbf{R}_0$. A numerical example to validate (18)-(20) is reported in Figures 1 and 2 showing three different resistivity patterns together with the eigenvalues for the three different cases (18)-(20).

We now discuss an inversion procedure that takes into account properties (18)-(20) to retrieve the unknown subset $V \subseteq D$ where the resistivity assumes the value η_i.

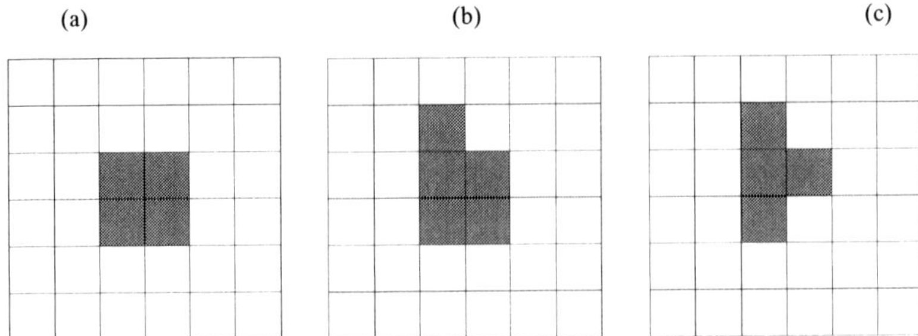

Figure 1. Three different resistivities profile on a gray scale map where white stays for 5Ωm and gray for 5kΩm for a two-dimensional example. The resistance matrix related to (a) is $\mathbf{R}_A$ whereas the resistance matrix related to (b) is $\mathbf{R}_B$ and the resistance matrix related to (c) is $\mathbf{R}_C$.

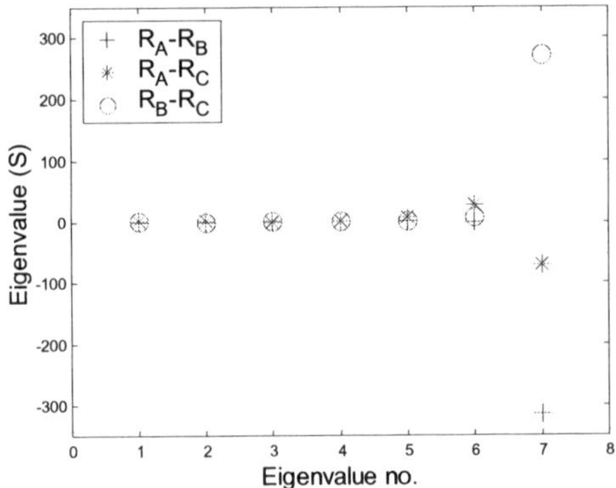

Figure 2. The eigenvalues for the three different cases (18)-(20) obtained by considering the resistance matrices $\mathbf{R}_A$, $\mathbf{R}_B$ and $\mathbf{R}_C$.

The main idea consists of partitioning the domain D in N small non-overlapping parts Ω_1, ..., Ω_N. Then by using (18)-(20) we check if Ω_k is contained in V by computing the eigenvalues of the matrix $\widetilde{\mathbf{R}} - \mathbf{R}_k$ where $\mathbf{R}_k$ is the numerically computed resistance matrix related to a resistivity η_k defined as:

$$\eta_k(\mathbf{r}) = \begin{cases} \eta_b \ \forall \mathbf{r} \in \mathbf{D} \setminus \Omega_k \\ \eta_i \ \forall \mathbf{r} \in \Omega_k \end{cases} \tag{21}$$

Then we set $\widetilde{V}$, the estimate of V, as the union of those Ω_k such that $\widetilde{\mathbf{R}} - \mathbf{R}_k$ is positive semi-definite. We notice that, from (18)-(20), it results that $V \subseteq \widetilde{V}$. Similarly, it is possible to build a set $\hat{V}$ s.t. $\hat{V} \subseteq V$.

It is worth noting that the computational cost of this procedure grows linearly with N, the number of Ω_k's. Moreover, the matrices $\mathbf{R}_k$ can be computed and stored thus the inversion of the data consists of computing only the eigenvalues of $\widetilde{\mathbf{R}} - \mathbf{R}_k$ for $k=1, ..., N$. Moreover, the size of the matrix $\widetilde{\mathbf{R}} - \mathbf{R}_k$ is modest being a $n \times n$ matrix where $n+1$ is the number of electrodes that, usually is not greater than few tenth.

The proposed procedure requires a slight modification when the matrix $\widetilde{\mathbf{R}}$ is affected by the measurement noise. Specifically, the noise affecting the matrix $\widetilde{\mathbf{R}}$ may corrupt the eigenvalues of the matrix $\widetilde{\mathbf{R}} - \mathbf{R}_k$ thus resulting with an undefined sign also when $\Omega_k \subseteq V$. To manage this situation, we compute a sign index s_k related to Ω_k defined as:

$$s_k = \frac{\sum_j \lambda_{k,j}}{\sum_j |\lambda_{k,j}|} \tag{22}$$

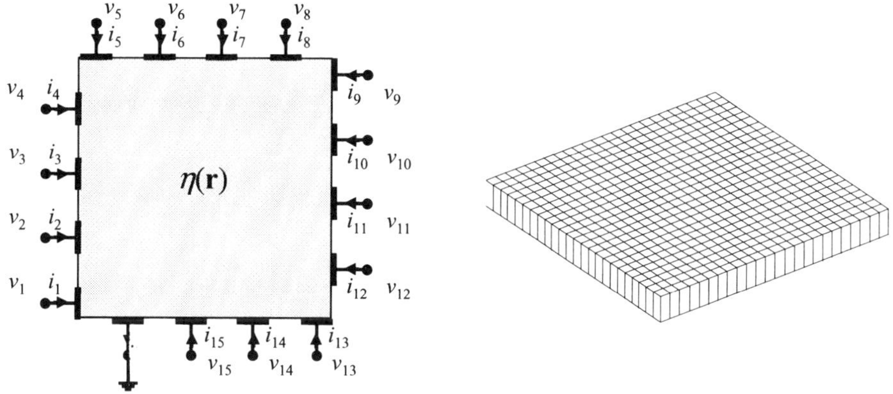

Figure 3. The reference geometry together with the finite element mesh.

where $\lambda_{k,j}$ is the j-th eigenvalue of the matrix $\tilde{\mathbf{R}} - \mathbf{R}_k$. The sign index s_k ranges from -1 when $\tilde{\mathbf{R}} - \mathbf{R}_k$ is a negative definite matrix (i.e. when $V_0 \subseteq \Omega_k$) to $+1$ when $\tilde{\mathbf{R}} - \mathbf{R}_k$ is a positive definite matrix (i.e. when $\Omega_k \subseteq V_0$). For values internal to the interval $(-1, +1)$ the sign of $\tilde{\mathbf{R}} - \mathbf{R}_k$ is not definite. The effect of noise consists of producing a sign index s_k different from $+1$ and -1 also when $\Omega_k \subseteq V_0$ or $V_0 \subseteq \Omega_k$. We define a reconstruction with threshold ε the union of those Ω_k with sign greater than ε, i.e.:

$$V_\varepsilon \triangleq \bigcup_{k \mid s_k > \varepsilon} \Omega_k \, . \tag{23}$$

The last problem is the selection of a proper value of the threshold ε. To this purpose, we solve the one parameter minimization problem:

$$\min_\varepsilon f(\varepsilon) \tag{24}$$

$$f(\varepsilon) \triangleq \left\| \tilde{\mathbf{R}} - \mathbf{R}_\varepsilon \right\|^2 \tag{25}$$

where the matrix $\mathbf{R}_\varepsilon$ is related to V_ε and $\|\cdot\|$ is a suitable matrix norm.

Finally, we notice that the inherit ill-posedness of this inverse problem directly affects the achievable resolution. In fact, as the dimension of Ω_k decreases, the matrix $\mathbf{R}_k$ tends to the resistance matrix related to an uniform resistivity equals to η_b and the noise affecting $\tilde{\mathbf{R}}$ produces a sign index s_k almost independent from k.

4. Numerical examples

The numerical example refers to a conducting body made by a square plate of 10cm×10cm×1cm and probed by using 16 electrodes on the boundary (see figure 3). Each electrode is 1.25 cm in length and 1cm in height; in addition the electrodes are equally spaced. The finite element mesh used for computing the matrices $\mathbf{R}_k$ and $\mathbf{R}_\varepsilon$ is made of 24×24×1 equal hexahedral elements (see figure 3), whereas the finite element mesh used for numerically computing the measurement consists of 48×48×1 elements. The resistivity is assumed to be independent from the z-coordinate (the axis z is orthogonal to the plate) inside the conductor; the unknown resistivity profile is reported in figure 4. We choose $\eta_b = 5\Omega\text{m}$ and $\eta_i = 5\text{k}\Omega\text{m}$ and we assume that the partitioning of the domain in the "small" sets Ω_k is made by a regular mesh of 12×12×1 hexahedral elements.

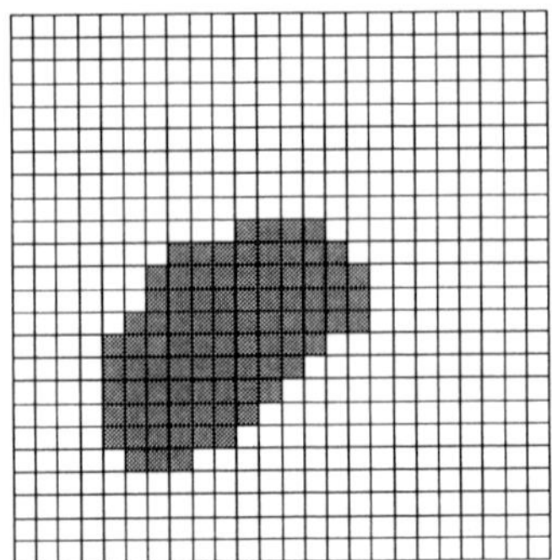

Figure 4. The unknown resistivity profile on a gray scale map. White stays for η_b and gray for η_i.

(a)   (b)

(c)   (d)

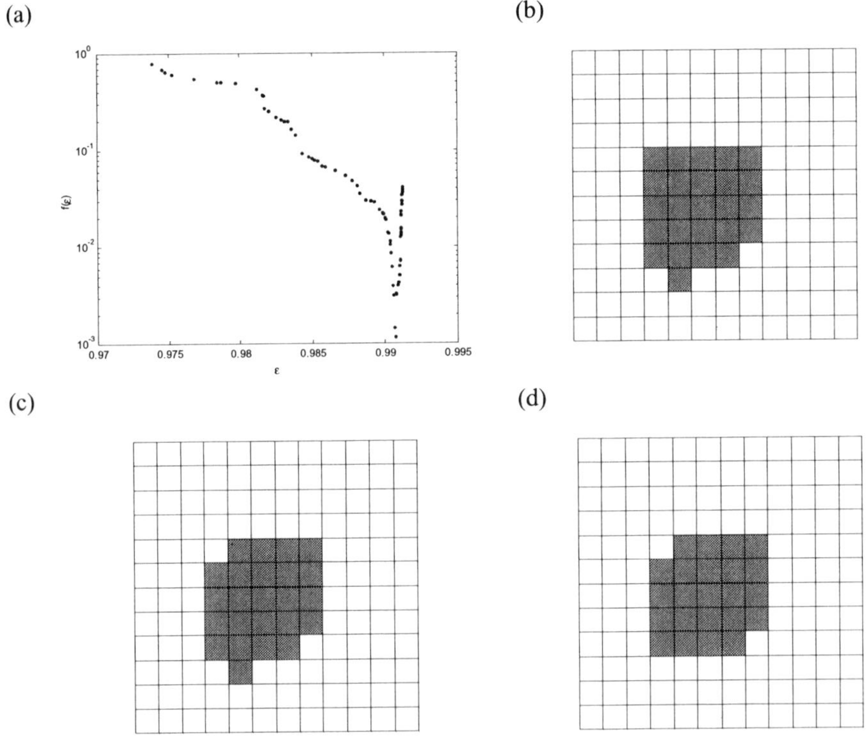

Figure 5. The (normalized) $f(\varepsilon)$ error functional (a), together with the reconstructions corresponding to $\varepsilon_-=0.99068$ (b) $\varepsilon_{\mathrm{opt}}=0.99075$ (c) and $\varepsilon_+=0.99079$ (d). White stays for η_b and gray for η_i.

Figure 5 reports the reconstruction obtained by the proposed method. Specifically, figure 5a reports the (normalized) error functional $f(\varepsilon)$ whereas figures 5b-5c report the reconstructions corresponding to $\varepsilon_-<\varepsilon_{\mathrm{opt}}<\varepsilon_+$, respectively, being $\varepsilon_{\mathrm{opt}}$ the minimizer of $f(\varepsilon)$ and ε_-, ε_+ the closest value to $\varepsilon_{\mathrm{opt}}$ giving different reconstructions.

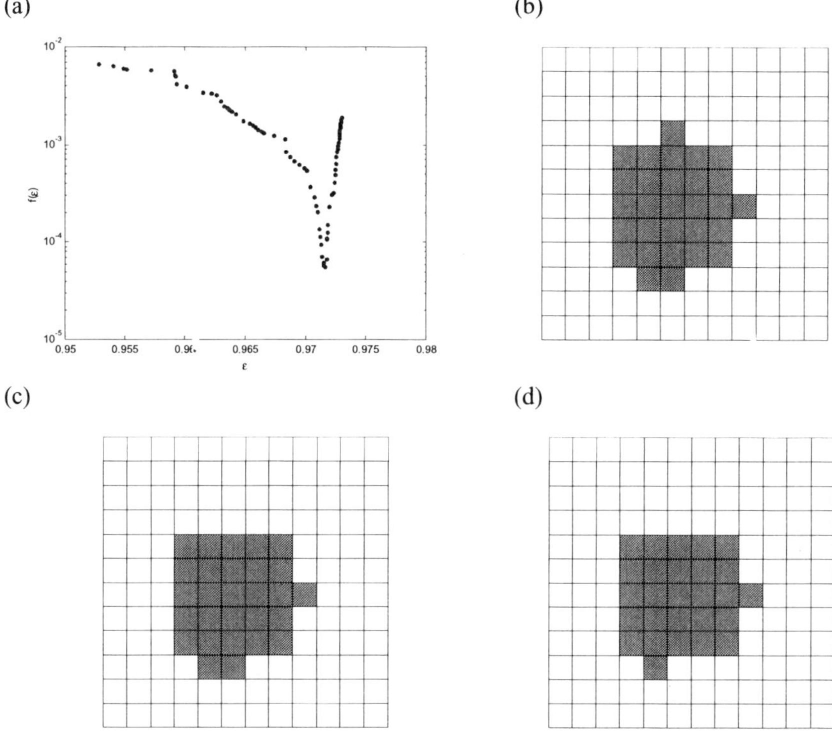

Figure 6. The (normalized) $f(\varepsilon)$ error functional (a), together with the reconstructions corresponding to $\varepsilon_-=0.9715$ (b) $\varepsilon_{opt}=0.9716$ (c) and $\varepsilon_+=0.9717$ (d). White stays for η_b and gray for η_i.

Figure 6 reports the results related to a case obtained by only changing η_i. Specifically, $\eta_i=7.5\Omega m$ is of the same order of magnitude of η_b. In this case the shape of the region of resistivity η_i has been identified worse than for the previous case since the contrast, i.e. the ratio η_i/η_b is close to the unity.

5. Conclusions

In this paper we have addressed the inverse problem of resistivity retrieval in Electrical Resistance Tomography. Specifically, we have shown some properties of the resistance and conductance matrices and proposed a novel method for solving the inverse problem of retrieving a resistivity distribution characterized by means of two different values. The key feature of the proposed method is its low computational cost that increases linearly with the number of regions used for partitioning the conductive domain under investigation.

Acknowledgements

This work was supported by the Italian MURST and by the Italian Space Agency.

Appendix

In this appendix we show that if **A** and **B** are two real, symmetric and invertible matrices, then

$$\mathbf{A} \geq \mathbf{B} \quad \Leftrightarrow \quad \mathbf{A}^{-1} \leq \mathbf{B}^{-1} \tag{A1}$$

In the following we use this straightforward property:

$$\mathbf{A} \geq \mathbf{B} \quad \Leftrightarrow \quad \mathbf{C}^{T}\mathbf{AC} \geq \mathbf{C}^{T}\mathbf{BC} \tag{A2}$$

if **C** is an invertible matrix.

In fact, thanks to the assumptions made, the matrix **B** can be factorized as:

$$\mathbf{B}=\mathbf{U}\,\Lambda\,\mathbf{U}^{T}$$

where **U** is an orthogonal matrix and Λ is a diagonal matrix with non zero entries on the main diagonal.

Then we notice that, thanks to (A2), it results that

$$\mathbf{A} \geq \mathbf{U}\,\Lambda\,\mathbf{U}^{T} \Leftrightarrow \Lambda^{-1/2}\,\mathbf{U}^{T}\,\mathbf{A}\,\mathbf{U}\,\Lambda^{-1/2} \geq \mathbf{I} \tag{A3}$$

where **I** is the identity matrix.

Since $\Lambda^{-1/2}\,\mathbf{U}^{T}\,\mathbf{A}\,\mathbf{U}\,\Lambda^{-1/2}$ is also a real, symmetric and invertible matrix the canonical problem becomes to show that

$$\mathbf{M} \geq \mathbf{I} \quad \Leftrightarrow \quad \mathbf{M}^{-1} \leq \mathbf{I} \tag{A4}$$

where **M** is a real, symmetric and invertible matrix.

In fact, if (A4) holds, then from (A3) it results that

$$\Lambda^{-1/2}\,\mathbf{U}^{T}\,\mathbf{A}\,\mathbf{U}\,\Lambda^{-1/2} \geq \mathbf{I}$$
$$\Leftrightarrow \Lambda^{1/2}\,\mathbf{U}^{T}\,\mathbf{A}^{-1}\,\mathbf{U}\,\Lambda^{1/2} \leq \mathbf{I}$$
$$\Leftrightarrow \mathbf{U}^{T}\,\mathbf{A}^{-1}\,\mathbf{U} \leq \Lambda^{-1}$$
$$\Leftrightarrow \mathbf{A}^{-1} \leq \mathbf{U}\,\Lambda^{-1}\,\mathbf{U}^{T} = \mathbf{B}^{-1}$$

To show (A4) we factorize the matrix **M** as $\mathbf{M}=\mathbf{V}\,\Sigma\,\mathbf{V}^{T}$; then

$$\mathbf{I} \leq \mathbf{V}\,\Sigma\,\mathbf{V}^{T} \Leftrightarrow \mathbf{I} \leq \Sigma \Leftrightarrow \mathbf{I} \geq \Sigma^{-1} \Leftrightarrow \mathbf{I} \geq \mathbf{V}\,\Sigma^{-1}\,\mathbf{V}^{T} = \mathbf{M}^{-1}.$$

References

[1] Williams R A and Beck M S (Eds.) 1995, Process Tomography-Principles, Techniques and Applications, Oxford: Butterworth-Heinemann, 1995

[2] A. Tamburrino, S. Ventre, G. Rubinacci, Electrical resistance tomography: complementarity and quadratic models Inverse Problems, vol. 16, 1585-1618, 2000.

[3] A. Bossavit, Computational electromagnetism, Variational formulations, Edge elements, Complementarity, Academic Press, Boston, 1998.

Electromagnetic Nondestructive Evaluation (VI)
F. Kojima et al. (Eds.)
IOS Press, 2002

A Numerical Method for Finding the Convex Hull of Inclusions using the Enclosure Method

Masaru IKEHATA[†] and Takashi OHE[‡]
† *Department of Mathematics, Faculty of Engineering, Gunma University,
Kiryu, Gunma 376-8515, Japan*
‡ *Department of Simulation Physics, Faculty of Informatics, Okayama University of Science,
Okayama, Okayama 700-0005, Japan*

Abstract. A numerical method for finding inclusions based on a formula in the enclosure method is discussed. We propose a numerical implementation of the formula by estimating the support function by the slope of the logarithm of the indicator function, and propose a rule to reject non-regular directions to avoid poor estimation. The effectiveness of our method is shown by numerical examples.

1 Introduction

Let $\Omega \subset \mathbf{R}^2$ be a simply connected bounded domain with smooth boundary, and D be an unknown domain with Lipschitz boundary satisfying $\overline{D} \subset \Omega$. Let $\gamma \in L^\infty(\Omega)$ be a function defined by

$$\gamma = \gamma_0 + (\gamma_1 - \gamma_0)\chi_D,$$

where γ_0, γ_1 are positive constants with $\gamma_0 \neq \gamma_1$, and χ_D is the characteristic function of D. Let $u \in H^1(\Omega \backslash \overline{D})$ be a non-constant solution of the equation

$$\nabla \cdot \gamma \nabla u = 0 \quad \text{in } \Omega. \tag{1}$$

Our problem is to extract some information about the location and shape of D from Cauchy data $(f, g) \equiv \left(u|_{\partial\Omega}, \left.\frac{\partial u}{\partial \nu}\right|_{\partial\Omega} \right)$ where ν denotes the outward unit normal vector to $\partial\Omega$. This problem is closely related to the electrical impedance tomography, and is widely discussed in mathematical, numerical and practical points of view (e. g. [1, 7, 8] and references cited therein).

The first author proposed a mathematical formula for finding the convex hull of D for the cases D consists of polygonal-shaped cavities [2], that is, $\gamma_1 = 0$ in D, and polygonal-shaped inclusions [3], that is, $\gamma_1 > 0$ in D. In [5], we implemented the formula numerically, and tested for the case that D is a cavity. However, it is an extreme case, and it is interesting to consider whether the method works well for more moderate case, that is, D is an inclusion.

In this paper, we proceed the numerical testings of our method for the case that D is an inclusion. Our main purpose and interest are to test the ability of the numerical method in [5] for cases that the jump of γ is small, and there are some inclusions that have different γ each other.

The contents of this paper are as follows. In section 2, we summarizes the enclosure method to find the convex hull of inclusions discussed in [3], and also summarize in section 3 a numerical implementation of the enclosure method proposed in [5]. Section 4 gives numerical experiments for some cases to examine the ability of our numerical method for finding the inclusions. In section 5, we describe some concluding remarks.

2 The Enclosure Method

In this section, we summarize the enclosure method [3] proposed by the first named author. Let $S^1 = \{\omega = (\omega_1, \omega_2), |\omega| = 1\}$ and define

$$h_D(\omega) = \sup_{\boldsymbol{x} \in D} \boldsymbol{x} \cdot \omega, \qquad \omega \in S^1. \tag{2}$$

The function h_D is called the support function of D. If we know the value of h_D on S^1, then the convex hull of D can be obtained by

$$\bigcap_{\omega \in S^1} \{\boldsymbol{x} \in \mathbf{R}^2 \,|\, \boldsymbol{x} \cdot \omega < h_D(\omega)\}.$$

Therefore, we call the method to determine the support function for finding the convex hull of D as the *enclosure method*.

We need three more definitions for the formulation of the reconstruction formula.

Definition (regular direction): *The direction $\omega \in S^1$ is regular with respect to D if the set $\{\boldsymbol{x} \in \mathbf{R}^2 \,|\, \boldsymbol{x} \cdot \omega = h_D(\omega)\} \cap \partial D$ consists of only one point.*

Definition (exponentially growing harmonic functions): *Let $\omega \in S^1$ and take $\omega^\perp \in S^1$ perpendicular to ω satisfying* $\det \begin{pmatrix} \omega^\perp \\ \omega \end{pmatrix} > 0$. *For $\tau > 0$, set*

$$v(\boldsymbol{x}; \tau, \omega) = \exp(\tau \boldsymbol{x} \cdot (\omega + i\omega^\perp)), \qquad \boldsymbol{x} \in \mathbf{R}^2,$$

where $i = \sqrt{-1}$.

Definition (Indicator function $I_\omega(\tau, t)$): *Let (f, g) be Cauchy data on $\partial\Omega$ of u that is a non-constant solution of (1). Define*

$$I_\omega(\tau, t) = e^{-\tau t} \int_{\partial\Omega} \left(gv - \frac{\partial v}{\partial \boldsymbol{\nu}} f \right) d\sigma, \quad \tau > 0, \quad t \in \mathbf{R}. \tag{3}$$

The reconstruction formula for the support function h_D is following [3]:

Theorem 1. *Assume that D is a polygon satisfying $\mathrm{diam}\, D < \mathrm{dis}(D, \partial\Omega)$, and that u is a non-constant solution of (1). Let ω be regular with respect to D. Then the formula*

$$\lim_{\tau \to \infty} \frac{\log |I_\omega(\tau, t)|}{\tau} = h_D(\omega) - t, \tag{4}$$

is valid.

Note that we need no information on the parameter γ_1 for the reconstruction formula. The same result has been obtained for the case where D is a cavity ($\gamma_1 = 0$), and a numerical method for this case is discussed in [5]. We also note that the formula (4) is valid for the case where $D \equiv D_1 \cup D_2 \cup \cdots \cup D_m, 1 \le m < \infty$, each D_j is a polygon, and $\overline{D_j} \cap \overline{D_k} = \emptyset$ if $j \ne k$ (see [4] for this arguments).

3 Numerical Implementation

3.1 Estimation of the support function

Now, let us consider a numerical implementation of the reconstruction formula (4). The implementation method is also the same as that of [5]. However, for readers' convenience, we explain the details of our method again.

The main difficulty in the numerical implementation is how to estimate asymptotic behavior of $\log|I_{\boldsymbol{\omega}}(\tau, t)|$ for large τ. In practical cases, observation points for Cauchy data are given only on discrete points because of some physical restrictions. Since $v(\boldsymbol{x}; \tau, \boldsymbol{\omega})$ oscillates violently for large τ and has exponentially growing property, it is very difficult to estimate $I_{\boldsymbol{\omega}}(\tau, t)$ with sufficient accuracy for large τ, and is also difficult to estimate $h_D(\boldsymbol{\omega})$ using the direct implementation of the formula (4).

The following lemma is a key point in the proof the reconstruction formula (4), and also in our numerical implementation:

Lemma 1. *Under the same assumption as Theorem 1, there exist* $\lambda = \lambda(\boldsymbol{\omega}, D) > 0$ *and* $A = A(\boldsymbol{\omega}, D) > 0$ *such that*

$$\lim_{\tau \to \infty} \tau^{\lambda} |I_{\boldsymbol{\omega}}(\tau, h_D(\boldsymbol{\omega}))| = A.$$

Using this lemma and the trivial identity

$$I_{\boldsymbol{\omega}}(\tau, t) = e^{\tau(h_D(\boldsymbol{\omega}) - t)} I_{\boldsymbol{\omega}}(\tau, h_D(\boldsymbol{\omega})),$$

we obtain the following estimate for $\log|I_{\boldsymbol{\omega}}(\tau, 0)|$:

$$\log|I_{\boldsymbol{\omega}}(\tau, 0)| \sim h_D(\boldsymbol{\omega})\tau - \lambda \log \tau + \log A \quad \text{for large } \tau. \tag{5}$$

Using this estimation, we choose an interval $[\tau_1, \tau_2] \subset \mathbf{R}^+$ in which $I_{\boldsymbol{\omega}}(\tau, 0)$ can be estimated with sufficient accuracy, and estimate $h_D(\boldsymbol{\omega})$ by the slope of $\log|I_{\boldsymbol{\omega}}(\tau. 0)|$ in $[\tau_1, \tau_2]$.

3.2 Rejection rule for non-regular directions

The another difficulty is how to distinguish regular directions $\boldsymbol{\omega}$. The reconstruction formula (4) is valid for regular directions with respect to D. However, it is impossible to give *a priori* information on regular directions.

Let $\boldsymbol{\omega}(\theta) = (\cos\theta, \sin\theta)$, and assume that $\boldsymbol{\omega}(\theta)$ is regular with respect to D in the interval $[\theta_1, \theta_2]$. Since D is a polygon, there exists only one point $\boldsymbol{x}_0 \in \partial D$ such that

$$\bigcap_{\theta \in [\theta_1, \theta_2]} \{\boldsymbol{x} \mid \boldsymbol{x} \cdot \boldsymbol{\omega}(\theta) = h_D(\boldsymbol{\omega}(\theta))\} = \{\boldsymbol{x}_0\}.$$

Let $\boldsymbol{x}_0 = (\rho_0 \cos\phi_0, \rho_0 \sin\phi_0)$, then $h_D(\boldsymbol{\omega}(\theta))$ is expressed by

$$h_D(\boldsymbol{\omega}(\theta)) - \rho_0 \cos(\theta - \phi_0), \quad \theta \subset [\theta_1, \theta_2].$$

Thus,

$$\frac{dh_D(\boldsymbol{\omega}(\theta))}{d\theta} = -\rho_0 \sin(\theta - \phi_0), \quad \theta \in (\theta_1, \theta_2) \tag{6}$$

is obtained. Equation (6) shows that $\left|\dfrac{dh_D(\boldsymbol{\omega}(\theta))}{d\theta}\right|$ do not exceed ρ_0. In some practical cases, we may assume that there exists a known domain $\widetilde{D} \subset \Omega$ that include the domain D. Then, the following rejection rule for non-regular ω is applicable:

Rejection rule

Let $R \equiv \max\limits_{\boldsymbol{x} \in \widetilde{D}} |\boldsymbol{x}|$, and $\hat{h}_D(\boldsymbol{\omega}(\theta))$ be the estimated value of $h_D(\boldsymbol{\omega}(\theta))$. If $\left|\dfrac{d\hat{h}_D(\boldsymbol{\omega}(\theta))}{d\theta}\right| > R$ at $\theta = \theta_0$, then reject $\boldsymbol{\omega}(\theta_0)$.

It seems very rare case that one comes across a non-regular direction. However, in the numerical computation, it may be considered that the estimation of $h_D(\boldsymbol{\omega})$ becomes worse when ω is near to some non-regular directions because of some numerical errors. The effectiveness of this rejection rule is shown in the next section.

4   Numerical Results

In this section, we show some numerical experiments for our method. We set the domain $\Omega = \{(x, y) \mid x^2 + y^2 < 1\}$ throughout this section.

First, we consider the case that the inclusion D consists of one polygon as shown in Figure 1 (a). For parameters γ_0 and γ_1, the following three cases are considered:

Case 1-A: $\gamma_0 < \gamma_1$ $(\gamma_0 = 1,\ \gamma_1 = 2.0)$

Case 1-B: $\gamma_0 > \gamma_1$ $(\gamma_0 = 1,\ \gamma_1 = 0.5)$

Case 1-C: $\gamma_0 \sim \gamma_1$ $(\gamma_0 = 1,\ \gamma_1 = 1.001)$

The case 1-C is given to test the behavior of our method for the case where the jump of γ is small.

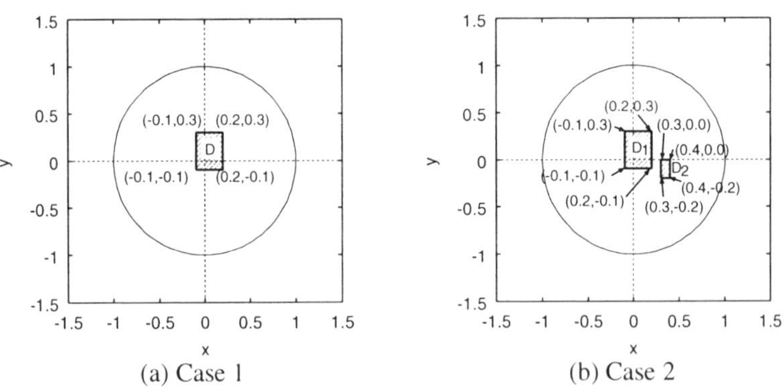

(a) Case 1   (b) Case 2

Figure 1: The domain Ω and inclusions D.

To give Cauchy data for our inverse problem, the following Dirichlet problem is considered:

$$\begin{cases} -\nabla \cdot \gamma \nabla u(x, y) = 0, & (x, y) \in \Omega, \\ u(x, y) = x, & (x, y) \in \partial\Omega. \end{cases} \tag{7}$$

We solve this *direct* problem by the finite element method, and give Cauchy data at 32 points uniformly on $\partial\Omega$. For the computation of $I_\omega(\tau, 0)$, we apply the trapezoidal rules.

Figure 2 shows the distribution of numerical values of $\log|I_\omega(\tau, 0)|$ for $\omega = \omega(\theta) = (\cos\theta, \sin\theta)$ with $\theta = 0, \pi/4, 2\pi/4, \cdots, 7\pi/4$ in case 1-A. Distributions of $|\log I_\omega(\tau, 0)|$ have corners at $\tau \simeq 8$. Since the slope value is nearly 1 for $\tau > 8$, numerical errors are dominant for $\tau > 8$, and a reliable interval of τ for the computation of $I_\omega(\tau, 0)$ is $[2, 6]$. Hence, we compute $I_\omega(\tau, 0)$ on $[2, 6]$ with step 0.25, and estimate $h_D(\omega)$ by the slope of $\log|I_\omega(\tau, 0)|$ using the least squares linear approximation at $\omega = \omega(\theta_k) = (\cos\theta_k, \sin\theta_k)$ with $\theta_k = k\pi/16$, $k = 0, 1, 2, \cdots, 31$. As *a priori* information for our rejection rule, we assume that unknown inclusions lie in the disc $\widetilde{D}_R = \{(x, y) \mid x^2 + y^2 < R\}$, and set $R = 0.8$. For the estimation of $\dfrac{\partial \hat{h}_D(\omega(\theta))}{\partial\theta}$, the forward finite differences are applied.

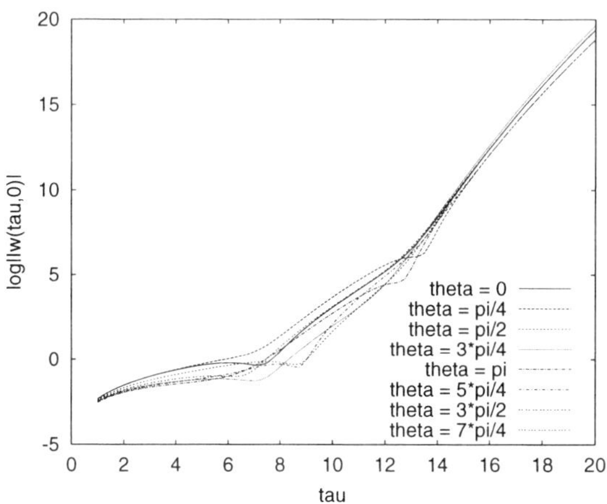

Figure 2: Distribution of numerical values of $\log|I_\omega(\tau, 0)|$.

Figure 3 shows lines $\{(x, y) \mid (x, y) \cdot \omega(\theta_k) = \hat{h}_D(\omega(\theta_k))\}$, $k = 0, 1, 2, \cdots, 31$ and estimated convex hull of unknown inclusions, where $\hat{h}_D(\omega)$ denotes the estimated value of $h_D(\omega)$. The results show that we can obtain good estimation for cases 1-A and 1-B. But for case 1-C, the estimated convex hull is rather small because the jump of γ is small. Note that the estimation result for case 1-C becomes rather worse if our rejection rule is not applied.

Next, we consider the case where the inclusion D consists of two polygons D_1 and D_2 as shown in Figure 1 (b), and γ is described by

$$\gamma = \gamma_0 + (\gamma_1 - \gamma_0)\chi_{D_1} + (\gamma_2 - \gamma_0)\chi_{D_2}. \tag{8}$$

For parameters in (8), the following three cases are considered:

Case 2-A: $\gamma_0 \neq \gamma_1 = \gamma_2$. ($\gamma_0 = 1.0$, $\gamma_1 = \gamma_2 = 2.0$)

Case 2-B: $\gamma_1 < \gamma_0 < \gamma_2$. ($\gamma_0 = 1.0$, $\gamma_1 = 0.5$, $\gamma_2 = 2.0$)

Case 2-C: $\gamma_0 < \gamma_2 \ll \gamma_1$ ($\gamma_0 = 1.0$, $\gamma_1 = 10.0$, $\gamma_2 = 1.1$)

To give Cauchy data for our inverse problem, we also consider the Dirichlet problem (7), and solve the problem by the finite element method. The Cauchy data is also given at 32 points

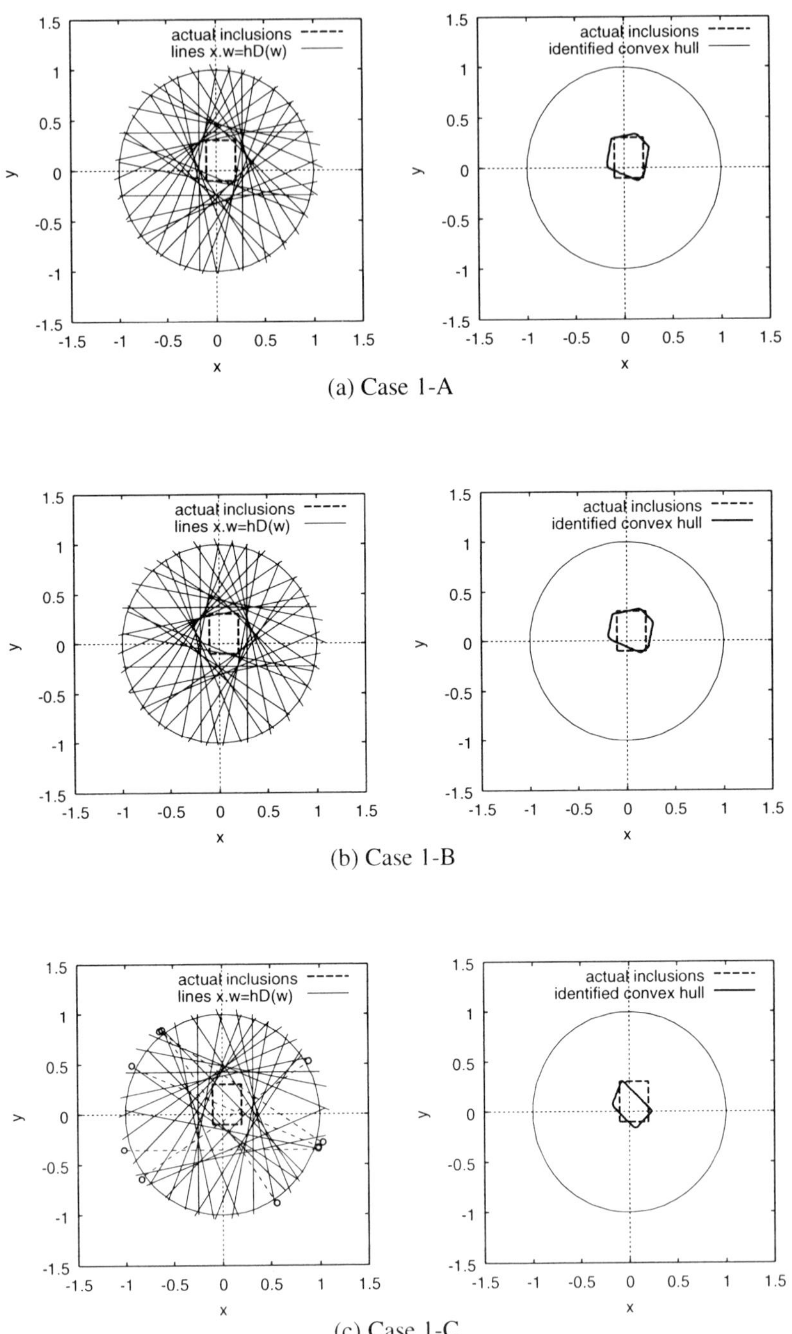

(a) Case 1-A

(b) Case 1-B

(c) Case 1-C

Figure 3: Results of lines $\{(x, y) \mid (x, y) \cdot \omega(\theta_k) = \hat{h}_D(\omega(\theta_k))\}$ (left hand side) and estimated convex hull of inclusions (right hand side) for single inclusion. Broken lines marked by circle are rejected by our rejection rule.

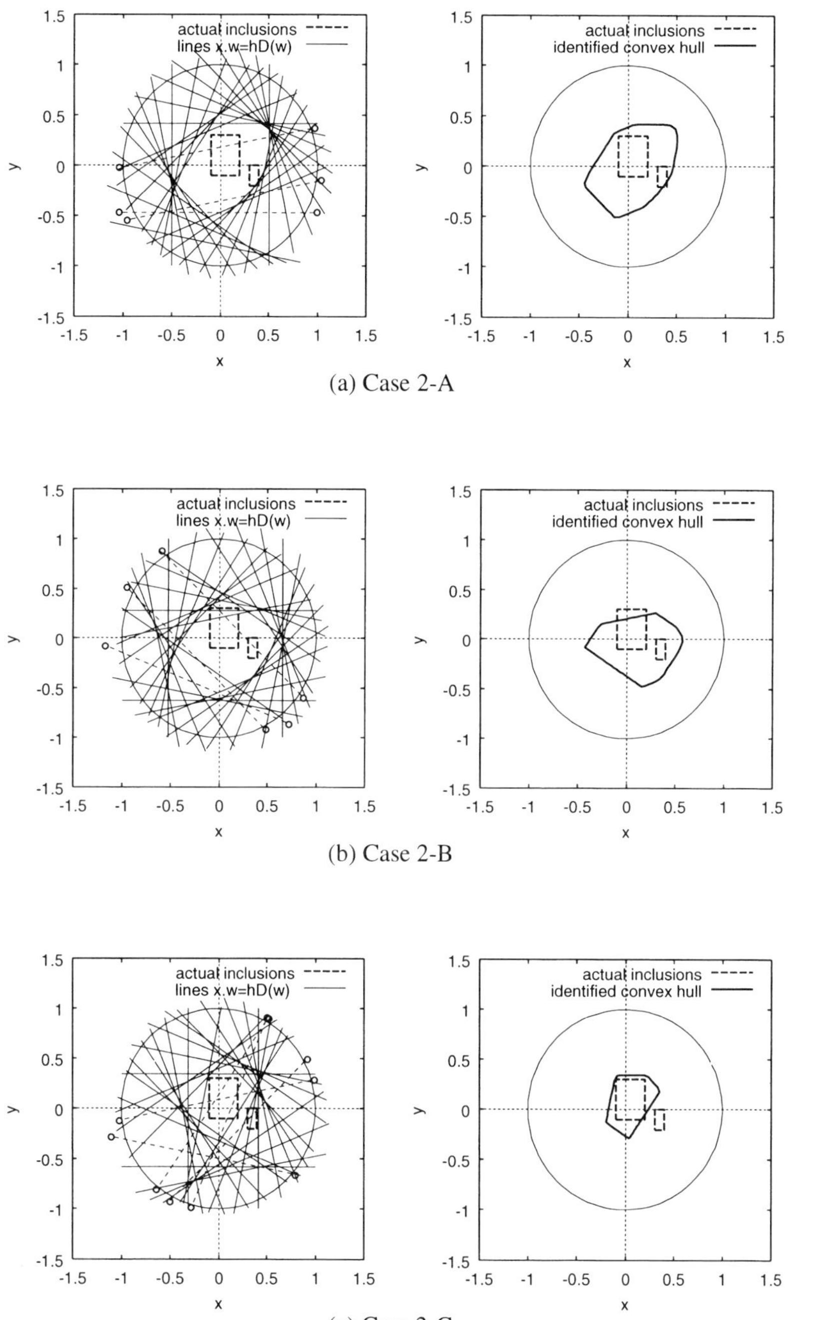

Figure 4: Results of lines $\{(x,y) \mid (x,y) \cdot \omega(\theta_k) = \hat{h}_D(\omega(\theta_k))\}$ (left hand side) and estimated convex hull of inclusions (right hand side) for two inclusions. Broken lines marked by circle are rejected by our rejection rule.

uniformly on $\partial\Omega$, and the trapezoidal rules are applied for the computation of $I_{\boldsymbol{\omega}}(\tau, 0)$. Before estimation of $h_D(\boldsymbol{\omega})$, we check the distribution of $\log|I_{\boldsymbol{\omega}}(\tau, 0)|$, and decide to compute $I_{\boldsymbol{\omega}}(\tau, 0)$ on $[3, 9]$ with step 0.25. The *a priori* parameter R in $\widetilde{D}_R$ for our rejection rule is chosen as 0.8.

Figure 4 shows lines $\{(x, y) \mid (x, y) \cdot \boldsymbol{\omega}(\theta_k) = \hat{h}_D(\boldsymbol{\omega}(\theta_k))\}$ and estimated convex hull of inclusions. The results for case 2-A and case 2-B show that we can obtain good estimations for the convex hull of two inclusions. However, in case 2-C, we can not find the smaller domain D_2 because the effects of the domain D_1 may be too large.

5   Conclusions

In this paper, we consider a numerical method for finding the convex hull of polygonal inclusions based on the enclosure method. In the numerical implementation, we have two difficulties. One is caused by characteristics of the function $v(\boldsymbol{x}; \tau, \boldsymbol{\omega})$, and the other one is because it is impossible to give *a priori* information on regular directions. To avoid these difficulties, we propose a method for the estimation of the support function $h_D(\boldsymbol{\omega})$ by the slope of $\log|I_{\boldsymbol{\omega}}(\tau, 0)|$, and a rejection rule for non-regular directions. Our numerical method is examined by some numerical examples. The numerical results show that our method is effective for finding the convex hull of polygonal inclusions.

Acknowledgement

This research was partially supported by Grant-in-Aid for Scientific Research (C) (2) (No. 13640152) of Japan Society for the Promotion of Science.

References

[1] M. Brühl and M. Hanke, Numerical implementation of two non-iterative methods for locating inclusions by impedance tomography, Inverse Problems 16 (2000) 1029–1042.

[2] M. Ikehata, Enclosing a polygonal cavity in a two-dimensional bounded domain from Cauchy data, Inverse Problems 15 (1999) 1231–1241.

[3] M. Ikehata, On reconstruction in the inverse conductivity problem with one measurement, Inverse Problems 16 (2000) 785–793.

[4] M. Ikehata, On reconstruction from a partial knowledge of the Neumann-to-Dirichlet operator, Inverse Problems 17 (2001) 45–51.

[5] M. Ikehata and T. Ohe, A numerical method for finding the convex hull of polygonal cavities using the enclosure method, in press in Inverse Problems.

[6] M. Ikehata and S. Siltanen, Numerical method for finding convex hull of an inclusion in conductivity from boundary measurements, Inverse Problems 16 (2000) 1043–1052.

[7] S. Siltanen, J. Mueller and D. Issacson, An implementation of the reconstruction algorithm of A. Nachman for the 2-D inverse conductivity problem, Inverse Problems 16 (2000) 681–699.

[8] Stavroulakis G E, Inverse and crack identification problems in engineering mechanics. Kluwer, Dordrecht, 2001.

Electromagnetic Nondestructive Evaluation (VI)
F. Kojima et al. (Eds.)
IOS Press, 2002

A Direct Reconstruction of Magnetic Charge for Non-Destructive Testing

Hajime Igarashi and Toshihisa Honma
Graduate School of Engineering, Hokkaido University
Kita 13, Nishi 8, Kita-ku, Sapporo 060, Japan

Abstract. The magnetic charge, which is due to martensite transformation around a crack in ferromagnetic materials, is reconstructed from the magnetic fields measured on a plane above the charge region. It is shown that both Tikhonov method with an appropriate regularization term and iterative regularization based on the CG method solving a modified normal equation yield satisfactory solutions. The regularization parameter for the Tikhonov method can be determined from the L-curve when the noise level is under 10% of measurement data. The residual tolerance of CG iteration, which gives great influence on the solution, can roughly be estimated from the discrepancy principle.

1 Introduction

Maintenance of aged plants requires non-destructive evaluation (NDE) for predicting initiation of cracks. The NDE based on measurement of passive leakage magnetic fields is one of the most promising methods so far as ferromagnetic materials are concerned. This method identifies the magnetization, due to martensite transformation around cracks, from the magnetic fields measured outside the material. Since this identification is a severely ill-posed problem, the system equation must be regularized by an appropriate method. Hitherto this ill-posed problem has been analyzed with the iterative methods such as an optimization method with the wavelet technique [1] and neural network method [8].

In the meantime, the authors have shown that the direct method based on the Tikhonov regularization and L-curve technique can provide successful identification of magnetization distribution in the permanent magnets used in copy machines and motors [4, 5, 6]. Moreover, this direct method has been applied to the reconstruction of magnetic charge for NDE, to show that the Tikhonov method with the appropriate regularization term gives accurate identification of magnetic charges, providing that there is no noise in measurement [7].

In this paper we report the feasibility of the direct method for the noisy measured magnetic fields. Moreover, an new iterative method based on the conjugate gradient (CG) method with a variable transformation is also applied to the problem to test its performance.

2 Formulation

Figure 1 shows two planar regions, that is, the charge region Ω_c which contains magnetic charge, and the measurement region Ω_m on which the magnetic fields generated by magnetic charges on Ω_c are measured.

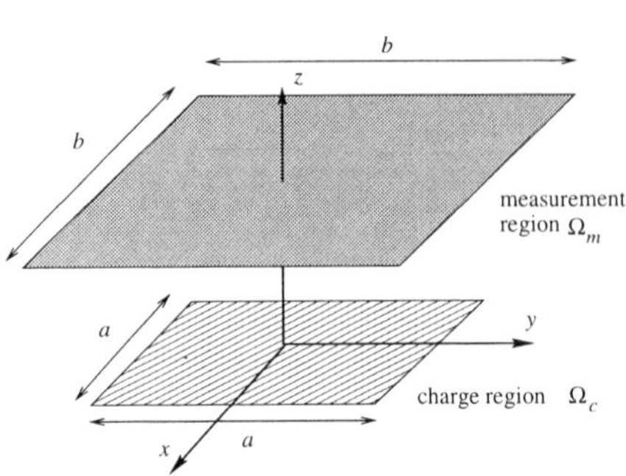

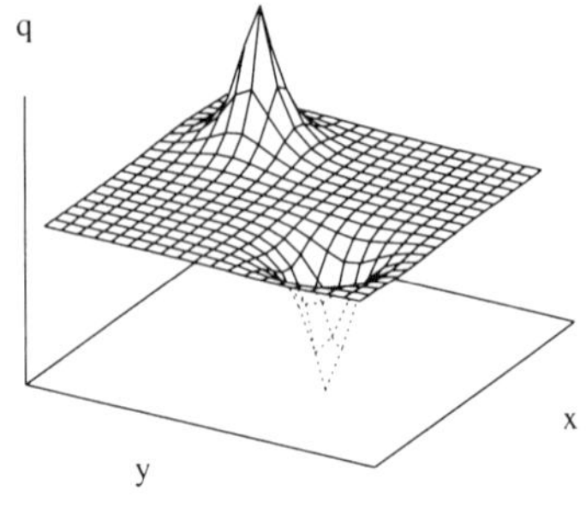

Figure 1: Charge and measurement regions	Figure 2: Original distribution

The magnetic field $\boldsymbol{H}$ on Ω_m is given by

$$\boldsymbol{H}(\boldsymbol{x}) = \frac{1}{4\pi}\int_{\Omega_c} q(\boldsymbol{x}')\frac{\boldsymbol{x}-\boldsymbol{x}'}{|\boldsymbol{x}-\boldsymbol{x}'|^3}\, dS', \tag{1}$$

where q denotes the surface magnetic charge density, which is defined by the divergence of magnetization. The surface magnetic charges along the fringe of Ω_c are neglected in (1) assuming that q concentrates near the cracks. In this work, only the z-components H of the magnetic field are used for reconstruction because the use of other components is expected to provide similar results [1].

The charge region Ω_c is discretized into small triangular elements Δ_k in which q is linearly interpolated with the shape function N_i, that is, $q = \sum_i N_i q_i$.

The Tikhonov regularization method corresponding to (1) is given by

$$\|[K]\{q\} - \{H\}\|^2 + \lambda\{q\}^t\{q\} \to \min., \tag{2}$$

where $\|\cdot\|$, $\{\cdot\}$ and $[\cdot]$ represent the 2-norm, column matrix and matrix, respectively, λ denotes the regularization parameter, and $[K]$ is the discrete counterpart of the integral operator in (1) whose entities are given by

$$K_{ij} = \frac{1}{4\pi}\sum_k \int_{\Delta_k} N_j\frac{\boldsymbol{x}_i-\boldsymbol{x}'}{|\boldsymbol{x}_i-\boldsymbol{x}'|^3}\, dS'. \tag{3}$$

It has been shown that (2) leads to solutions polluted with numerical noise even if there is no measurement noise [7]. This is due to the fact that the singular vectors of $[K]$ have undesirable ripples with short wavelength. Note here that the solution to (2) is expressed in the form [2]:

$$\{q\} = \sum_i f_i\frac{\{u_i\}^t\{H\}}{\sigma_i}\{v_i\} \tag{4}$$

where $\{u_i\}$ and $\{v_i\}$ are the left and right singular vectors of $[K]$, respectively, σ_i the ith singular value, and f_i the filter function defied by $f_i = \sigma_i^2/(\sigma_i^2 + \lambda^2)$.

It is known that the iterative solution to the normal equation

$$[K]^t[K]\{q\} = [K]^t\{H\}, \tag{5}$$

can also be expanded in the form of (4). The solution to (5) is, therefore, polluted by numerical noise.

Instead of (2), we consider

$$\|[K]\{q\} - \{H\}\|^2 + \lambda\{q\}^t[M]\{q\} \to \min., \tag{6}$$

where $[M] = \int_{\Omega_c}\{N\}^t\{N\}\,dS$. Note that the regularization term $\{q\}^t[M]\{q\}$ corresponds to the discrete counterpart of $\int_{\Omega_c} q^2\,dS$. The problem (6) is equivalent to

$$([K]^t[K] + \lambda[M])\{q\} = [K]^t\{H\}, \tag{7}$$

which is solved to obtain $\{q\}$.

The solution to (7) can be expressed by the linear combination of the generalized singular vectors $\{x_i\}$ of the pair $([K],[L])$, that is [2],

$$\{q\} = \sum_i g_i \frac{\{u_i\}^t\{H\}}{\sigma_i}\{x_i\} \tag{8}$$

where $[L]$ denotes the upper triangular matrix obtained by the Cholesky factorization of $[M]$ which satisfies $[L]^t[L] = [M]$, g_i the filter function defined by $g_i = \gamma_i^2/(\gamma_i^2 + \lambda^2)$, γ_i the generalized singular value and $\sigma_i = \gamma_i/\sqrt{1+\gamma_i^2}$. The generalized singular vectors $\{x_i\}$ has been shown to be smooth in contrast to the singular vectors of $[K]$ [7]. Hence the solution to (7) is adequately smooth.

The above fact suggests that the iterative solutions to the modified normal equation

$$[L]^{-t}[K]^t[K][L]^{-1}\{\tilde{q}\} = [L]^{-t}[K]^t\{H\}, \tag{9}$$

where $\{\tilde{q}\} = [L]\{q\}$, are also smooth because they would be expressed in the form of (8). In this paper (9) is iteratively solved by the CG method with a controlled stopping criterion for regularization. In the next section, the effects of noise in measurement on the solutions to (7) and (9) will be discussed.

3   Numerical Results

The magnetic charge q is reconstructed assuming the original distribution shown in Fig.2. The white noise, whose level n is defined by $n = \|n_a\|/\|\{H\}\|$, where n_a denotes the amplitude of the noise, is superposed to $\{H\}$. The sizes of Ω_c and Ω_m are set as $a = 20$ and $b = 30$. The domain Ω_c is uniformly discretized into triangular elements with 400 unknowns at the nodal points. The lift-off is 2.0, and the noise level n is set to 0.05 unless notified.

The solutions to (7) for different values of λ are shown in Figs.3-5. The solution seems adequate when $\lambda \simeq 0.1$. Figure 6 shows the L-curves [3] for different noise levels. The corner of the L-curve for $n = 0.05$ was given when $\lambda = 0.01$ while the error becomes minimum for $\lambda = 0.1$, as shown in Fig.7. Hence the L-curve method yields a bit too small estimate of the optimal value of λ. Moreover there would exist no clear corner of the L-curves when n is over 0.1. Thus the L-curve method cannot be used for such highly noisy cases.

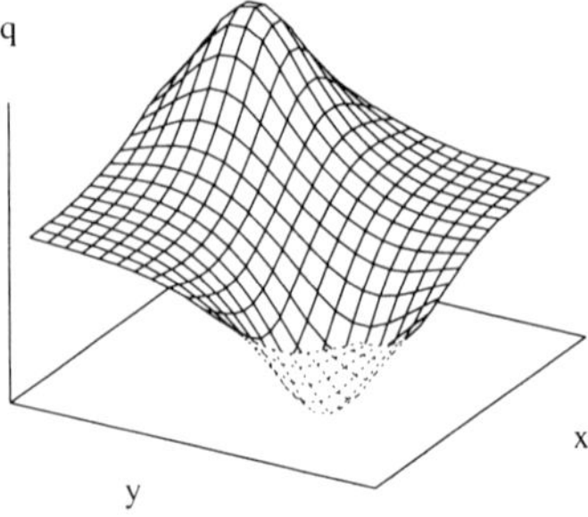

Figure 3: Solution to (7), $\lambda = 100$

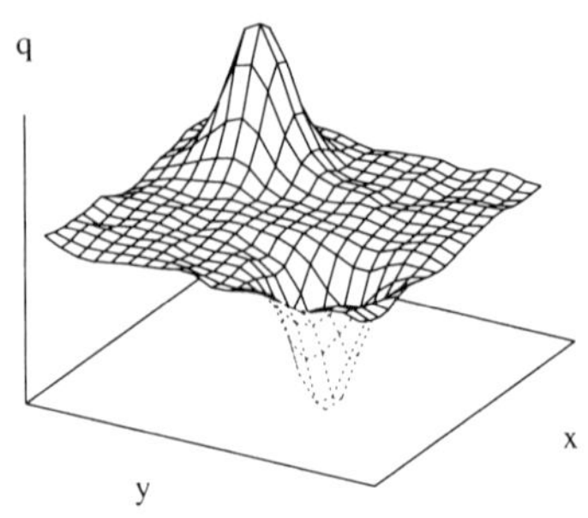

Figure 4: Solution to (7), $\lambda = 0.1$

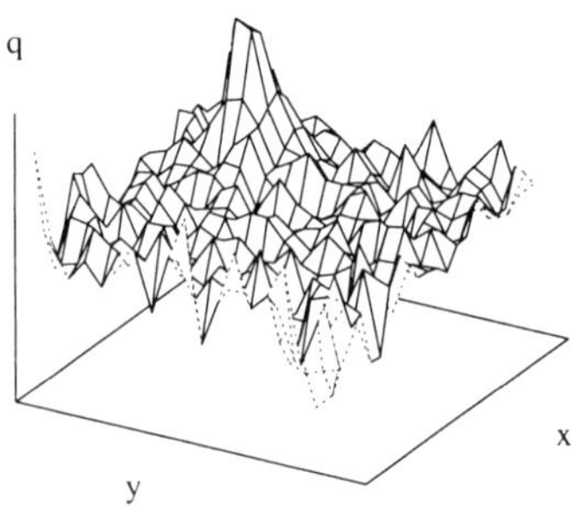

Figure 5: Solution to (7), $\lambda = 0.001$

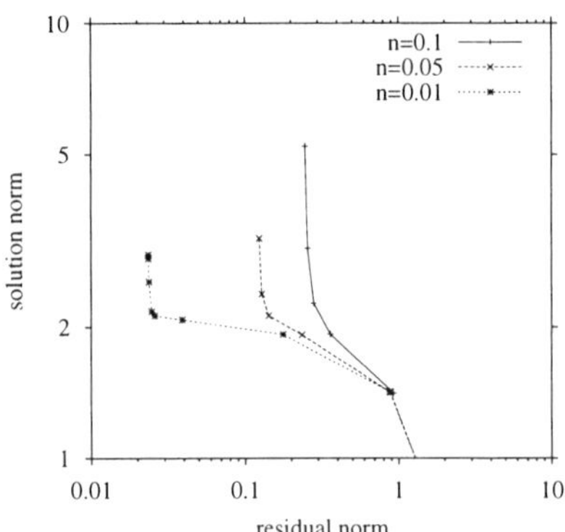

Figure 6: L-curves for different noise levels n

Figures 8-10 show the solutions to (9) obtained by the CG method, whose iteration processes are stopped when the normalized residue of a linear system $\|\delta\|/\|\{H\}\|$ is smaller than the tolerance ϵ. The solutions in Figs. 8 and 10 seem over- and underdamping, respectively, as are in Figs. 3 and 5, whereas that in Fig.9 seems adequate. The CG solution to (5), in which numerical noise is observed, is shown in Fig.11 for reference.

The dependence of the solution errors on ϵ for different noise levels is shown in Fig.12. The discrepancy principle states that the appropriate value of ϵ equals to the noise level [2]. For instance, the adequate value of ϵ is 0.05 for $n = 0.05$. The error, however, becomes minimum when $\epsilon = 0.01$ when $n = 0.05$. The discrepancy principle, therefore, yields a rough estimate of the optimal value of ϵ.

We conclude that both Tikhonov method and iterative regularization method based on the CG method in conjunction with the L-curve method and the discrepancy principle, respectively, can be promising methods for the NDE.

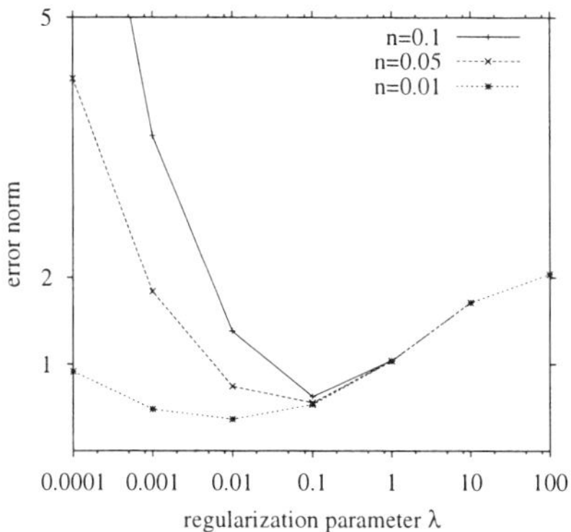

Figure 7: Dependence of errors on regularization parameter λ

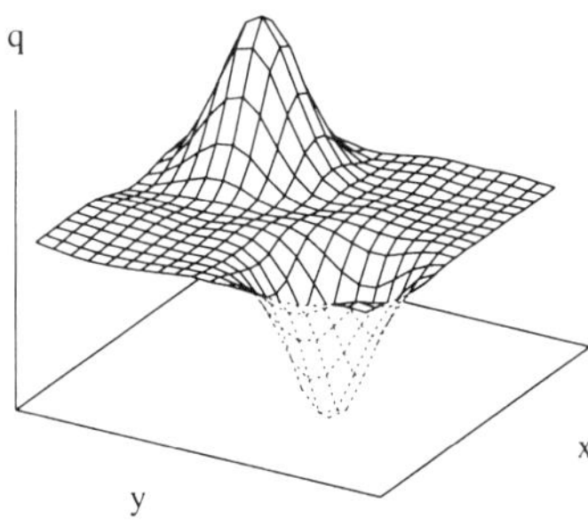

Figure 8: Solution to (9), $\epsilon = 0.1$

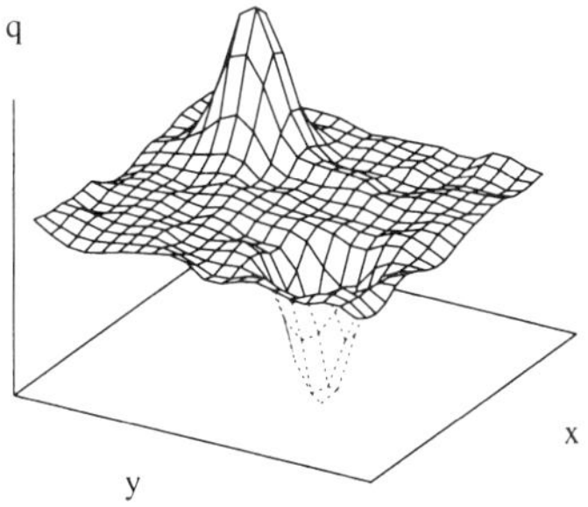

Figure 9: Solution to (9), $\epsilon = 0.01$

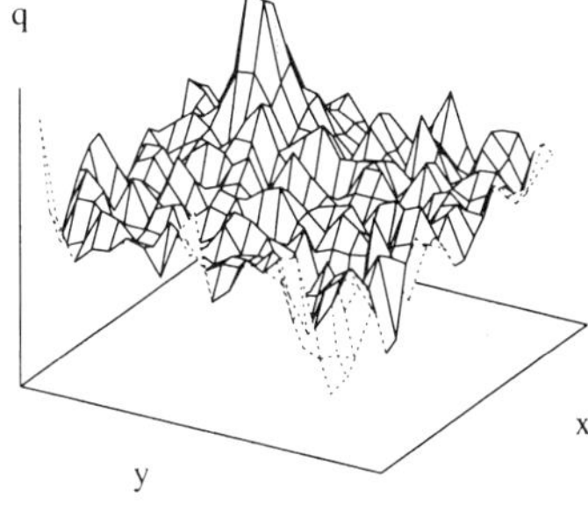

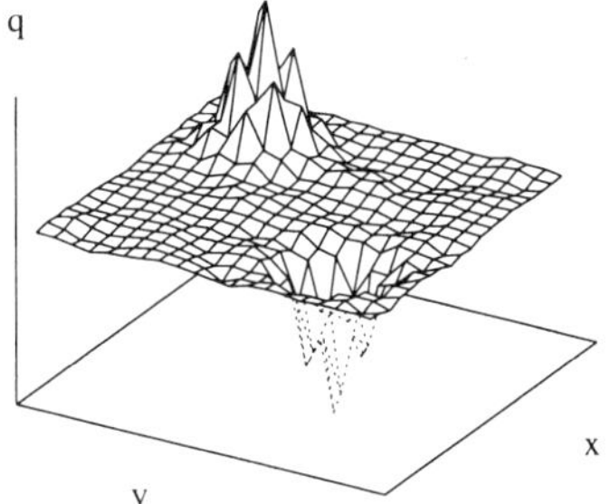

Figure 10: Solution to (9), $\epsilon = 0.001$

Figure 11: Solution to (5), $\epsilon = 0.01$

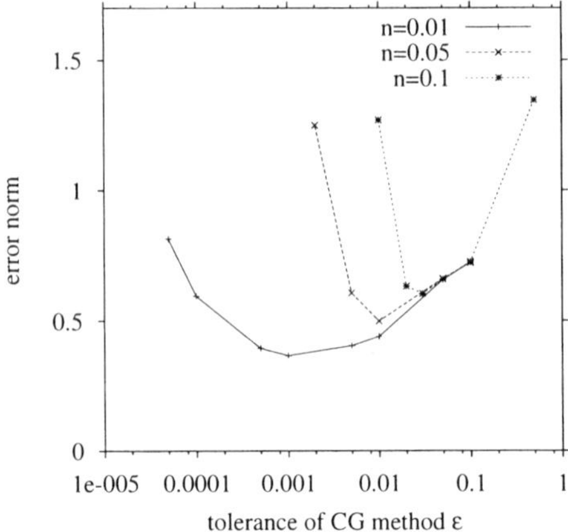

Figure 12: Dependence of errors on CG tolerance ϵ

4　Conclusion

In this paper, the Tikhonov method as well as the new regularization method based on the CG method in which the modified normal equation is solved have been applied to the NDE problem with noisy measurement data. The L-curve method and discrepancy principle yield roughly optimal estimate of the regularization parameters. Both regularization methods are thought to be promising for real NDE.

References

[1] Z. Chen, K.Aoto and S. Kato, Reconstruction of magnetic charges using an optimization method and wavelet *J. JSAEM* **8**(3) (2000) 363-371.

[2] P.C.Hansen, *Rank-Deficient and Discrete Ill-posed Problems*, SIAM, 1998.

[3] P.C.Hansen and D.P.O'leary, The use of the L-Curve in the Regularization of discretization of discrete ill-posed problems, *SIAM J. Sci. Comput.*, **14** (1993) 1487-1503.

[4] H.Igarashi, Forward and inverse analyses of cylindrical permanent magnets, *J. IEE Japan*, **vol.120-A** (2000) 791-797 (in Japanese).

[5] H.Igarashi, T.Honma and A.Kost, Inverse inference of magnetization distribution in cylindrical permanent magnets, *IEEE Trans. Magn.*, **36** (2000) 1168-1171.

[6] H. Igarashi, On the reconstruction of magnetic source in cylindrical permanent magnets, Inverse Problems in Engineering Mechanics II (eds. M. Tanaka and G.S. Dulikravich), Elsevier, Amsterdam (2000) 467-476.

[7] H. Igarashi and T.Honma, Reconstruction of Planar Magnetic Charge Distribution, Proc. ISEM-Tokyo, 621-622, extended version will appear in *Int. J. Applied Electromagnetics and Mechanics*.

[8] G. Preda, S. Takaya, K. Demachi and K. Miya, Reconstruction of magnetic moments distribution from 2D scan data using neural networks Proc. Symposium on Electromagnetics and Dynamics (1999) 618-621.

Electromagnetic Nondestructive Evaluation (VI)
F. Kojima et al. (Eds.)
IOS Press, 2002

STABILITY AND UNIQUENESS IN DETERMINING A CURRENT SOURCE IN QUASISTATIC MAXWELL EQUATIONS

[1] MASAHIRO YAMAMOTO AND [2] JUN ZOU

[1] *Department of Mathematical Sciences*
The University of Tokyo
3-8-1 Komaba Meguro Tokyo 153 Japan
tel: +81-3-5465-8328
fax: +81-3-5465-7017
e-mail : myama@ms.u-tokyo.ac.jp
[2] *Department of Mathematics*
The Chinese University of Hong Kong
Shatin, Hong Kong, China
fax: +852-2603-5154
e-mail: zou@math.cuhk.edu.hk

ABSTRACT. We consider an inverse problem of determining the strength depending only on one direction of a current source in the quasistatic Maxwell equations and prove the uniqueness and stability for this inverse problem. We also stress the importance of such a mathematical analysis in the reconstruction problem.

§1. Introduction.

Let $\Omega \subset \mathbb{R}^3$ be a bounded domain whose boundary $\partial\Omega$ is of class C^1. In Ω, we consider the quasistatic Maxwell equations:

$$-\mu\frac{\partial H}{\partial t} = \operatorname{rot} E, \quad x = (x_1, x_2, x_3) \in \Omega, \, t > 0. \tag{1.1}$$

Key words and phrases. quasistatic Maxwell equations, current source, inverse problem, uniqueness, stability.

$$\mathrm{rot}\, H = J(x,t) + \sigma E(x,t), \qquad x \in \Omega,\, t > 0. \tag{1.2}$$

$$\mathrm{div}\, H(x,t) = 0, \qquad x \in \Omega,\, t > 0. \tag{1.3}$$

Here $E = (E_1, E_2, E_3)$, $H = (H_1, H_2, H_3)$ are the electric field and the magnetic field respectively, and we assume that the permeability $\mu > 0$ and the conductivity $\sigma > 0$ are constants. Throughout this paper, we further assume that functions under consideration are sufficiently smooth, in other words, all the functions are sufficiently many times continuously differentiable with respect to x_1, x_2, x_3, t.

Here $J = J(x,t)$ is an impressed current density and is considered as source term. We discuss an inverse problem of determining J from overdetermining measurements of H. In this paper, we mainly consider the following form of J:

$$J(x,t) = f(x_1)\lambda(x_2, x_3, t), \qquad x \in \Omega,\, t > 0, \tag{1.4}$$

where $\lambda = \lambda(x_2, x_3, t) = (\lambda_1(x_2, x_3, t), \lambda_2(x_2, x_3, t), \lambda_3(x_2, x_3, t))$ is a given function and the strength of the current source is assumed to depend only on x_1. The function $f(x_1)$ is an unknown function to be determined. For general accounts of radiation, we refer to Kong [6] for example.

We set

$$\Omega_1 = \{x_1 \in \mathbb{R};\, (x_1, x_2, x_3) \in \Omega \quad \text{for some } x_2, x_3 \in \mathbb{R}\}. \tag{1.5}$$

In other words, Ω_1 is the projection of Ω on the x_1-axis.

We consider

Inverse Problem. Let $\omega \subset \Omega$ be an arbitrarily fixed subdomain and $\theta \in (0, T)$ be arbitrary. Then determine $f(x_1)$, $x_1 \in \Omega_1$ from

$$\begin{cases} H_1(x,t), & x \in \partial\Omega,\, 0 < t < T, \\ H_1(x,\theta), & x \in \Omega, \\ H_1(x,t), & x \in \omega,\, 0 < t < T. \end{cases} \tag{1.6}$$

More precisely, for this inverse problem, we will discuss the theoretical subjects:

Uniqueness. Do the measurements (1.6) determine f uniquely? That is, let $\widetilde{H} = (\widetilde{H}_1, \widetilde{H}_2, \widetilde{H}_3)$ and $\widetilde{E} = (\widetilde{E}_1, \widetilde{E}_2, \widetilde{E}_3)$ satisfy (1.1) - (1.3), where $J(x,t) = f(x_1)\lambda(x_2, x_3, t)$ is replaced by $\widetilde{J}(x,t) = \widetilde{f}(x_1)(x_2, x_3, t)$. Then we are required to prove that

$$\begin{cases} H_1(x,t) = \widetilde{H}_1(x,t), & x \in \partial\Omega, \ 0 < t < T, \\ H_1(x,\theta) = \widetilde{H}_1(x,\theta), & x \in \Omega, \\ H_1(x,t) = \widetilde{H}_1(x,t), & x \in \omega, \ 0 < t < T \end{cases} \tag{1.7}$$

implies $f(x_1) = \widetilde{f}(x_1)$, $x_1 \in \Omega_1$.

Stability. Is the mapping from the measuements (1.6) to f_1 continuous by suitable norms?

Needless to say, if $\lambda \equiv 0$ in (1.4), then we cannot expect the uniqueness, so that some conditions on the given λ are necessary. Moreover, for the stability, we have to specify the norms for estimating $f = f(x_1)$ and measurement data (1.6). Henceforth, for simplicity, we fix the boundary data $H_1(x,t)$, $x \in \partial\Omega$, $0 < t < T$:

$$H_1(x,t) = \widetilde{H}_1(x,t), \qquad x \in \partial\Omega, \ 0 < t < T. \tag{1.8}$$

In this paper, we exclusively discuss the theoretical issue, and postpone execution of numerical analysis. To the authors' knowledge, there are very few theoretical results for inverse problems for the Maxwell equations, so that many trials of numerical computations are not well grounded. In this paper, we intend to provide theoretically least backgrounds for our inverse problem which has been even never done. The necessity of such theoretical backgrounds for guaranteeing a reasonable reconstruction result, is discussed in Section 2. Moreover, after establishing the uniqueness and the stability, we can apply our numerical algorithm (Yamamoto and Zou [11]) based on multigrid techniques for the optimization, and the numerical analysis will be done in a succeeding paper.

Now we state our main result.

Theorem. *Let $0 < \theta < T$. We assume that*

$$\frac{\partial \lambda_3}{\partial x_2}(x_2, x_3, \theta) \neq \frac{\partial \lambda_2}{\partial x_3}(x_2, x_3, \theta) \quad \textit{for all } x_2, x_3. \tag{1.9}$$

Then there exists a constant $C > 0$, depending on $\Omega, T, \omega, \lambda, \theta, \mu$, such that

$$\|f - \widetilde{f}\|_{L^2(\Omega_1)} \leq C \Bigg(\|(H_1 - \widetilde{H_1})(\cdot, \theta)\|_{H^2(\Omega)}$$

$$+ \left\| \frac{\partial}{\partial t}(H_1 - \widetilde{H_1}) \right\|_{L^2(\omega \times (0,T))} + \|H_1 - \widetilde{H_1}\|_{L^2(\omega \times (0,T))} \Bigg).$$

Here we set

$$\|f - \widetilde{f}\|_{L^2(\Omega_1)} = \left(\int_{\Omega_1} |(f - \widetilde{f})(x_1)|^2 dx_1 \right)^{\frac{1}{2}}$$

and

$$\|(H_1 - \widetilde{H_1})(\cdot, \theta)\|_{H^2(\Omega)} = \Bigg\{ \int_{\Omega} \Bigg(|(H_1 - \widetilde{H_1})(x, \theta)|^2$$

$$+ |\nabla(H_1 - \widetilde{H_1})(x, \theta)|^2 + \sum_{i,j=1}^{3} \left| \frac{\partial^2}{\partial x_i \partial x_j}(H_1 - \widetilde{H_1})(x, \theta) \right|^2 \Bigg) dx \Bigg\}^{\frac{1}{2}}$$

and

$$\|H_1 - \widetilde{H_1}\|_{L^2(\omega \times (0,T))} = \left(\int_0^T \int_\omega |(H_1 - \widetilde{H_1})(x, t)|^2 dx dt \right)^{\frac{1}{2}}.$$

The theorem automatically guarantees the uniqueness in our inverse problem. The condition (1.9) is restrictive but we do not know any uniqueness and stability results on the inverse problem without this condition.

As for other inverse problems for the Maxwell equations, we refer to Neittaanmäki, Rudnicki and Savini [7], Romanov and Kabanikhin [8], Yamamoto [9], [10]. Moreover the readers can consult Banks and Kunisch [1], Isakov [5], for example, for more general accounts of inverse problems.

§2. Why should we discuss the uniqueness?

The uniqueness in the inverse problem is a basic theoretical issue, and should be established also for numerical computations. In this section, we will explain a rôle of the uniqueness result for the Tikhonov regularization by the following simple example: For the heat equation:

$$\frac{\partial u}{\partial t}(x_1, t) = \frac{\partial^2 u}{\partial x_1^2}(x_1, t), \qquad 0 < x_1 < 1, \, t > 0 \tag{2.1}$$

and

$$u(0, t) = u(1, t) = 0, \qquad t > 0, \tag{2.2}$$

we consider an inverse problem of determining an initial value from the measurements at $x_1 = \frac{1}{2}$:

$$u\left(\frac{1}{2}, t\right), \, 0 < t < T \longrightarrow u(x_1, 0), \, 0 < x_1 < 1. \tag{2.3}$$

As it is easily seen, our measurement $u\left(\frac{1}{2}, t\right)$, $0 < t < T$, cannot determine an initial value $u(\cdot, 0)$ uniquely. In fact, let $N \in \mathbb{N}$ be fixed. Then two solutions

$$\widetilde{u}(x_1, t) = \sum_{k=1}^{N} e^{-4k^2 \pi^2 t} \sin 2k\pi x_1, \quad u(x_1, t) = 0, \quad 0 < x_1 < 1, \, t > 0,$$
$$\tag{2.4}$$

satisfy (2.1), (2.2) and yield the same observation:

$$\widetilde{u}\left(\frac{1}{2}, t\right) = u\left(\frac{1}{2}, t\right) = 0, \qquad t > 0.$$

Next let us reconstruct the initial value $\sum_{k=1}^{N} \sin 2k\pi x_1$ from observation data with errors. For this, the Tikhonov regularization is widely used, which is formulated as follows: let $u(a) = u(a)(x_1, t)$ be the solution to (2.1) and (2.2) with $u(x_1, 0) = a(x_1)$, $0 < x_1 < 1$. We take

$$H_0^1(0, 1) = \left\{ a; a(0) = a(1) = 0, \quad \int_0^1 \left|\frac{da}{dx_1}(x_1)\right|^2 dx_1 < \infty \right\}$$

as an admissible set of a. Then, as approximate solutions, we look for a minimizer of the functional

$$J(a) = \int_0^T \left| u(a)\left(\frac{1}{2}, t\right) - \eta(t) \right|^2 dt + \gamma \int_0^1 \left|\frac{da}{dx_1}(x_1)\right|^2 dx_1 \tag{2.5}$$

over $a \in H_0^1(0,1)$. Here $\gamma > 0$ is a parameter and $\eta = \eta(t)$ is an observation with noises. This is the Tikhonov regularization. If we can make exact measurement for the reconstruction of $\sum_{k=1}^{N} \sin 2k\pi x_1$, then $\eta(t) = 0$, $0 < t < T$. However, in practise, errors in observations must be taken into consideration and we can expect only that practical observation data are near the data without any errors:

$$\left(\int_0^T |\eta(t)|^2 dt \right)^{\frac{1}{2}} < \delta, \tag{2.6}$$

where $\delta > 0$ is a noise level. Thus, in the Tikhonov regularization, one important task is the choice of a parameter $\gamma > 0$ in (2.5) so that one minimizer $a(\delta, \gamma)$ of J converges to $\sum_{k=1}^{N} \sin 2k\pi x_1$, $0 < x_1 < 1$. By the general theory of the regularization (e.g. Groetsch [3]), however, all that we can expect after the optimal choice of γ depending on $\delta > 0$, is that $a(\delta, \gamma)$ converges to the "minimum-norm" solution a_0 in a suitable norm as $\delta \to 0$, where $a_0 = a_0(x_1)$ minimizes $\int_0^1 |a(x_1)|^2 dx_1$ among a set of a such that $u(a) = u(a)(x_1, t)$ satisfies (2.1), (2.2) and $u(a)\left(\frac{1}{2}, t\right) = 0$, $0 < t < T$. By (2.4), our measurement cannot distinguish $a = 0$ from $\sum_{k=1}^{N} \sin 2k\pi x_1$, and the minimum-norm solution a_0 is the zero function: $a_0(x_1) = 0$, $0 < x_1 < 1$. Thus, even if we want to reconstruct the oscillating initial function $\sum_{k=1}^{N} \sin 2k\pi x_1$, we can never reconstruct such an initial value by our adopted observation $u\left(\frac{1}{2}, t\right)$, $0 < t < T$, but only trivial approximation to $a_0 = 0$ can be obtained. Thus it is necessary to verify that the adopted observation guarantees the uniqueness in the corresponding formulation of the inverse problem, and this is the motivation for the mathematical analysis concerning the uniqueness in the inverse problem.

Finally we add that the stability in the inverse problem is essential for proving the convergence rate of approximating solutions by the Tikhonov regularization (e.g. Cheng and Yamamoto [2]).

§3. Sketch of the proof of Theorem.

By (1.2) and (1.4), we have

$$E(x,t) = \frac{1}{\sigma} \operatorname{rot} H(x,t) - \frac{1}{\sigma} f(x_1)\lambda(x_2, x_3, t),$$

so that (1.1) and (1.3) imply

$$-\mu\frac{\partial H}{\partial t}(x,t) = \frac{1}{\sigma}\operatorname{rot}\operatorname{rot}H(x,t) - \frac{1}{\sigma}\operatorname{rot}\left(f(x_1)\lambda(x_2,x_3,t)\right)$$

$$= -\frac{1}{\sigma}\Delta H(x,t) - \frac{1}{\sigma}\operatorname{rot}\left(f(x_1)\lambda(x_2,x_3,t)\right).$$

Therefore, taking the first component of the above equation, we obtain

$$\frac{\partial H_1}{\partial t}(x,t) = \frac{1}{\sigma\mu}\Delta H_1(x,t) + \frac{1}{\sigma\mu}f(x_1)R(x_2,x_3,t),$$

$$x = (x_1,x_2,x_3) \in \Omega,\, 0 < t < T, \tag{3.1}$$

where $R(x_2,x_3,t) = \frac{\partial\lambda_3}{\partial x_2}(x_2,x_3,t) - \frac{\partial\lambda_2}{\partial x_3}(x_2,x_3,t)$. For the proof, setting $y = H_1 - \widetilde{H_1}$ and $p = f - \widetilde{f}$, it is sufficient to consider

$$\frac{\partial y}{\partial t}(x,t) = \frac{1}{\sigma\mu}\Delta y(x,t) + \frac{1}{\sigma\mu}p(x_1)R(x_2,x_3,t), \quad x \in \Omega,\, 0 < t < T \tag{3.2}$$

and

$$y(x,t) = 0, \qquad x \in \partial\Omega,\, 0 < t < T \tag{3.3}$$

with given $y(x,t)$ for $(x,t) \in \omega \times (0,T)$ and $y(x,\theta)$ for $x \in \Omega$. Therefore the argument in Imanuvilov and Yamamoto [4] can complete the proof of our theorem.

References

[1]. H.T. Banks and K. Kunisch, *Estimation Techniques for Distributed Parameter Systems*, Birkhäuser Verlag, Boston, 1989.

[2]. J. Cheng and M. Yamamoto, *One new strategy for a priori choice of regularizing parameters in Tikhonov's regularization*, Inverse Problems **16** (2000), L31–L38.

[3]. C.W. Groetsch, *Inverse Problems in the Mathematical Sciences*, Vieweg, Braunschweig, 1993.

[4]. O. Yu. Imanuvilov and M. Yamamoto, *Lipschitz stability in inverse parabolic problems by the Carleman estimate*, Inverse Problems **14** (1998), 1229–1245.

[5]. V. Isakov, *Inverse Problems for Partial Differential Equations*, Springer-Verlag, Berlin, 1998.

[6]. J.A. Kong, *Electromagnetic Wave Theory*, John-Wiley & Sons, New York, 1990.

[7]. P. Neittaanmäki, M. Rudnicki and A. Savini, *Inverse Problems and Optimal Design in Electricity and Magnetism*, Clarendon Press, Oxford, 1996.

[8]. V.G. Romanov and S.I. Kabanikhin, *Inverse Problems for Maxwell's Equations*, VSP, Utrecht, 1994.

[9]. M. Yamamoto, *A mathematical aspect of inverse problems for non-stationary Maxwell's equations*, Int. J. of Applied Electromagnetics and Mechanics **8** (1997), 77–98.

[10]. M. Yamamoto, *On an inverse problem of determining source terms in Maxwell's equations with a single measurement*, in Inverse Problems, Tomography, and Image Processing (1998), Plenum Press, New York, 241–256.

[11]. M. Yamamoto and J. Zou, *Simultaneous reconstruction of the initial temperature and heat radiative coefficient*, to appear in Inverse Problems **17** (2001).

Direct Problems

Electromagnetic Nondestructive Evaluation (VI)
F. Kojima et al. (Eds.)
IOS Press, 2002

Adaptive Multigrid Techniques in Electromagnetic Field Computation

Ronald H.W. Hoppe

Institute of Mathematics , University of Augsburg, D-86159 Augsburg, Germany

Abstract. The use of standard nodal finite elements in the numerical solution of Maxwell's equations is marred by the occurrence of spurious modes causing severe stability problems, if no proper gauging is performed. Moreover, in case of corner or edge singularitites there might be solutions that due to a lack of regularity cannot be approximated by nodal finite elements at all. On the other hand, it is well-known that curl-conforming edge elements avoid such difficulties, since they are much closer to the variational formulation of boundary and initial-boundary value problems in electromagnetics.
In this contribution, we are concerned with efficient numerical solution techniques in terms of adaptive multigrid methods based on edge element discretizations with respect to adaptively generated hierarchies of triangulations of the computational domain.
In particular, we deal with multigrid algorithms whose basic ingredients are hybrid or distributive smoothing processes that take care of the nontrivial kernel of the discrete curl-operator. The characteristic feature is an additional defect correction on the subspace of irrotational vector fields. Adaptive grid refinement can be performed by means of efficient and reliable a posteriori error estimators. We present residual estimators based on an appropriate Helmholtz decomposition of the error into an irrotational and weakly solenoidal part.
The performance of the multigrid algorithm and the a posteriori error estimator is illustrated by some numerical results.

1 Introduction

Physical processes based on electromagnetic phenomena can be adequately described by Maxwell's equations which in differential form are given as follows

$$\operatorname{curl} \mathbf{E} \;=\; -\frac{\partial \mathbf{B}}{\partial t} \;\;,\quad \operatorname{div} \mathbf{B} \;=\; 0 \;, \tag{1}$$

$$\operatorname{curl} \mathbf{H} \;=\; \mathbf{J} + \frac{\partial \mathbf{D}}{\partial t} \;\;,\quad \operatorname{div} \mathbf{D} \;=\; \rho \tag{2}$$

where $\mathbf{E}$ and $\mathbf{H}$ are the electric and the magnetic field, $\mathbf{D}$ and $\mathbf{B}$ are the dielectric and magnetic induction, $\mathbf{J}$ is the electric current density and ρ stands for the space charge density. These equations have to be supplemented by material dependent constitutive equations which in case of linear and isotropic materials are of the form

$$\mathbf{D} \;=\; \varepsilon \mathbf{E} \;\;,\quad \mathbf{B} \;=\; \mu \mathbf{H} \;\;,\quad \mathbf{J} \;=\; \sigma \mathbf{E} + \mathbf{J}_i \tag{3}$$

where ε and μ are the electric and magnetic permeabilities, σ is the electric conductivity, and $\mathbf{J}_i$ refers to an intrinsic current density.

For slowly varying fields the time derivative of the dielectric induction is usually neglected resulting in the so-called eddy current equations that are of parabolic type

$$\sigma \frac{\partial \mathbf{E}}{\partial t} + \mathbf{curl}\, \mu^{-1} \mathbf{curl}\, \mathbf{E} = -\frac{\partial \mathbf{J}_i}{\partial t} \ . \tag{4}$$

Discretizing implicitly in time by the backward Euler scheme results in an elliptic boundaty value problem for the double curl-operator of the form

$$\mathbf{curl}\, \alpha\, \mathbf{curl}\, \mathbf{E} + \beta\, \mathbf{E} = \mathbf{F} \tag{5}$$

with appropriate boundary conditions on the boundary of the computational domain.

In this paper, we focus on efficient adaptive multilevel techniques based on edge element discretizations with respect to an adaptively generated hierarchy of simplicial triangulations. In section 2, we provide the variational setting for stationary second order boundary value problems for the double curl operator and give an introduction to curl-conforming edge element discretizations, whereas section 3 is devoted to the presentation of a multigrid algorithm whose characteristic feature is a hybrid smoothing process that takes particular care of the nontrivial kernel of the discrete curl operator and to an efficient and reliable residual type a posteriori error estimator that may serve as the basis for adaptive grid refinement strategies.

2 Curl-conforming edge element discretizations

We assume $\Omega \subset \mathbf{R}^3$ to be a bounded polyhedral domain with boundary $\Gamma = \partial\Omega$. We further denote by $L^2(\Omega)$ and $L^2(\Gamma)$ the Hilbert spaces of square integrable functions on Ω and Γ with inner products $(\cdot,\cdot)_{0,D}$ and norms $\|\cdot\|_{0,D}$, $D = \Omega$ or $D = \Gamma$ and refer to $H^r(\Omega)$ and $H^s(\Gamma)$, $r, s \in \mathbf{R}$, as the Sobolev spaces with norms $\|\cdot\|_{r,\Omega}$ and $\|\cdot\|_{s,\Gamma}$, respectively (cf., e.g., [3]). For vector fields we adopt the notations $\mathbf{H}^r(\Omega) := (H^r(\Omega))^3$ and $\mathbf{H}^s(\Gamma) := (H^s(\Gamma))^3$.

The variational formulation of boundary value problems for Maxwell's equations in 3D involving the double curl operator naturally gives rise to the space

$$\mathbf{H}(\mathbf{curl}; \Omega) := \{\, \mathbf{q} \in \mathbf{L}^2(\Omega) \mid \mathbf{curl}\, \mathbf{q} \in \mathbf{L}^2(\Omega) \,\}$$

which is a Hilbert space with respect to the graph norm

$$\|\mathbf{q}\|_{curl,\Omega} := (\, \|\mathbf{q}\|_{0,\Omega}^2 + \|\mathbf{curl}\, \mathbf{q}\|_{0,\Omega}^2\,)^{1/2} \ .$$

Finally, we refer to $\mathbf{H}_0(\mathbf{curl}; \Omega)$ as the subspace

$$\mathbf{H}_0(\mathbf{curl}; \Omega) := \{\, \mathbf{q} \in \mathbf{H}(\mathbf{curl}; \Omega) \mid \mathbf{q} \wedge \mathbf{n} = 0 \text{ on } \Gamma \,\}$$

of vector fields with vanishing tangential trace on Γ and to $\mathbf{H}^0(\mathbf{curl}; \Omega)$ as the subspace

$$\mathbf{H}^0(\mathbf{curl}; \Omega) \ := \ \{\, \mathbf{q} \in \mathbf{H}(\mathbf{curl}; \Omega) \mid \mathbf{curl}\, \mathbf{q} = 0 \,\}$$

of irrotational vector fields and set $\mathbf{H}_0^0(\mathbf{curl}; \Omega) := \mathbf{H}_0(\mathbf{curl}; \Omega) \cap \mathbf{H}^0(\mathbf{curl}; \Omega)$. We shall be concerned with the numerical solution of edge element discretized variational problems of the form:
Find $\mathbf{j} \in \mathbf{H}_0(\mathbf{curl}; \Omega)$ such that

$$a(\mathbf{j}, \mathbf{q}) \ = \ \ell(\mathbf{q}) \quad , \quad \mathbf{q} \in \mathbf{H}_0(\mathbf{curl}; \Omega) \tag{6}$$

where

$$a(\mathbf{j}, \mathbf{q}) \ := \ \int_\Omega (\alpha\, \mathbf{curl}\, \mathbf{j} \cdot \mathbf{curl}\, \mathbf{q} \ + \ \beta\, \mathbf{j} \cdot \mathbf{q})\, dx \quad , \quad \ell(\mathbf{q}) \ := \ \int_\Omega \mathbf{f} \cdot \mathbf{q}\, dx \ .$$

Here, α is supposed to be a bounded, symmetric matrix-valued function that is uniformly positive definite on $\bar{\Omega}$, and β is a bounded positive function on $\bar{\Omega}$.

For the efficient numerical solution of (6) we aim at constructing a multilevel iterative solver based on edge element discretizations with respect to a hierarchy $(\mathcal{T}_k)_{k=0}^\ell$ of grids generated, for instance, by an adaptive grid refinement strategy relying on the residual type a posteriori error estimator that will be described in the following section.

Given a simplicial triangulation $\mathcal{T}_h$ of the computational domain $\Omega \subset \mathbf{R}^\mathbf{d}$, d=2 or d=3, for $D \subset \Omega$ we denote by $\mathcal{N}_h(D), \mathcal{E}_h(D)$, and $\mathcal{F}_h(D)$ the sets of vertices, edges, and faces of $\mathcal{T}_h$ in D and by $P_\ell(D)$ resp. $\tilde{P}_\ell(D)$ the set of polynomials resp. homogeneous polynomials of degree $\ell \in \mathbf{N}_0$ on D. For $T \in \mathcal{T}_h$ and $k \in \mathbf{N}$ Nédélec's lowest order curl-conforming edge element is given by

$$\mathbf{Nd}_1(T) := (P_0(T))^d \ + \ \mathbf{S}_1(T) \ , \tag{7}$$

where $\mathbf{S}_1(T) := \{\mathbf{q} \in (\tilde{P}_1(T))^d \mid \mathbf{q}(\mathbf{x}) \cdot \mathbf{x} = 0 \, , \, \mathbf{x} \in T\}$. Any $\mathbf{q} \in \mathbf{Nd}_1(T)$, $T \in \mathcal{T}_h$, $1 \leq \nu \leq 2$, is uniquely determined by the following degrees of freedom

$$\int_E \mathbf{t}_E \cdot \mathbf{q}\, ds \quad , \quad E \in \mathcal{E}_h(T) \ ,$$

where $\mathbf{t}_E$ is the tangential unit vector with respect to the edge $E \in \mathcal{E}_h(T)$ (cf., e.g., [5]).
Specifying the basis fields accordingly, we are thus led to the H(curl)-conforming edge element spaces

$$\mathbf{Nd}_1(\Omega; \mathcal{T}_h) \ := \ \{\mathbf{q} \in H(\mathbf{curl}; \Omega) \mid \mathbf{q}\,|_T \in \mathbf{Nd}_1(T)\, , \, T \in \mathcal{T}_h\} \ . \tag{8}$$

In particular, using the lowest order $\mathbf{H}(\mathbf{curl})$-conforming edge elements $\mathbf{Nd}_1(T)$, $T \in \mathcal{T}_k$, the discrete problem on the finest grid reads as follows: Find $\mathbf{j_h} \in \mathbf{Nd}_{1,0}(\Omega; \mathcal{T}_\ell)$ such that

$$a(\mathbf{j_h}, \mathbf{q_h}) = \ell(\mathbf{q_h}) \quad , \quad \mathbf{q_h} \in \mathbf{Nd}_{1,0}(\Omega; \mathcal{T}_\ell) \tag{9}$$

where $\mathbf{Nd}_{1,0}(\Omega; \mathcal{T}_\ell) := \{\mathbf{q_h} \in \mathbf{Nd}_1(\Omega; \mathcal{T}_\ell) \mid \mathbf{q_h} \wedge \mathbf{n} = 0 \text{ on } \Gamma\}$.

3 Adaptive multigrid iterative solvers

The basic ingredients of a multigrid method are the smoother and the intergrid transfer operators (prolongations and restrictions). Whereas the intergrid transfers can be done canonically (cf., e.g., [1,4]), the design of the smoother requires particular attention due to the nontrivial kernel of the discrete curl-operator. Indeed, considering the discrete Helmholtz type decomposition

$$\mathbf{Nd}_{1,0}(\Omega; \mathcal{T}_k) = \mathbf{Nd}_{1,0}^0(\Omega; \mathcal{T}_k) \oplus \mathbf{Nd}_{1,0}^\perp(\Omega; \mathcal{T}_k)$$

into the subspace $\mathbf{Nd}_{1,0}^0(\Omega; \mathcal{T}_k) := \{\mathbf{q_h} \in \mathbf{Nd}_{1,0}(\Omega; \mathcal{T}_k) \mid \mathbf{curl}\, \mathbf{q_h} = 0\}$ and its L^2-orthogonal complement $\mathbf{Nd}_{1,0}^\perp(\Omega; \mathcal{T}_k)$, a standard smoother like the Gauss-Seidel iteration, applied to the edge element discretized problem, performs well on the subspace $\mathbf{Nd}_{1,0}^\perp(\Omega; \mathcal{T}_k)$ of weakly solenoidal discrete vectorfields, but insufficiently on the subspace $\mathbf{Nd}_{1,0}^0(\Omega; \mathcal{T}_k)$ of irrotational discrete vectorfields. Therefore, the idea is to apply a hybrid smoothing process that is characterized by an additional defect correction step on the subspace of irrotational vectorfields. Taking into account $\mathbf{Nd}_{1,0}^0(\Omega; \mathcal{T}_k) = \mathbf{grad}\, S_{1,0}(\Omega; \mathcal{T}_k)$, where $S_{1,0}(\Omega; \mathcal{T}_k)$ is the standard FE-space of continuous, piecewise linear finite elements, on level ℓ the hybrid smoother works as follows:

Step 1: Gauss-Seidel sweeps (edge element discretized problem)

Given an iterate $\mathbf{j}_\ell^{(m)}$, perform $\kappa > 0$ Gauss-Seidel iterations on the edge element discretized problem (9) resulting in the smoothed iterate $\bar{\mathbf{j}}_\ell^{(m)}$.

Step 2: Correctional Gauss-Seidel sweeps (irrotational part)

Compute the residual

$$r(\mathbf{grad}\, v_h) := \ell(\mathbf{grad}\, v_h) - a(\bar{\mathbf{j}}_\ell^{(m)}, \mathbf{grad}\, v_h) \quad , \quad v_h \in S_{1,0}(\Omega; \mathcal{T}_h)$$

and determine $u_h \in S_{1,0}(\Omega; \mathcal{T}_h)$ by $\kappa > 0$ Gauss-Seidel iterations on the defect correction equation

$$\int_\Omega c\, \mathbf{grad}\, u_h \cdot \mathbf{grad}\, v_h \, d\mathbf{x} = r(\mathbf{grad}\, v_h) \quad , \quad v_h \in S_{1,0}(\Omega; \mathcal{T}_h)$$

using $u_h^{(0)} = 0$ as a startiterate.

For an eddy currents problem with $\alpha = 1$ and $\Omega = (0,1)^3$, Table 1 displays the convergence rates $\rho = \rho_{MG}$ of multigrid V(1,1)-cycles. For comparison, on each level the convergence rate $\bar{\rho} = \bar{\rho}_{MG}$ for V(1,1)-cycles without smoothing on the irrotational part (Step 2 of the smoother) are also listed.

ℓ	2		3		4		5		6	
	ρ	$\bar{\rho}$	ρ	$\bar{\rho}$	ρ	$\bar{\rho}$	ρ	$\bar{\rho}$	ρ	$\bar{\rho}$
$\beta = 0.100$	**0.15**	0.98	**0.16**	0.98	**0.16**	0.99	**0.16**	0.99	**0.16**	0.99
$\beta = 0.500$	**0.15**	0.97	**0.16**	0.97	**0.16**	0.98	**0.16**	0.98	**0.16**	0.99
$\beta = 1.000$	**0.15**	0.96	**0.16**	0.97	**0.16**	0.97	**0.16**	0.97	**0.16**	0.98
$\beta = 2.000$	**0.15**	0.96	**0.16**	0.96	**0.16**	0.96	**0.16**	0.97	**0.16**	0.98
$\beta = 10.00$	**0.14**	0.88	**0.16**	0.93	**0.16**	0.96	**0.16**	0.97	**0.16**	0.97
$\beta = 100.0$	**0.10**	0.75	**0.13**	0.91	**0.15**	0.92	**0.16**	0.95	**0.16**	0.96

Table 1: Multigrid convergence rates (effect of hybrid smoothing)

A posteriori error estimation is a prerequisite for grid adaptation techniques. Here, we follow [2] and sketch a residual type a posteriori error estimator for edge element approximations of boundary value problems of the form

$$\mathbf{curl}\,\alpha\,\mathbf{curl}\,\mathbf{j} + \beta\mathbf{j} = \mathbf{f} \quad \text{in } \Omega \subset \mathbf{R}^3 , \tag{10}$$

$$\mathbf{j} \wedge \mathbf{n} = \mathbf{g} \quad \text{on } \partial\Omega \tag{11}$$

discretized by means of the lowest order $\mathbf{H}(\mathbf{curl})$-conforming edge elements of the first family.

The residual based error estimator, which will be given in terms of the energy norm

$$||| \mathbf{j} |||^2 := (a\,\mathbf{curl}\,\mathbf{j}, \mathbf{curl}\,\mathbf{j})_{0,\Omega} + (c\mathbf{j}, \mathbf{j})_{0,\Omega} ,$$

relies on an appropriate evaluation of the residual with respect to an approximation $\tilde{\mathbf{j}}_{\mathbf{h}} \in \mathbf{Nd}_1(\Omega; \mathcal{T}_h)$ of (10),(11). Assuming for simplicity $\tilde{\mathbf{j}}_{\mathbf{h}} \wedge \mathbf{n} = \mathbf{g}$ on $\partial\Omega$, it is easy to see that the total error $\mathbf{e}_{\mathbf{h}} := \mathbf{j} - \tilde{\mathbf{j}}_{\mathbf{h}}$ is in $\mathbf{H}_0(\mathbf{curl}; \Omega)$ and satisfies the defect equation

$$(a\,\mathbf{curl}\,\mathbf{e}_{\mathbf{h}}, \mathbf{curl}\,\mathbf{q})_{0,\Omega} + (c\,\mathbf{e}_{\mathbf{h}}, \mathbf{q})_{0,\Omega} = r(\mathbf{q}) \quad , \quad \mathbf{q} \in \mathbf{H}_0(\mathbf{curl}; \Omega)$$

with the residual $r \in \mathbf{H}_0(\mathbf{curl}; \Omega)^*$ given by

$$r(\mathbf{q}) := (\mathbf{f}, \mathbf{q})_{0,\Omega} - (a\,\mathbf{curl}\,\tilde{\mathbf{j}}_{\mathbf{h}}, \mathbf{curl}\,\mathbf{q})_{0,\Omega} - (c\tilde{\mathbf{j}}_{\mathbf{h}}, \mathbf{q})_{0,\Omega} \quad , \quad \mathbf{q} \in \mathbf{H}_0(\mathbf{curl}; \Omega) .$$

The error analysis can be substantially facilitated by the Helmholtz type decomposition

$$\mathbf{H}_0(\mathbf{curl}; \Omega) = \mathbf{H}_0^0(\mathbf{curl}; \Omega) \oplus \mathbf{H}_0^\perp(\mathbf{curl}; \Omega) \tag{12}$$

where $\mathbf{H}_0^\perp(\mathbf{curl}; \Omega)$ is the orthogonal complement of $\mathbf{H}_0^0(\mathbf{curl}; \Omega)$ with respect to the L^2-inner product $(c\cdot, \cdot)_{0,\Omega}$, i.e.,

$$\mathbf{q}^\perp \in \mathbf{H}_0^\perp(\mathbf{curl}; \Omega) \iff (c\mathbf{q}^\perp, \mathbf{q}^0)_{0,\Omega} = 0 \quad , \quad \mathbf{q}^0 \in \mathbf{H}_0^0(\mathbf{curl}; \Omega) .$$

Note that $\mathbf{H}_0^0(\mathbf{curl}; \Omega)$ represents the subspace of irrotational vectorfields whereas $\mathbf{H}_0^\perp(\mathbf{curl}; \Omega)$ can be interpreted as a subspace of weakly solenoidal vectorfields.

In the sequel we assume that a decomposition (12) can be found such that $\mathbf{H}_0^\perp(\mathbf{curl}; \Omega)$ is continuously embedded in $\mathbf{H}^1(\Omega) \cap \mathbf{H}_0(\mathbf{curl}; \Omega)$ and

$$| \mathbf{q}^\perp |_{1,\Omega} \le C(\Omega) \|\mathbf{curl}\, \mathbf{q}^\perp\|_{0,\Omega} \ , \ \mathbf{q}^\perp \in \mathbf{H}_0^\perp(\mathbf{curl}; \Omega) \ . \tag{13}$$

Now, splitting the error according to $\mathbf{e_h} = \mathbf{e_h^0} + \mathbf{e_h^\perp}$, where $\mathbf{e_h^0} \in \mathbf{H}_0^0(\mathbf{curl}; \Omega)$ and $\mathbf{e_h^\perp} \in \mathbf{H}_0^\perp(\mathbf{curl}; \Omega)$, it follows readily that $\mathbf{e_h^0}$ and $\mathbf{e_h^\perp}$ satisfy

$$(c\,\mathbf{e_h^0}, \mathbf{q^0})_{0,\Omega} = r(\mathbf{q}) \ , \ \mathbf{q^0} \in \mathbf{H}_0^0(\mathbf{curl}; \Omega) \ , \tag{14}$$

$$(a\,\mathbf{curl}\,\mathbf{e_h^\perp}, \mathbf{curl}\,\mathbf{q}^\perp)_{0,\Omega} + (c\,\mathbf{e_h^\perp}, \mathbf{q}^\perp)_{0,\Omega} = r(\mathbf{q}) \ , \ \mathbf{q}^\perp \in \mathbf{H}_0^\perp(\mathbf{curl}; \Omega) \tag{15}$$

The advantage of the Helmholtz decomposition (12) is that the irrotational and weakly solenoidal part of the error can be estimated separately (cf. sections 4,5 in [2]). In particular, the bounds for $\mathbf{e_h^0}$ involve the error term

$$\delta^{(0)} := \Big(\sum_{T \in \mathcal{T}_h} (\delta_T^{(0)})^2 \Big)^{1/2} + \Big(\sum_{F \in \mathcal{F}_h(\bar{\Omega})} (\delta_F^{(0)})^2 \Big)^{1/2}$$

with the local contributions $\delta_T^{(0)}$ and $\delta_F^{(0)}$ given by

$$\delta_T^{(0)} := h_T \|\mathrm{div}\,(c^{1/2}\tilde{\mathbf{j_h}})\|_{0,T} \ , \ \delta_F^{(0)} := h_F^{1/2} \|c_A^{-1/2}[\mathbf{n} \cdot c\tilde{\mathbf{j_h}}]_J\|_{0,F} \ ,$$

where c_A is the arithmetic average of c on F with respect to the adjacent elements, $[\mathbf{n} \cdot c\tilde{\mathbf{j_h}}]_J$ denotes the jump of the normal component of $c\tilde{\mathbf{j_h}}$ across F, and $h_T := \mathrm{diam}\,T$, $h_F := \mathrm{diam}\,F$.

On the other hand, the estimation of $\mathbf{e_h^\perp}$ gives rise to

$$\delta^{(1)} := \Big(\sum_{T \in \mathcal{T}_h} (\delta_T^{(1)})^2 \Big)^{1/2} + \Big(\sum_{F \in \mathcal{F}_h(\Omega)} (\delta_F^{(1)})^2 \Big)^{1/2} \ , \ \delta^{(2)} := \Big(\sum_{T \in \mathcal{T}_h} (\delta_T^{(2)})^2 \Big)^{1/2}$$

with the local contributions $\delta_T^{(\nu)}$, $1 \le \nu \le 2$, and $\delta_F^{(1)}$ given by

$$\delta_T^{(1)} := h_T \|a^{-1/2}\,(\mathbf{f_h} - \mathbf{curl}\,a\,\mathbf{curl}\,\tilde{\mathbf{j_h}} - c\tilde{\mathbf{j_h}})\|_{0,T} \ ,$$

$$\delta_T^{(2)} := h_T \|a^{-1/2}\,(\mathbf{f} - \mathbf{f_h})\|_{0,T} \ , \ \delta_F^{(1)} := h_F^{1/2} \|a_A^{-1/2}[\mathbf{n} \wedge a\,\mathbf{curl}\,\tilde{\mathbf{j_h}}]_J\|_{0,F} \ .$$

Here, $\mathbf{f_h}$ stands for the L^2-projection of $\mathbf{f}$ onto $\prod_{T \in \mathcal{T}_h}(P_1(T))^3$, a_A is the average of a on F, and $[\mathbf{n} \wedge a\,\mathbf{curl}\,\tilde{\mathbf{j_h}}]_J$ refers to the jump of $\mathbf{n} \wedge a\,\mathbf{curl}\,\tilde{\mathbf{j_h}}$ across F.

Denoting the iteration error by $\delta^{iter} := ||| \mathbf{j_h} - \tilde{\mathbf{j_h}} |||$ and setting $\delta^{(3)} := \delta^{(0)} + \delta^{(1)}$, we obtain (cf. [2]; Thm 3.3):

Theorem 5.1 Under the assumption (13) there exist constants γ_ν, Γ_ν, $1 \le \nu \le 2$, depending only on Ω, a, c and on the local geometry of $\mathcal{T}_h$ such that

$$\gamma_1 \delta^{(3)} - \gamma_2 \delta^{(2)} \le ||| \mathbf{e_h} ||| \le \Gamma_1 (\delta^{(3)} + \delta^2) + \Gamma_2 \delta^{iter} \ . \tag{16}$$

We remark that due to the asymptotic optimality of the multigrid iterative scheme the iteration error δ^{iter} is under control and that $\delta^{(2)}$ is a higher order term provided the right-hand side $\mathbf{f}$ is smooth enough.
The following Table 2 contains the computed effectivity indices, i.e., the ratio of the energy norms of the computed and the exact error.

Level	0	1	2	3	4
$\beta = 10^{-4}$	4.21	4.70	4.90	4.98	5.01
$\beta = 10^{-2}$	4.21	4.70	4.90	4.98	5.01
$\beta = 1.00$	4.21	4.70	4.89	4.98	5.01
$\beta = 10^{+2}$	4.22	4.68	4.87	4.96	5.01
$\beta = 10^{+4}$	4.22	4.66	4.84	4.92	4.96

Table 2: Effectivity indices

For further examples and a detailed discussion and documentation of quality measures for the error estimator we refer to [1,2].

Acknowledgments. The work has been supported by grants from the Federal Ministry for Education and Research (BMBF) under Grant No. 03HO7AU1-8 and Grant No. 03HOM3A1.

References

1. R. Beck, P. Deuflhard, R. Hiptmair, R.H.W. Hoppe, and B. Wohlmuth, "Adaptive multilevel methods for edge element discretizations of Maxwell's equations", Surveys of Math. in Industry, Vol. **8**, pp. 271–312, (1999)
2. R. Beck, R. Hiptmair, R.H.W. Hoppe, and B. Wohlmuth, "Residual based a posteriori error estimators for eddy current computation", M^{2}AN Math. Modelling and Numer. Anal. **34**, 159-182, (2000)
3. P. Grisvard, "Elliptic Problems in Nonsmooth Domains", Pitman, Boston, 1985
4. R. Hiptmair, "Multigrid method for Maxwell's equations", SIAM J. Numer. Anal. **36**, 204-225, (1999)
5. J.-C. Nédélec, "Mixed finite elements in $\mathbf{R}^3$", Numer. Math. **35**, 315-341, (1980)

Electromagnetic Nondestructive Evaluation (VI)
F. Kojima et al. (Eds.)
IOS Press, 2002

Eddy-current evaluation of 3-D defects in a metal plate: a first analysis of a contrast-source gradient method

D. Dos Reis, M. Lambert, D. Lesselier

Département de Recherche en Électromagnétisme – Laboratoire des Signaux et Systèmes
CNRS-SUPÉLEC-UPS, Plateau de Moulon, 3 rue Joliot Curie,
91192 Gif-sur-Yvette Cedex, France, http://www.lss.supelec.fr

Abstract Retrieval of voluminous defects in a non-magnetic metal plate from maps of time-harmonic anomalous magnetic fields in air due to a source nearby is performed by a contrast-source, gradient-type technique which uses the binary aspect of the sought conductivity distributions. The functional analysis within a vector integral formulation of the diffusive fields is sketched and the efficiency of the approach is discussed in the light of preliminary inversion results obtained with synthetic data.

1 Introduction

We investigate the evaluation of 3-D bounded (voluminous) defects of known conductivity (voids, inclusions) inside a flat, horizontal non-magnetic metal plate from anomalous magnetic fields observed when a low-frequency source is operated nearby. The source consists of an air-core coil probe positioned above the expectedly damaged zone. The data consist of a 2-D map of one single component (the vertical one) or of all three components of the anomalous magnetic field at discrete locations in a planar surface above the plate.

Our main aim is not to introduce fully novel modeling tools of, and solution methods to, that demanding 3-D shape inversion problem, but to study whether one could tackle this problem by combining a number of recently introduced methods, when as now the embedding environment is a planar (conductive) layer, when the scatterers are three-dimensionally bounded, when illumination and observation are made from above the layer (aspect-limited data), and even when a single component of the vector magnetic field is collected on a limited, coarse mesh. These methods have been developed in order

- to calculate the eddy-current pattern in a planarly-layered conductive environment affected by a localized inhomogeneity of arbitrary geometry using exact solutions as well as extended-Born approximations developed within a rigorous, *full-wave* vector domain integral formulation of the diffusive fields inside and outside the affected sample [1, 2];

- to extract the distribution of conductivity of a 3-D, possibly multiply-connected voluminous object buried in an infinite conductive space from a magnetic field observed nearby, using a contrast-source-based gradient-type method [3, 4] in an Earth's subsurface imaging context;

- to use the prior information that the sought defects are of prescribed conductivity contrast with respect to their environment, their 3-D maps ideally consisting of distributions of black and white voxels (in the exterior and in the interior of the defects) as is the case already in [5] and as it has been much studied in 2-D geometries since the early approach [6] of binary-specialized modified gradient methods.

Previous investigations [1, 2] have shown that anomalous magnetic fields are modeled with fair accuracy –in manageable computational time on a standard work-station– by the aforementioned integral formulation, and we do not come back to this topic here. We mostly outline below how this *hybrid* inversion method can be properly developed, and give a few typical results. Test cases at this still preliminary stage come from synthetic data calculated on canonical 3-D models of mm-sized parallelepipeded void defects in an Inconel 600 plate probed in the 150 – 300 kHz frequency range.

2   The algorithm

Let us refer to the configuration sketched in Fig. 1. A possibly multiply-connected 3-D defect Ω of constant conductivity σ (it will be 0 in the numerical examples) is embedded within a non-magnetic linear isotropic metal plate of conductivity σ_2. Eddy currents are generated in the plate using a time-harmonic (circular frequency ω, time-dependence $\exp\left(-\mathrm{j}\omega t\right)$) coil source carrying electrical currents J_C at fixed location in air (permeability μ_0, permittivity ϵ_0). Perturbation of the eddy current pattern produces an anomalous magnetic field that is observed in a prescribed planar surface S in air.

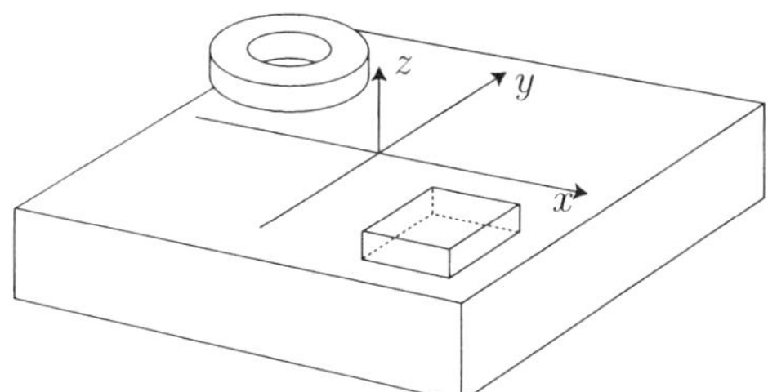

Figure 1: Configuration of study: a pancake-type source probe is set at fixed location above the damaged metal plate, here sketched as a void parallelepiped, the time-harmonic magnetic field being sampled in a plane of observation at fixed height above the plate.

The electromagnetic field $\mathbf{E}, \mathbf{H}$ at any $\mathbf{r}$ in the plate (corresponding quantities will be marked by subscript 2) and in air (using subscript 1 if above the plate, and 3 if below) are cast into a rigorous vector domain (contrast-source) integral formulation from application of the Green theorem onto the quasi-static Maxwell's PDE –involving the Green dyads of the layered environment. So, the field modeling requires the calculation of a fictitious contrast source $\mathbf{J}_2\left(\mathbf{r}\right) = \sigma_2\chi\left(\mathbf{r}\right)\mathbf{E}_2\left(\mathbf{r}\right)$, where $\chi\left(\mathbf{r}\right)$, equal to 0 outside Ω and to $\chi_c = \sigma/\sigma_2 - 1$ inside Ω, is the electrical contrast between the metal and the defect Ω (for a void, $\chi_c = -1$). For any $\mathbf{r}$ in Ω,

$$\mathbf{J}_2\left(\mathbf{r}\right) = \mathbf{J}_{20}\left(\mathbf{r}\right) + \mathrm{j}\omega\mu_0\sigma_2\chi\left(\mathbf{r}\right)\int_\Omega \underline{\mathbf{G}}_{i2}^{ee}\left(\mathbf{r},\mathbf{r}'\right)\mathbf{J}_2\left(\mathbf{r}'\right)d\mathbf{r}', \tag{1}$$

whilst the anomalous magnetic field $\mathbf{H}_1^{\mathcal{S}}$ in $\mathcal{S}$ follows from

$$\mathbf{H}_1^{\mathcal{S}}(\mathbf{r}) = \int_{\Omega} \underline{\mathbf{G}}_{12}^{me}(\mathbf{r},\mathbf{r}')\,\mathbf{J}_2(\mathbf{r}')\,d\mathbf{r}'. \tag{2}$$

In the above $\mathbf{J}_{20} = \sigma_2\chi\mathbf{E}_{20}$, where $\mathbf{E}_{20}(\mathbf{r}) = \mathrm{j}\omega\mu_0\int_{\mathcal{S}}\underline{\mathbf{G}}_{21}^{ee}(\mathbf{r},\mathbf{r}')\,\mathbf{J}_C(\mathbf{r}')\,d\mathbf{r}'$ represents primary currents associated to the electrical current $\mathbf{J}_C$ in the coil source C in medium 1; $\underline{\mathbf{G}}_{ij}^{ee}(\mathbf{r},\mathbf{r}')$ and $\underline{\mathbf{G}}_{ij}^{me}(\mathbf{r},\mathbf{r}')$ are the electric-electric and magnetic-electric Green dyads made of the electric and magnetic fields at $\mathbf{r}$ in medium i due to a Dirac electrical current at $\mathbf{r}'$ in medium j, respectively.

Upon iteratively solving (1) by a Conjugate-Gradient Fast-Fourier-Transform Method of Moments using pre-calculated samples of the Green dyads the anomalous field follows from (2) (extended-Born approximations enabling us to bypass the solution of (1) via the introduction of a depolarization dyad independent of the field) [1, 2, 7]. Correspondingly, the inversion problem requires us to determine some conductivity contrast within each voxel of a prescribed 3-D box $\mathcal{D}$ containing the unknown defect of support Ω in order to produce an anomalous magnetic field close enough in some sense to data ζ collected in $\mathcal{S}$, being assumed that the true distribution is binary.

As already indicated, binary-specialized modified gradient methods [5, 6] and contrast-source gradient-type methods [3, 4] have been investigated for nonlinearized wave field inversion. Lack of place precludes us to go into their respective machineries. Combining both methods by specializing the contrast-source approach to a binary contrast –in original fashion and with promising results as shown in section 3 by numerical experimentation– can be done as follows.

The formulation of interest is written into shorthand operator form using Green dyadic operators: $\underline{\mathcal{G}}_{22}^{ee}\mathbf{J}_2 = \mathrm{j}\omega\mu_0\int_{\mathcal{D}}\underline{\mathbf{G}}_{22}^{ee}(\mathbf{r},\mathbf{r}')\,\mathbf{J}_2(\mathbf{r}')\,d\mathbf{r}'$ and $\underline{\mathcal{G}}_{12}^{me}\mathbf{J}_2 = \int_{\mathcal{D}}\underline{\mathbf{G}}_{12}^{me}(\mathbf{r},\mathbf{r}')\,\mathbf{J}_2(\mathbf{r}')\,d\mathbf{r}'$. Two equations are of interest: the data equation

$$\rho = \zeta - \underline{\mathcal{G}}_{12}^{me}\,\mathbf{J}_2^n \tag{3}$$

tells us that some contrast source $\mathbf{J}_2$ in $\mathcal{D}$ radiates a certain magnetic field in $\mathcal{S}$ whose discrepancy with data ζ is measured by vector residual ρ^n; the state equation

$$\epsilon = \sigma_2\,\chi\,\mathbf{E}_2 - \mathbf{J}_2 \tag{4}$$

tells us that this contrast source is consistent with a contrast distribution χ (here, purely real) in $\mathcal{D}$, the degree of consistency being appraised by vector residual ϵ, letting $\mathbf{E}_2 = \mathbf{E}_{20} + \underline{\mathcal{G}}_{22}^{ee}\,\mathbf{J}_2$.

As for the sought binary aspect of the contrast distribution (valued to χ_c or to 0 at any $\mathbf{r}$ in $\mathcal{D}$), and in order to use gradient methods that require us to restore differentiability with respect to this distribution, it is enforced via the nonlinear transformation

$$\chi = \chi_c\,\Psi(\tau)\,,\,\Psi(\tau) = (1 + \exp(-\tau/\theta))^{-1}, \tag{5}$$

where θ is a strictly positive, real-valued tuning parameter which is controlling the slope of the strictly monotonous real function Ψ (bounded between 0 and 1 and varying from 0 to 1 when τ is increased), the $\mathbf{r}$-dependence being implied. In practice, decreasing θ yields an electrical contrast χ closer to χ_c (resp., to zero) for a given positive (resp., negative) τ at same $\mathbf{r}$.

The solution itself now relies on the construction of two sequences of τ^n and $\mathbf{J}_2^n$ (correspondingly, of $\mathbf{E}_2^n$) that are both function of space $\mathbf{r}$ (implied) in $\mathcal{D}$ –once chosen proper first guesses at $n = 1$– so as a suitable cost functional $\mathbf{F}$ is decreased from one iteration to the next. Using $\|.\|_\mathcal{S}$ and $\|.\|_\mathcal{D}$ as norms on $L_2\left(\mathcal{S}\right)$ et $L_2\left(\mathcal{D}\right)$ and (later on) corresponding scalar products $\langle,\rangle_\mathcal{S}$ and $\langle,\rangle_\mathcal{D}$, the general expression of $\mathbf{F}$ reads as

$$\mathbf{F} = \eta_S \|\boldsymbol{\rho}\|_\mathcal{S}^2 + \eta_\mathcal{D} \|\boldsymbol{\epsilon}\|_\mathcal{D}^2 , \tag{6}$$

where weight $\eta_S = \|\boldsymbol{\zeta}\|_\mathcal{S}^{-2}$ is constant for given data $\boldsymbol{\zeta}$ and where weight $\eta_\mathcal{D} = \|\sigma_2\,\chi\,\mathbf{E}_{20}\|_\mathcal{D}^2$ is contrast-dependent.

The determination of the contrast source $\mathbf{J}_2^n$ is carried out first. At iteration n, one simply chooses $\mathbf{J}_2^n = \mathbf{J}_2^{n-1} + \alpha^n \mathbf{v}^n$, where the complex-valued, r-dependent (3-component) vector $\mathbf{v}^n$ is a conjugate-gradient direction of displacement of the Polak-Ribière type and where the complex-valued constant parameter α^n is a coefficient of displacement. One sets for $n > 1$

$$\mathbf{v}^n = \mathbf{g}^{v,n} + \frac{\langle \mathbf{g}^{v,n}, \mathbf{g}^{v,n} - \mathbf{g}^{v,n-1}\rangle_\mathcal{D}}{\langle \mathbf{g}^{v,n-1}, \mathbf{g}^{v,n-1}\rangle_\mathcal{D}} \tag{7}$$

(at $n = 1$, $\mathbf{v}^1 = \mathbf{g}^{v,1}$). $\mathbf{g}^{v,n}$ is the complex-valued gradient (Fréchet derivative) of $\mathbf{F}$ with respect to the contrast source $\mathbf{J}_2$ at $\mathbf{J}_2^{n-1}$, the contrast being kept fixed at χ^{n-1} in the calculation (i.e., τ^{n-1} is kept fixed). Formal derivation in $\mathbf{F}$ of the squared norms of the residuals that are given in operator form in (3) and (4) being carried out in line with earlier works, e.g., [3, 4, 5, 6] and many references therein, one can show that this gradient reads as

$$\mathbf{g}^{v,n} = -\eta_S\,\underline{\mathcal{G}}_{12}^{me*}\,\boldsymbol{\rho}^{n-1} - \eta_\mathcal{D}^{n-1}\left(\boldsymbol{\epsilon}^{n-1} - \underline{\mathcal{G}}_{22}^{ee*}\,\sigma_2\,\chi^{n-1}\,\boldsymbol{\epsilon}^{n-1}\right) , \tag{8}$$

where * denotes the adjoint operation, weight $\eta_\mathcal{D}^{n-1}$ and residuals $\boldsymbol{\rho}^{n-1}$ and $\boldsymbol{\epsilon}^{n-1}$ being those associated with the iterates of order $n - 1$ (since one is interested into the gradient of $\mathbf{F}$ at $\mathbf{J}_2^{n-1}$). Accordingly, minimizer α^n of cost functional $\mathbf{F}\left(\mathbf{J}_2^{n-1} + \alpha^n\,\mathbf{v}^n\right)$ for the just determined set of directions $\mathbf{v}^n$ is written in closed form as

$$\alpha^n = \frac{\eta_S\langle\boldsymbol{\rho}^{n-1}, \underline{\mathcal{G}}_{12}^{me}\,\mathbf{v}^n\rangle_\mathcal{S} + \eta_\mathcal{D}^{n-1}\langle\boldsymbol{\epsilon}^{n-1}, \mathbf{v}^n - \sigma_2\,\chi^{n-1}\,\underline{\mathcal{G}}_{22}^{ee}\,\mathbf{v}^n\rangle_\mathcal{D}}{\eta_S\left\|\underline{\mathcal{G}}_{12}^{me}\,\mathbf{v}^n\right\|_\mathcal{S}^2 + \eta_\mathcal{D}^{n-1}\left\|\mathbf{v}^n - \sigma_2\,\chi^{n-1}\,\underline{\mathcal{G}}_{22}^{ee}\,\mathbf{v}^n\right\|_\mathcal{D}^2} . \tag{9}$$

According to the above, contrast source and field are now updated to new values $\mathbf{J}_2^n$ and $\mathbf{E}_2^n = \mathbf{E}_2^{n-1} + \alpha^n\,\underline{\mathcal{G}}_{22}^{ee}\,\mathbf{v}^n$. However, the residual of the *updated* state equation, written as $\boldsymbol{\epsilon} = \sigma_2\,\chi\,\mathbf{E}_2^n - \mathbf{J}_2^n$, valued at $\widetilde{\boldsymbol{\epsilon}}^n$ when the contrast is the present iterate $\chi^{n-1} = \chi_c\,\Psi\left(\tau^{n-1}\right)$, might not be close enough to zero in order to be satisfied with. So, proper updating of the contrast comes by letting $\tau^n = \tau^{n-1} + \beta^n d^n$, where real-valued r-dependent scalar d^n is again of the Polak-Ribière type and where real-valued parameter β^n is a displacement coefficient. For $n > 1$

$$d^n = g^{d,n} + \frac{\langle g^{d,n}, g^{d,n} - g^{d,n-1}\rangle_\mathcal{D}}{\langle g^{d,n-1}, g^{d,n-1}\rangle_\mathcal{D}} \tag{10}$$

(at $n = 1$, $d^1 = g^{d,1}$). $g^{d,n}$ is the real-valued gradient (Fréchet derivative) at τ^{n-1}, both contrast source $\mathbf{J}_2^n$ and field $\mathbf{E}_2^n$ being now kept fixed, of a weighted-in squared norm $\mathbf{F}_\mathcal{D}$ of the above state residual $\boldsymbol{\epsilon}$; in conveniently expanded form, $\mathbf{F}_\mathcal{D}$ reads as

$$\mathbf{F}_\mathcal{D} = \eta_\mathcal{D}^{n-1}\left\|\sigma_2\,\chi_c\,\Psi\left(\tau\right)\mathbf{E}_2^n - \mathbf{J}_2^n\right\|_\mathcal{D}^2 , \tag{11}$$

where weight $\eta_D^{n-1} = \left\| \sigma_2 \, \chi_c \, \Psi \left(\tau^{n-1} \right) \mathbf{E}_{20} \right\|_D^{-2}$.

Proceeding as it was the case with the gradient of $\mathbf{F}$ in the above, one can show that

$$g^{d,n} = -2 \, \eta_D^{n-1} \, \Psi' \left(\tau^{n-1} \right) \Re \left(\sigma_2 \, \chi_c \, \overline{\mathbf{E}_2^n} \cdot \widetilde{\boldsymbol{\epsilon}}^n \right), \tag{12}$$

where Ψ' is the derivative of Ψ with respect to τ, where the upper bar denotes complex conjugation, where $\Re$ denotes the real part, and where residual $\widetilde{\boldsymbol{\epsilon}}^n$ at τ^{n-1} has been defined before; here, $(\cdot)$ represents the usual scalar product of (3-component) vectors at given location. (Notice that the weight η_D^{n-1} has been kept constant for simplicity, which means that one uses it simply as a normalization factor, and that one is differentiating the factor $\left\| \sigma_2 \, \chi_c \, \Psi \left(\tau \right) \mathbf{E}_2^n - \mathbf{J}_2^n \right\|_D^2$ only.) As for β^n, the minimizer of $\mathbf{F}_D^n \left(\tau^{n-1} + \beta^n \, d^n \right)$ into the set of directions d^n, it is obtained in closed form as

$$\beta^n = -\frac{\Re \langle d^n \, \sigma_2 \, \chi_c \, \Psi' \left(\tau^{n-1} \right) \mathbf{E}_2^n, \widetilde{\boldsymbol{\epsilon}}^n \rangle_D}{\left\| d^n \, \sigma_2 \, \chi_c \, \Psi' \left(\tau^{n-1} \right) \mathbf{E}_2^n \right\|_D^2}. \tag{13}$$

The algorithm now proceeds by alternating an update of contrast source $\mathbf{J}_2$ via equations (6-9) and an update of contrast τ via equations (10-13) inside each voxel of the prescribed search domain $\mathcal{D}$ until the value reached by the cost functional $\mathbf{F}$ is small enough, or until some plateau is reached. In the latter case, the tuning parameter θ might be reduced in order to push contrast values χ either to 0 or to χ_c, and equivalently to re-distribute voxels into darker ones (at positive τ) and lighter ones (at negative τ), from which the construction of the two sequences is then started again.

3   Numerical results

A number of results has been obtained by the above inversion algorithm. Those shown in figures 2 and 3 consist of gray-level 2-D cross-sectional maps at successive depths (0.1 mm step) of the retrieved 3-D distributions of the contrast χ (after about 2500 iterations, when the cost functional has notably decreased) and of evolution curves of the cost functional $\mathbf{F}$ as a function of the number of iterations.

The example is the one of a 2 mm thick slab of conductivity $\sigma_2 = 1$ MS/m where voluminous void defects are sought. A thick circular coil (0.6 and 1.6 mm inner and outer radii, 0.8 mm height) is centered 0.9 mm above the slab at $x = -0.75$ mm and $y = 0$ to avoid undue symmetries with respect to the search domain $\mathcal{D}$ (the latter is centered on the z axis and discretized into $52 \times 52 \times 20$ cubical voxels of 0.1 mm^3 volume). The frequency of operation is 150 kHz, which is corresponding with a plane-wave skin depth δ in metal of about 1.3 mm. One considers in figure 2 and 3 two void parallelepipeds opening in air (the top surface of the slab), both of 1.1×1.1 mm^2 squared horizontal cross-section, the first one 1 mm deep and centered at $x = -1.1$ mm, $y = -1.1$ mm, the other 0.5 mm deep and symmetrically located at $x = 1.1$ mm, $y = 1.1$ mm; one considers a single void opening in air in figure 4, which is still of 1.1×1.1 mm^2 squared horizontal cross-section and 1 mm depth, but its axis is now inclined by $22.5°$.

Data consist of the three components of the anomalous magnetic (Figs. 2-4) or of its single vertical (z) component (Fig. 4) collected at nodes 0.1 mm apart in a planar surface $\mathcal{S}$ of overall area 10 mm^2 and centered 1.55 mm above $\mathcal{D}$. The inversion is started from a two-level contrast map valued at $\chi = 0.5 \, \chi_c$ for the three shallower layers of voxels of $\mathcal{D}$ and at $0.01 \, \chi_c$ for the deeper ones, or from a contrast map that evolves with depth as $\chi = 0.5 \, \chi_c \exp \left(-z/\delta \right)$

(Fig.4). No refreshment (decrease of θ) has been enforced at this stage of the study, a proper θ having been equated to 1 by numerical experimentation.

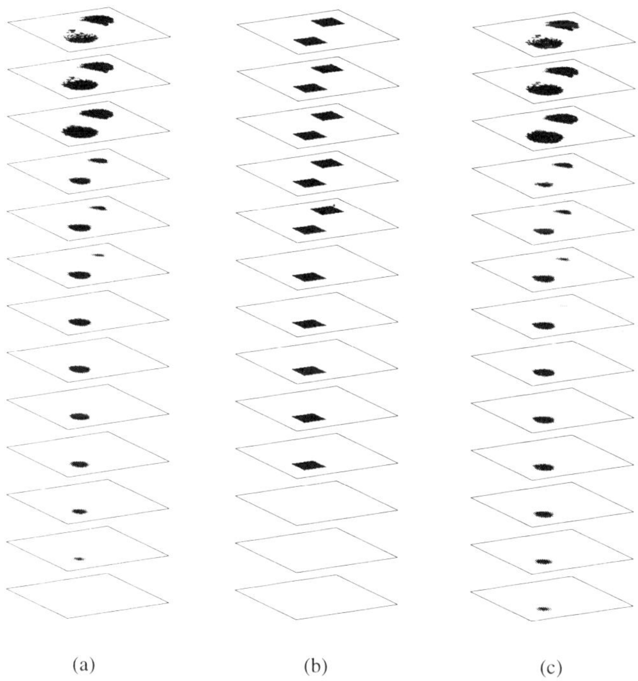

(a)   (b)   (c)

Figure 2: Retrieval of two voids of different depths in a metal slab (see text for details). Cross-sectional gray-level maps of the conductivity contrast χ at successive 0.1 mm-stepped depths using all three components of the anomalous magnetic field (a), and the vertical component only (c). Exact maps are given in (b) for comparison.

From those results retrievals appear almost identical (and the cost functionals evolve similarly) when all three components of the field are used and when the vertical one is the only one used. This is rather comforting in terms of application to real measurement configurations, e.g., by displacing a small coil probe throughout a plane of observation parallel with the probed metal plate. However, only the shallower parts of the voids, down to about $2/3$ of a skin depth, are fairly well retrieved, the volume of those shallower parts being increased with respect to the exact one and those of the deeper parts being decreased. Such shadowing and blurring due to skin effect are expected phenomena that are faced independently of the inversion algorithm used.

Nevertheless, the results displayed here show that the proposed hybrid technique is rather effective. This is illustrated by the retrieval of two close voids (Fig. 2-3) that appear both fairly located and well discriminated, the inclination of a single one (Fig.4) being well reproduced. Obviously one needs an appropriate initial guess to do so. An initial contrast that exponentially decreases as a function of the depth-to-skin-depth ratio from the usual mid value of $\chi_c/2$, does not lead to a better retrieval than an initial contrast that is set to $\chi_c/2$, in the three first layers of voxels and to almost 0 below (the cost functional reaches a value one-third larger in the first case than in the second case).

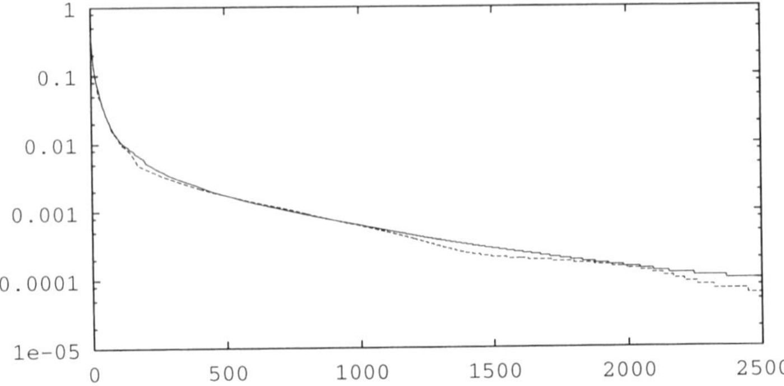

Figure 3: Evolution of the cost functional **F** as a function of the number of iterations for the two voids as studied in Fig. 2. Dotted line, case (c); solid line, case (a).

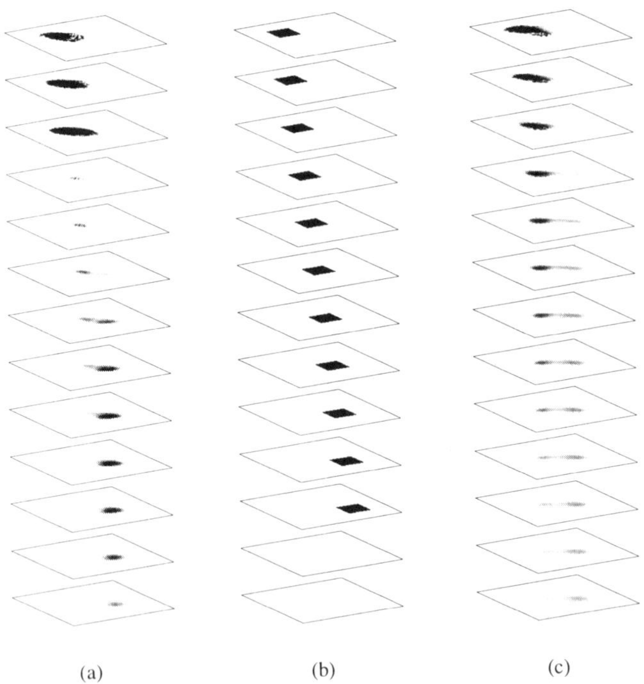

(a)   (b)   (c)

Figure 4: Retrieval of an inclined void in a metal slab (see text for details). Cross-sectional gray-level maps of the conductivity contrast χ at successive 0.1 mm-stepped depths (using all three components of the anomalous magnetic field). The initial contrast χ is chosen as the two-level one in (a), and is chosen as the one exponentially decreasing with depth in (c). Exact maps are given in (b) for comparison.

4   Concluding remarks

So far, data have been assumed to be collected throughout a planar surface (not only along a probing line within a user-prescribed symmetry plane of a defect as is often assumed); this appears to be a pre-requisite when mapping geometrically unknown 3-D objects. Two configurations, other than this multistatic configuration (fixed source – moving field probe), could be attacked similarly, e.g., when source and probe devices are moved together (a reference case would be a SQUID-based magnetometer [7]), or when the variation of the impedance of the probe itself is observed [8]. Work is presently in progress for such two configurations.

However the difficulty in dealing with such new configurations does not lie in the functional analysis, which is carried out as the one sketched here, but in the numerical burden, (one has to construct induced sources for each location of the source device) and in the lack of information (in the second situation, one has to extract pertinent information from a quantity expected to vary more smoothly than the anomalous field).

In addition it remains to study in further detail the influence of the initial guess and of refreshment, and the advantages of frequency-hopping (using data at successive frequencies, from a low one to a high one), whereas the consequence of data noise and model errors should be appraised closely, the synthetic data here having been modeled using the same discretization as the one used in the inversion itself.

References

[1] D. Dos Reis *et al.*, Extended Born domain integral models of diffusive fields, Proc. COMPUMAG'01 Evian, 2001, IV-74-75.

[2] D. Dos Reis *et al.*, On the modeling of 3-D inclusions in conductive media using extended Born models in the diffusive regime, Int. J. Appl. Electromagn. Mechan. (2002) in press.

[3] A. Abubakar and P. M. van den Berg, Three-dimensional inverse scattering applied to cross-well induction sensors, IEEE Trans. Geosci. Remote Sensing 38 (2000) 1669-81.

[4] A. Abubakar *et al.*, A conjugate gradient contrast source technique for 3D profile inversion, IEICE Trans. Electron. E83-C (2000) 1864-1874.

[5] V. Monebhurrun *et al.*, 3-D inversion of eddy current data for Nondestructive Evaluation of steam generator tubes, Inverse Problems 14 (1998) 707-24.

[6] L. Souriau *et al.*, Modified gradient approach to inverse scattering for binary objects in stratified media, Inverse Problems 12 (1996) 463-81.

[7] A. Ruosi *et al.*, High Tc SQUIDS and eddy-current NDE: a comprehensive investigation from real data to modelling, Meas. Sci. Technol. 12 (2000) 1639-48.

[8] T. Takagi *et al.*, ECT research activities in JSAEM Benchmark models of eddy current testing for steam generator tube, Parts 1 and 2. In: R. Collins *et al.* (ed.), Nondestructive Testing of Materials. IOS Press, Amsterdam, 1995, pp. 253-64 and 313-20.

Electromagnetic Nondestructive Evaluation (VI)
F. Kojima et al. (Eds.)
IOS Press, 2002

Numerical Analysis of Ultrasonic Inspection using Electromagnetic Acoustic Transducer

Takahiro MITSUDA and Eiji MATSUMOTO
Department of Energy Conversion Science, Kyoto University, Kyoto 606-8501, Japan

Abstract. This paper analyses ultrasonic inspection of defects using an Electro-Magnetic Acoustic Transducer of Lorentz type. This type of EMATs can be applied to many structural materials without contact, and they can easily excite various wave modes by arranging the coils and the magnets. The eddy current induced by a coil generates the Lorentz force under the static bias magnetic field by permanent magnets. The excited ultrasonic wave is reflected from the boundary or the defects and reaches the boundary surface faced to the EMAT. The eddy current induced by the reflected wave under the bias magnetic field generates the dynamic magnetic field, which is received by the coil through the electromagnetic induction. This paper simulates such transmission and reception processes by the EMAT and attempts to identify an inside crack by scanning the probe over the surface of the specimen.

1. Introduction

In many techniques of Non-Destructive Evaluations, ultrasonics are most fluently used on site because in spite of portability they can obtain the important information from the interior of the object. Ultrasonic NDE is applied not only to inspection of defects such as cracks, voids, delaminations of composite materials but also to evaluation of material properties such as the density and the elastic coefficients. Objectives are widely spread from architectural materials to semiconductors in electromagnetic devices. In general, transmission and reception of ultrasonic waves are achieved by probes which have transformation functions between the electric and the mechanical signals. An EMAT consists of permanent magnets and coils, and utilises the electromagnetic forces and electromagnetic induction in transmission and reception processes. Thus it can be applied to almost metals without contact, even when their surfaces are covered with rust or membranes and they are moving or melted at high temperature, cf. [1]. In the case of piezoelectric or magnetostrictive transducers, the surface of the objective material should be polished or coupling materials are required. On the other hand, EMAT does not need such processes and can exite many wave modes such as longitudinal, transverse and surface waves are by setting arrangement of magnets and coils. It is not difficult to design EMAT suitable for the assigned shape and the size of the object because it utilises non-contact forces.

Two types of EMATs have been developed, i.e., the Lorentz and the magnetostrictive types. When an electric current is supplied to the coil of an EMAT, the induced dynamic magnetic field gives rise to the magnetostriction in the ferromagnetic material, which generates the local wave motion of the object in the magnetostrictive type. The dynamic magnetic field also gives rise to the eddy current near the surface of the conductive material and the Lorentz type of EMATs utilise the Lorentz force due to the eddy current under the static magnetic field by the magnet, cf. [2[. Compared with the magnetostrictive type, the allowable lift-off of the Lorentz type probe is smaller, but by scanning the frequency of the transmission or reception, it can obtain much more information from the object. Ludwig et al [3] and Thompson [4] studied the basic principle of the Lorentz type of EMATs, and Hagi et al [5] numerically analysed transmission of ultrasonic waves by such an EMAT on the basis of FEM.

In this paper, we analyse inspection by an EMAT of the Lorentz type, including reception process of reflected waves from the defect and the boundary of the object. In the next section we derive the basic equations describing the transmission process by the EMAT. In section 3, we lay down the basic equations for wave propagation exited by the Lorenz force and the reception process by the EMAT. In section 4, we simulate ultrasonic inspection and attempt to identify an inside crack by scanning the EMAT probe. We employ FEM for analysis of the electromagnetic fields and the finite difference method for analysis of wave propagation. In the last section, we summarise the obtained results.

2. Electromagnetic Fields around EMAT

Figure 1 shows the sketch of the cross section of a simple EMAT of the Lorentz type. A coil of the track shape is located between the specimen and a couple of permanent magnets with opposite poles. When a high frequency current is supplied to the coil, the induced dynamic magnetic field generates the eddy current near the surface of the conductive specimen. The free charges in the eddy current are subjected to the Lorentz force under the static bias magnetic field by the permanent magnets. The free charges collide to the neutral moleculars and ions, which generates the local motion of the objective material. The ions and the free charges in the conductive material are also subjected to the Coulomb force in the parallel and the opposite directions of the electric field corresponding to the eddy current. It is known that the local motion due to the Coulomb force to the ions and the free charges are cancelled in the conductive material. In a result, only the local motion in the direction of the Lorentz force takes place. It is also known that when the frequency of the driving current is less than 100 MHz, the wavelength of the exited motion is much lager than the mean free path of the charged particles in the room temperature, cf. [4]. This fact means that the above external force by the Lorentz force can be regarded as a continuous field in the space and the time regions, i.e., it can be identified with the Lorentz force determined by the vector fields of the current and the magnetic flux density.

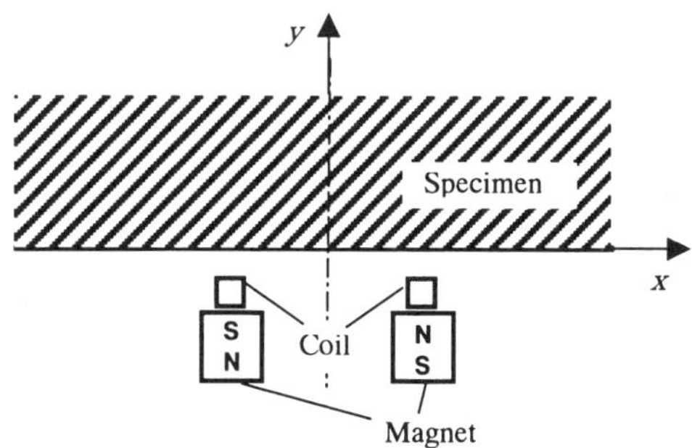

Fig.1 Two-dimensional sketch of EMAT

The electromagnetic fields around the EMAT are governed by the Maxwell equations

$$\nabla \times H = J, \quad \nabla \times E + \frac{\partial B}{\partial t} = 0, \quad \nabla \cdot B = 0, \tag{1}$$

where H is the magnetic field, J the current density, E the electric field, and B the magnetic flux density. Note that the electric displacement and the induced current by it can be neglected in the frequency range of the EMAT. The magnetic and the electric constitutive equations of an electromagnetic material are given by

$$B = \mu H, \quad J = \sigma E,$$ (2)

where μ is the magnetic permeability and σ the electric conductivity of the material. Introducing the vector and the scalar potentials A and ϕ, we can write

$$B = \nabla \times A, \quad E = -\frac{\partial A}{\partial t} - \nabla\phi.$$ (3)

If we divide the current density into the forced current density J_0 and the eddy current density J_e such that $J=J_0+J_e$, we can derive the differential equations for the potentials around the EMAT

$$\Delta A - \mu\sigma\left(\nabla\phi + \frac{\partial A}{\partial t}\right) - \mu J_0 = 0,$$ (4)

$$\nabla \cdot J_e = \nabla \cdot -\sigma\left(\nabla\phi + \frac{\partial A}{\partial t}\right) = 0.$$ (5)

In the following discussions, we assume that the electromagnetic fields are two-dimensional, and the boundary conditions are the forced current density in the coil, the magnetic flux density in the permanent magnets, and the electric conductivities in the coil and the specimen. We also assume that the material is non-magnetic. For efficiency and accuracy, we divide the electromagnetic fields into the static fields due to the permanent magnets and the small dynamic fields due to the driving coil.

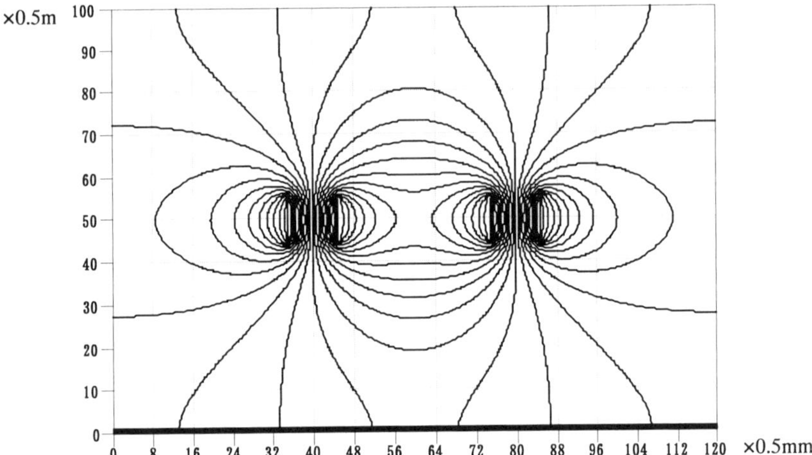

Fig. 2 Equipotential lines of static magnetic field around permanent magnets

We shall show the numerical result for the static and the dynamic electromagnetic fields. When x and y are the two-dimensional coordinates as shown in Fig.1, the vector potential and the current take the forms

$$A = (0,0,A), \quad J = (0,0,J), \quad J_e = (0,0,J_e).$$ (6)

For analysis of the static magnetic field by the permanent magnets, we employ FEM with first order rectangular elements. The length of the minimum edge of a rectangular element is 0.5mm. Each permanent magnet has a square cross section with 5mm edges, and two magnets are located at 15mm distance. The flux density in the permanent magnets is 1.0T. We analyze the magnetic field in the 60mm×50mm quadrangle region around the magnets. The magnetic field is assigned to be normal to the boundary of the analysis region. Since there is no current and the scalar potential vanishes in this case, from (4) the component A of the vector potential is found to be governed by the Laplace's equation. The calculated equipotential lines of A is shown in Fig. 2.

We next consider the dynamic magnetic field generated by the coil of the EMAT. In this case, the component A and the scalar potential ϕ are governed by the system of partial differential equations (4) and (5). We use similar finite elements to the static analysis in the region containing the specimen of aluminum with quadrangle cross section 30mm×10mm. The driving coil is made of copper wire of the truck shape with quadrangle cross section 1mm×1mm and located over the center of the specimen surface with 1.0mm lift-off. The distance between the centers of two wires is 7mm. The material parameters used in the calculation are as follows. The permeabilities of the environment (the air), aluminum and copper are assumed to equal to the one of the vacuum 1.26×10^{-6}H/m, the conductivities of aluminum and copper are, respectively, 3.64×10^{7}S/m and 5.99×10^{7}S/m, and the environment is assumed to be an insulator. We supply one cycle of sinusoidal current to the coil such that the direction of the current in each wire is opposite to each other. The current density has the amplitude 1.00×10^{7}A/m^2 and the frequency 100kHz. Figure 3 indicates the distribution of the induced eddy current density in the specimen at the instance of its maximum. We see that the eddy current is locally distributed near the surface of the specimen faced to the EMAT coil.

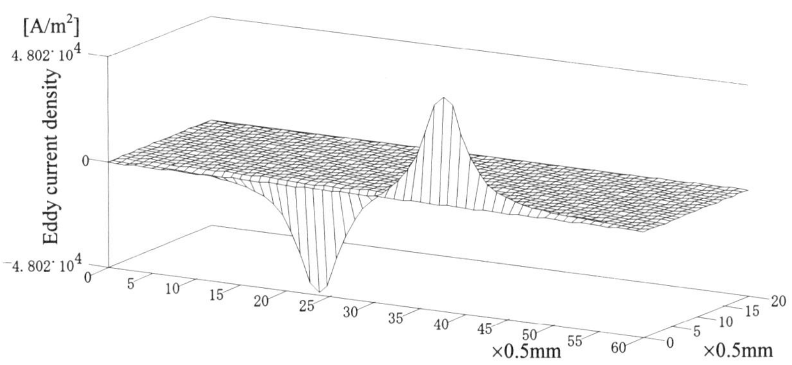

Fig.3 Eddy current distribution in conductive material by EMAT coil

3. Ultrasonic Wave Propagation and Reception Process by EMAT

If the magnetic flux density and the eddy current density are obtained, the Lorentz force per unit volume is given by

$$F = J_e \times B. \tag{7}$$

Since we are concerned with the two-dimensional problem discussed in the previous section, from (6) the magnetic flux density $\boldsymbol{B}$ and the Lorentz force $\boldsymbol{F}$ take the forms

$$\boldsymbol{B} = \left(B_x, B_y, 0\right), \quad \boldsymbol{F} = \left(F_x, F_y, 0\right) = \left(-J_e B_y, J_e B_x, 0\right). \tag{8}$$

Thus the Lorentz force lies on the x–y plane, and its distribution has symmetry with respect to the y-axis from the location of the coil and the magnets, refer to Fig.1. That is, since J_e and B_y are antisymmetric and B_x is symmetric with respect to the y-axis, from (8) F_x and F_y become, respectively, symmetric and antisymmetric. Figure 4 shows the simulated components of the Lorentz force near the surface of the specimen. Here the forced current, the specimen and the coil are the same as in the previous section. Each of the permanent magnets has 3mm×3mm square cross sections and the flux density 1.0T, and they are located just below of each wire of the coil with common center lines. It is found that each components of the Lorentz force has two peaks corresponding to each pair of wire and magnet and the maximum strength of the horizontal component F_x is larger than that of the vertical one F_y. In other words, this type of EMAT mainly excites the transverse wave. Since the eddy current and the Lorentz force are localized within the skin depth of the dynamic magnetic field, an elastic wave is excited in the neighborhood of the boundary faced to the EMAT.

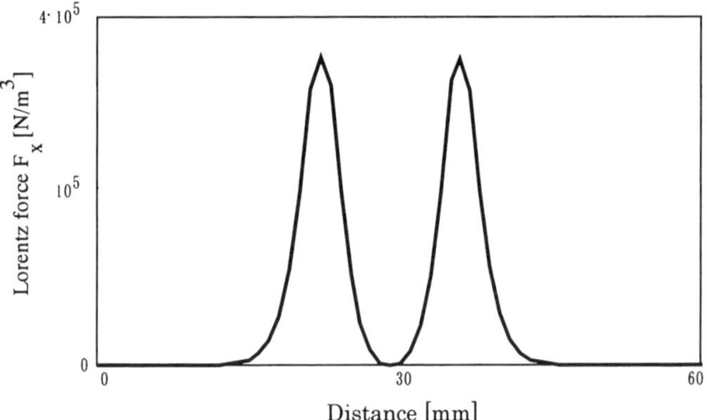

Fig. 4(a) Lorentz force parallel to specimen surface

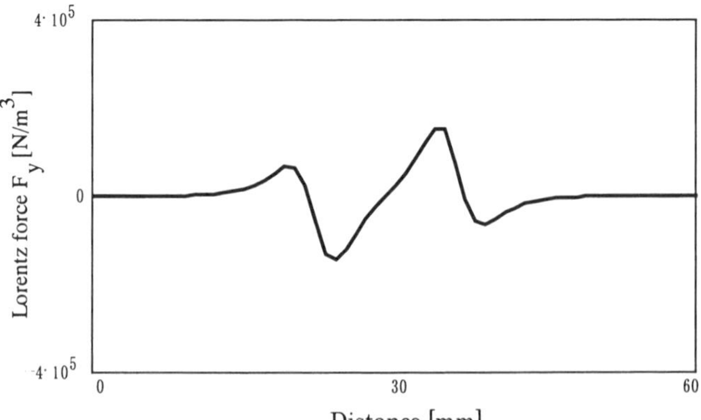

Fig. 4(b) Lorentz force normal to specimen surface

We next assume that the specimen is an isotropic homogeneous elastic material. Propagation of elastic waves in the material is governed by the balance of linear momentum

$$\rho \frac{\partial^2 u_i}{\partial t^2} = \frac{\partial \sigma_{ij}}{\partial x_j} + F_i \tag{9}$$

and the constitutive equation

$$\sigma_{ij} = \lambda \cdot \varepsilon_{kk} \delta_{ij} + 2\mu \varepsilon_{ij}, \tag{10}$$

where double indices in each term are summed from 1 to 3, u is the displacement, ρ the material density, σ the stress tensor, and λ and μ the Lame's elastic constants. The strain tensor ε is defined by

$$\varepsilon_{ij} = \frac{1}{2}\left(\frac{\partial u_i}{\partial x_j} + \frac{\partial u_j}{\partial x_i} \right). \tag{11}$$

The external body force in (9) is given by the Lorentz force (8). The mechanical fields are expressed in terms of the material coordinate system, i.e., in the Lagrangian form, so that the velocity and the acceleration of a material point are denoted by the partial derivatives with respect to time.

The elastic wave excited near the boundary surface by the Lorentz force is reflected from other boundary surfaces and the flaws such as cracks and inclusions. When a material point is moving under the static magnetic field B_0 by the permanent magnets, $v \times B_0$ term appears in the effective electric field. Thus the induced current takes the form

$$J = \sigma\left(E + \frac{\partial u}{\partial t} \times B_0 \right). \tag{12}$$

The current (12) plays a similar role to the forced current in the coil for the transmission process. That is, for the reception process of the EMAT, instead of (4) and (5) the vector and the scalar potentials are governed by

$$\Delta A - \left[\mu\sigma\left(\nabla\phi + \frac{\partial A}{\partial t} + \frac{\partial u}{\partial t} \times B_0 \right) \right] = 0, \tag{13}$$

$$\nabla \cdot \left[\sigma\left(\nabla\phi + \frac{\partial A}{\partial t} + \frac{\partial u}{\partial t} \times B_0 \right) \right] = 0. \tag{14}$$

Then the waveform of the reflected wave is given by the voltage variation in the coil generated by the electromagnetic induction due to the perturbed magnetic field.

Similarly to the case of the electromagnetic fields, we assume that the mechanical fields are also two-dimensional. Then the displacement induced by the Lorentz force (8) always lie on the x–y plane, so that the induced $v \times B_0$ term is in the direction of the z–axis. Thus also in the reception process of the EMAT, the electromagnetic fields are two-dimensional such that the magnetic field lie on the x–y plane and the electric fields are directed to the z–axis. It is easy to see that similarly to the transmission process, the EMAT coil is more sensitive to the horizontal motion of the specimen.

4. Simulation of Ultrasonic Inspection by EMAT

In this section, we first simulate the total process of the transmission and the reception of the elastic wave by the EMAT. As we have discussed so far, the transverse and the longitudinal waves are exited near the boundary surface of the specimen. For the numerical calculation of the elastic waves, we employ the finite difference method such that a difference formula of higher order is applied to the time domain. Using the same EMAT as in the previous section, we transmit the elastic wave into the specimen. In this case we set the amplitude of the current density in the coil $1.00\times10^7\text{A/m}^2$ and the frequency 1.0MHz. The specimen is aluminum with quadrangle cross section 300mm×50mm and a crack with 20mm×2mm cross section exists at the centre of the specimen. Figure 5 shows the wave fields near the crack at time 12μs. The longitudinal and the transverse components of the wave field are given by $\nabla\cdot\boldsymbol{u}$ and $\nabla\times\boldsymbol{u}$, respectively. We see that the longitudinal wave is reflected from the crack corners and the transverse wave is reflected from the crack surface and diffracted beyond the crack. This implies that a mode transformation from the transverse to the

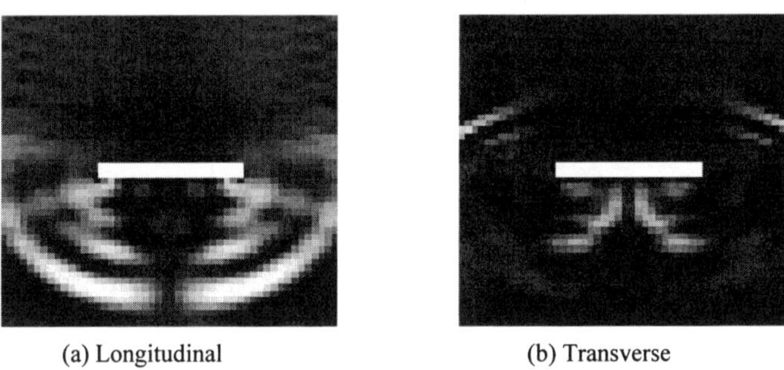

(a) Longitudinal   (b) Transverse

Fig.5 Wave fields around a crack with 20mm×2mm cross section

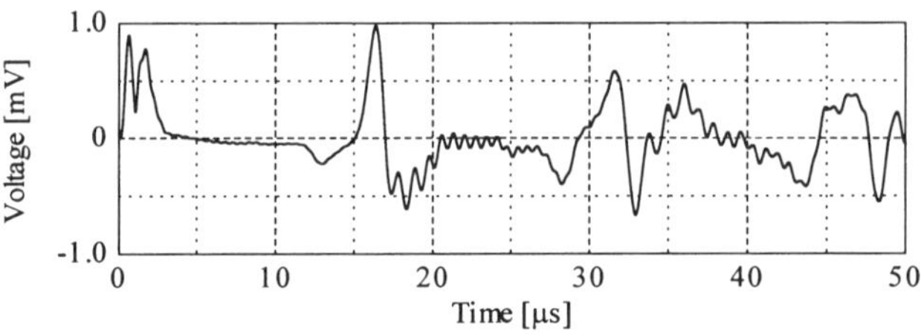

Fig.6 Received waveform by EMAT

longitudinal waves takes place at the corner of the crack. The received waveform by the coil of the EMAT is given in Fig.6. The first peak indicates the incident wave and the second one at 16.5 μs and the third one at 33.0 μs indicate the reflected transverse waves from the crack surface and the opposite boundary surface, respectively. As shown in Fig.4(a) and Fig.4(b), the EMAT also excites the longitudinal wave by the component F_y of the Lorentz force. Although the amplitude of the exited longitudinal wave is not negligibly small compared with the one of the transverse wave, the signal of the longitudinal wave is not clear in Fig. 6. This fact comes from that in the reception process the EMAT is not sensitive to the longitudinal motion of the specimen similarly to the transmission process. This characteristic is convenient to ultrasonic inspection by means of excluding the effects of other waves with different speeds bringing errors to the estimation. From Fig.4(a) the source of the transverse motion does not have a uniform distribution and has two peaks. We see however that the wave front transmitted by the EMAT is almost straight near the center of the path from Huygens principle, which means that the wave is nearly a plane wave. This property of the exited waves is also convenient to the inspection.

We next scan the EMAT probe linearly from the original center position in the right direction by keeping the constant lift-off. Figure 7 shows the distribution of the flying time of the first reflected wave for each position of the probe. By converting the half of the flying time into the distance by means of the speed of the transverse wave, we can identify the location and the width the crack. We see that the estimated position is somewhat far from the boundary on the EMAT side. The cause may be that we have not taken into account the transmission time of the incident wave 1μs and the observed flying time of the wave was longer. The estimated crack edge is not sharp because the cylindrical waves including the longitudinal wave are reflected from the crack edge by the collision of the plane transverse wave, which makes the peak of the pure reflected transverse wave ambiguous. Although taking into account these points, we can state that the EMAT in consideration can detect the inside crack in certain accuracy.

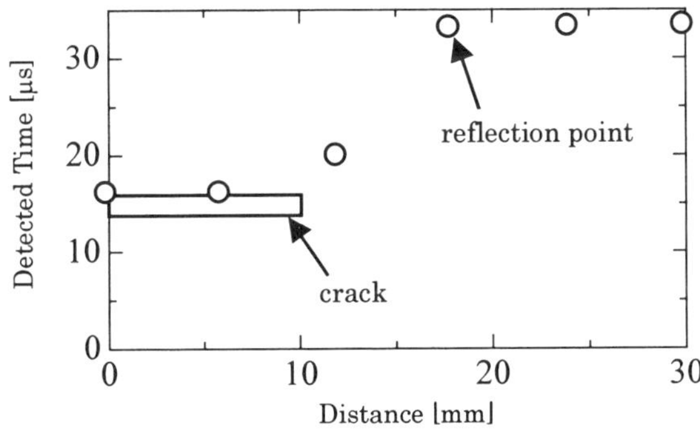

Fig. 7 Identification of crack by EMAT

5. Conclusions

Throughout the paper, we obtained the following conclusions.
1. Governing equations are derived for the transmission and the reception of elastic waves by EMAT.
2. Distribution of the Lorentz force by a typical type of EMAT is obtained.
3. The coupled electromagnetic and mechanical equations for dynamic motions caused by the EMAT are numerically solved.
4. The simulated results imply that the EMAT is suitable for the transmission and the reception of transverse waves and that simple linear scanning of the EMAT probe is effective to detect an inside crack in the conductive specimen.

References

[1] H. Ogi, M. Hirao and H. Fukuoka, Acoustoelastic Stress Measurement on Railroad Rails Using Electromagnetic Acoustic Transducer, *Trans. JSME* **60A** (1994) 881-887.
[2] H. Ogi and M. Hirao, Line-Focusing of Ultrasonic SV Wave by Electromagnetic Acoustic Transducer, *J. Acoust. Soc. Am.* **103** (1998) 2411-2415.
[3] R. Ludwig and X-W. Dai, Numerical Simulation of Electromagnetic Acoustic Transducer in the Time Domain, *J. Appl. Phys.* **69** (1991) 89-98.
[4] R. B. Thompson, Physical Acoustics 19. Academic Press, New York, 1988.
[5] H. Ogi, M. Hirao, K. Minoura and H. Fukuoka, Quasi-Nonlinear Analysis of Lorentz-Type EMAT by Finite Element Method, *Trans. JSME* **61A** (1995) 638-645.

Electromagnetic Nondestructive Evaluation (VI)
F. Kojima et al. (Eds.)
IOS Press, 2002

Finite Element Model for Crack with Narrow Gap in ECT Problem

Motoo TANAKA[1], Hajime TSUBOI[1], Kenichi OOSHIMA[2] and Mitsuo HASHIMO[2]

1) Department of Information Processing Engineering, Fukuyama University,
Gakuen-cho, Fukuyama, Hiroshima 729-0292, Japan.
E-mail: tanaka@fuip.fukuyama-u.ac.jp,   tsuboi@fuip.fukuyama-u.ac.jp

2) Department of Electrical Engineering, Polytechnic University,
4-1-1, Hashimotodai, Sagamihara, Knagawa 229-1196, Japan.
E-mail: hasimoto@uitec.ac.jp

Abstract. It is important to develop an efficient approximation model for the natural crack and the narrow gap in finite element analysis of ECT problems. In this paper, a finite element model for approximation of the crack in the ECT problems is proposed. In the proposed model, the anisotropic conductivity is introduced for the finite element method based on the hexahedral edge element. Numerical results and experimental results of the ECT problems are shown, and the validity of the proposed model is verified by comparing both the numerical results and the experimental results.

1. Introduction

Various types of numerical methods in eddy current testing (ECT) problems have been proposed and the validities have been verified. However, the conventional numerical method, such as finite element method (FEM) and boundary element method, have a problem on practical use in the analysis of the natural crack and the crack with narrow gaps, because a large amount of computer resources are necessary in generating the accurate computation model. The number of unknowns for the final simultaneous equations for the crack model increases in proportion to the complexity of the crack shape.

On the other hand, the aim of the eddy current testing is to detect the crack from ECT probe response caused by the change of the eddy current due to the crack. The ECT probe response is the change of the coil impedance and it is obtained from the magnetic flux linkage of the probe coil. Accordingly, the response is given by the integral equation taking account of the vector potential. It is more important to calculate the impedance of the coil than a detailed analysis of the eddy current distribution in the ECT problem. Therefore, the approximation methods [1,2] of the crack were proposed in the numerical analysis.

In this paper, a finite element model to approximate the crack in the ECT problems is proposed. In the proposed model, the anisotropic conductivity is introduced to the finite element method based on the hexahedral edge element. The proposed finite element model is applied to an ECT problem with a narrow gap crack. The computational results and the experimental results of the ECT problem are compared and the validity of the proposed model is verified. The applicability of the proposed model as the approximation method of the natural crack is shown by the comparison with the experimental results for a fatigue crack with several contact conditions.

2. ECT problem

In this section, the calculated ECT signals are compared with experimental results to estimate the conductivity for the finite element model. An ECT problem consists of two conductor blocks and a coil as shown in Fig. 1. There is a gap between the two blocks in the model, and the gap is assumed a crack. The pancake-type coil moves parallel to x-axis. The lift off is 0.1mm and the internal diameter 1.1mm, the external diameter 2.2mm. The frequency of the exciting current is 100 kHz. The coil moves from -3.5mm to 3.5mm with steps of 0.5mm.

The finite element code [3,4] using hexahedral edge elements was applied to the model. The influence of the conductivity on the ECT signal was investigated. The ECT signals for the gap of 0.05mm are shown in Fig 2. In the case of 2.5×10^7 S/m, the behavior of the computation results is similar to the experimental results near 0mm. Accordingly, the conductivity of the computation model was assumed to be 2.5×10^7 S/m. The computational results of the ECT signal agree with the experimental results for each gap as shown in Fig. 3.

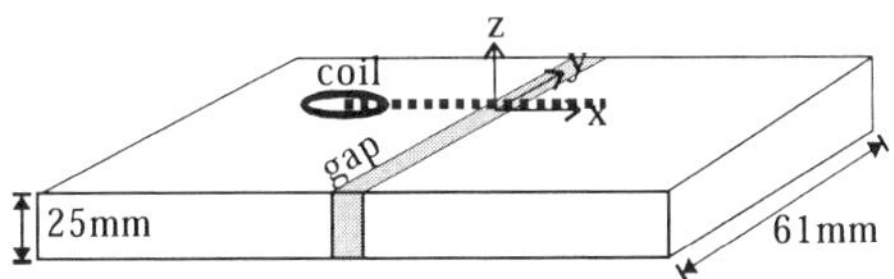

Fig.1 ECT problem.

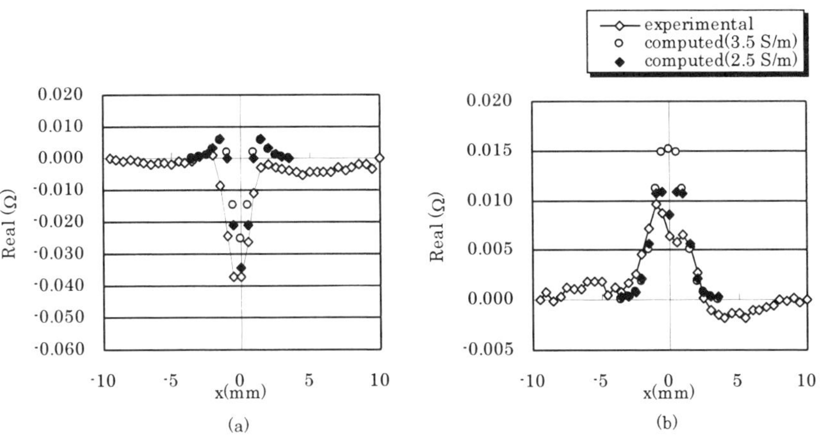

Fig. 2. Impedance change and coil position, (a) gap:0.5 mm, (b) gap:0.05 mm.

3. Novel Finite Element Model

The Galerkin's weighted residual equation for the finite element method using the edge element in the eddy current problem is given by the following equation.

$$\iiint_V \{\frac{1}{\mu} \nabla \times N \cdot \nabla \times A + j\omega\sigma A \cdot N\}dV = 0 \tag{1}$$

where μ is the permiablity, σ the conductivity, ω the angular frequency, A the magnetic vector potential, N the interpolation function and V the region to be analysed.

The computational resources for the eddy current analysis around the crack increase with the complexity of the shape of the crack, because many finite elements are necessary to generate an accurate model of the crack. A special element for the crack is necessary to complete the eddy current analysis in practical time.

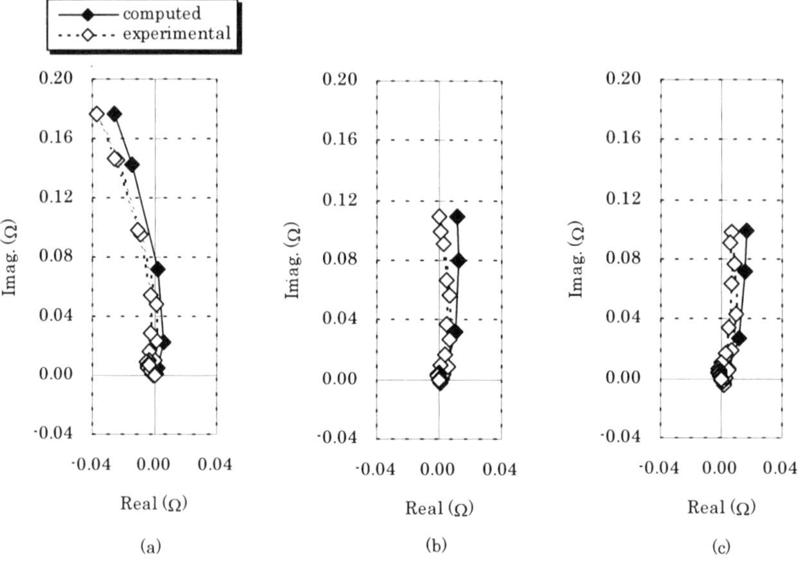

Fig. 3. ECT signals and gap, (a) gap: 0.5 mm, (b) gap: 0.1 mm, (c) gap: 0.05 mm.

Fig. 4 shows eddy current vectors in a finite element containing the crack. The eddy current depends on the crack shape and the contact conditions of the crack in the element. Each component of the eddy current vector can be expressed by

$$J_x = -j\omega C_x \sigma A_x \tag{2}$$

$$J_y = -j\omega C_y \sigma A_y \tag{3}$$

$$J_z = -j\omega C_z \sigma A_z \tag{4}$$

where σ is the conductivity of the conductor, J is the eddy current. C is a constant by which the change of the equivalent conductivity for each direction in the element is expressed. The constant C for each element is determined in the relation to the crack shape and the contact conditions. The value changes form 0% to 100%.

When the approximation in (2)-(4) is introduced to (1), the following equations are obtained.

$$\iiint_V \{\frac{1}{\mu} \nabla \times N \cdot \nabla \times A + j\omega(\sigma^* A) \cdot N\} dV = 0 \tag{5}$$

$$\sigma^* = \begin{pmatrix} C_x\sigma & & 0 \\ & C_y\sigma & \\ 0 & & C_z\sigma \end{pmatrix} \tag{6}$$

Where σ^* is the matrix which expresses the anisotropic conductivity due to the crack.

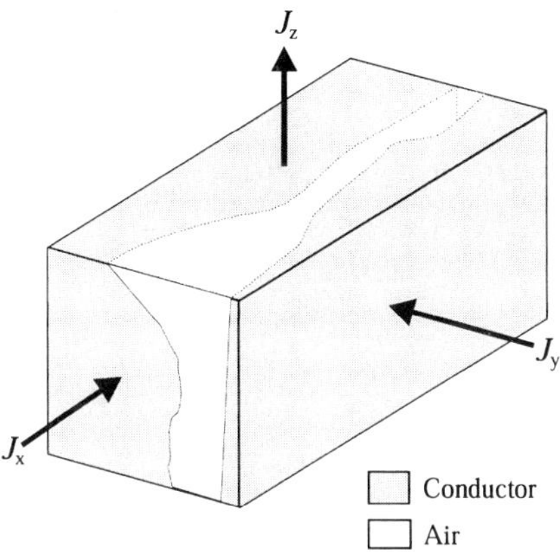

Fig. 4. Finite element containing conductor and a crack.

3. Numerical results

The ECT model in Fig. 1 was used to investigate the validity of the proposed model. The maximum values of the ECT signals are plotted in Fig. 5. The computational results of the conventional model and the experimental results are same as Fig 3.

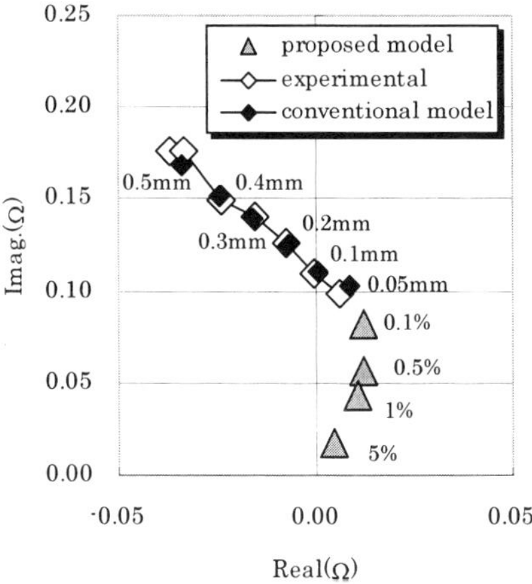

Fig. 5. Computational results of the proposed finite element model.

The proposed finite element model was applied to the gap in Fig. 1. The thickness of the finite element was 0.025mm. The labels (0.1%, .., 5%) in Fig. 5 denote the values of Cx in (2). In these cases, the finite element includes the thinner crack than the thickness of the finite element. When the Cx decreases, the calculated values move continuously to zero from the calculated results of 0.05 mm. We can verify the proposed finite element model is able to simulate the thinner crack than the traditional finite element.

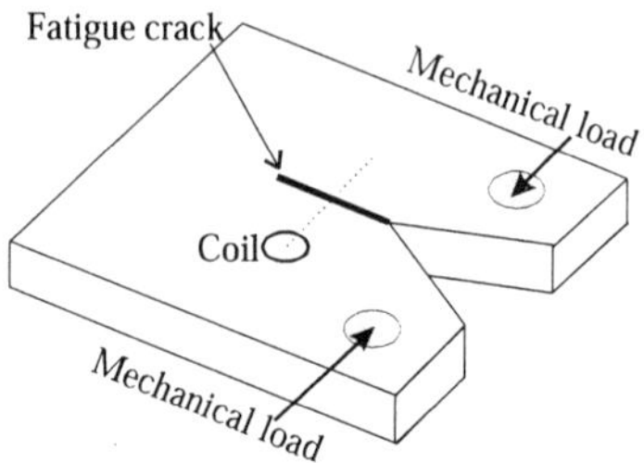

Fig. 6. Fatigue crack model.

The applicability of the proposed method to approximate the natural crack was investigated by the comparison between the computational results of the finite element method and the experimental results of fatigue crack model. Fig. 6 shows the fatigue crack model for the experiment. Both sides of the crack was pressed or expanded by the mechanical load. Therefore, the contact condition is changed by the change in the mechanical load.

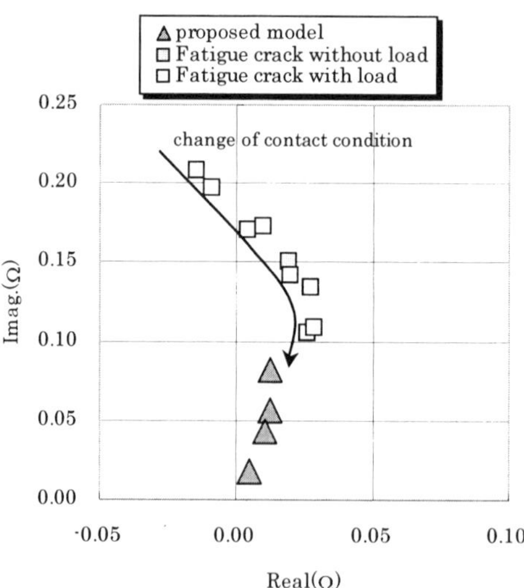

Fig. 7. Computational results of the proposed finite element model and Experimental results of the fatigue crack model.

Fig. 7 shows the comparison between the computation results of the proposed finite element model and the experimental results. The maximum values of the impedance change for each load are plotted in Fig. 7. In this case, the width of the fatigue crack without the load is from 5 μm to 10 μm, and the coil was moved along the dotted line. The computation results are same as Fig. 5. The ECT signals of the experimental results change in proportion to change of the contact condition of the fatigue crack. The ECT signals of the proposed finite element model are small compared with the experimental results, and the ECT signals have the similar tendency to the experimental results and move continuously to zero from the experimental results. Therefore, it seems that the computation model simulated the contact conditions that both sides of the crack is strongly pressed than the experiment.

From these results, the possibility of the proposed method as the approximation method of the fatigue crack was verified, when the parameter σ^* in (6) can be given by any functions.

4. Conclusions

A finite element model for the crack in the analysis of the ECT problem was proposed. The proposed finite element model was applied to the ECT problem and the computational results and the experimental results were compared.

(1) The finite element model which conductivity is approximated by the anisotropic conductivity was proposed.

(2) The finite element code based on the edge finite element using the proposed finite element model was applied to the ECT problem with a narrow gap. Both the conventional finite element model and the proposed finite element model were compared, and it was verified that the proposed finite element model is able to model the narrower crack than the conventional finite element model.

(3) The computational results of the proposed finite element model and the experimental results of the ECT model with the fatigue crack model were compared, and the applicability of the proposed model as the approximation method of the natural crack was shown.

References

[1] H.Fukutomi, H.Huang, T.Takagi and J.Tani, Identification of Crack Depths from Eddy Current Testing Signal, IEEE Transactions on Magnetics, Vol. 34, No.5,pp.2893-2896,1988.

[2] M.Tanaka and H.Tsuboi, Finite Element Model of Natural Crack in Eddy Current Testing Problem, Proceedings of IEEE CEFC2000, Milwaukee, USA, June 2000, p.209.

[3] M.Tanaka, M.Tsuboi et al, Computational Results of the Benchmark Problem Step 5-3: Conductor Plate with a Support Plate, Electromagnetic Nondestructive Evaluation (IV), S.SS. Upda et al.(Eds.), IOSS Press, 2000.

[4] M.Tanaka, K.Ikeda and H.Tsuboi, Fast Simulation Method for Eddy Current Testing, IEEE Transactions on Magnetics, Vol. 36, No. 4, July 2000.

Electromagnetic Nondestructive Evaluation (VI)
F. Kojima et al. (Eds.)
IOS Press, 2002

Eddy Current Analysis by Integral Equation Method for ECT

Kazuhisa ISHIBASHI
Mechanical Engineering, Faculty of Engineering, Tokai University
2-28-4 Tomigaya, Shibuya-ku, Tokyo, 151 Japan
E-mail: isibasi@keyaki.cc.u-tokai.ac.jp

Abstract. In eddy current analysis for the eddy current testing (ECT), it is essential to estimate the electromagnetic fields around cracks accurately because the fields contain information as to the cracks. In order to get accurate solutions, a method of the eddy current analysis for ECT is proposed by using the surface integral equation method. The unknowns of the surface integral equations are the loop electric and surface magnetic currents. The loop electric current satisfies automatically the condition that the divergence of the surface electric current is zero. The magnetic surface current satisfies the condition that the total magnetic charge is zero. In modeling the cracks, a crack element introduced. In order to check the effectiveness and adequacy of the proposed method, we solve the benchmark models of ECT.

1. Introduction

Eddy current testing (ECT) is a type of non-destructive methods for detecting a crack in a conductor. In eddy current analysis for ECT, we have to evaluate the electromagnetic fields accurately for detecting the crack. In the case of employing the integral equation method for ECT analysis [1, 2], there exist two methods. One is a direct method. We determine the electromagnetic fields directly by solving the surface integral equations. The other is an indirect method. We determine the electromagnetic fields by introducing the dipole current replacing the crack. In this paper, we discuss how to get the accurate solution by the direct method formulated by the surface integral equation method.

Applying Green's theorem, we can obtain integral representations of electromagnetic fields from Maxwell's equation [3]. Obtaining the electromagnetic fields on the surface of the conductor and enforcing the continuity conditions of the tangential and normal components of the electromagnetic fields, we get surface integral equations whose unknowns are the surface electric and magnetic currents, J_s and K_s [4]. Here, we call the method employing the surface integral equations with the surface electric and magnetic currents as the unknowns "conventional SIEM". At the sharp edge, the normal and tangential directions can not be defined and the continuity conditions become void. Therefore, the conventional SIEM is employed for solving eddy current problems when the skin depth is small and the surface of the conductor is flat [5]. In ECT analysis, we have to solve accurately eddy currents near the crack including sharp edges and corners. The skin depth for ECT is large for finding the crack in the conductor. In addition, as the crack width becomes narrower compared with the depth, the surface integral equations become ill conditioned and so we cannot get accurate computed results. For ECT analysis by the surface integral equations, a crack element for formulating the eddy current with the narrow crack width has been introduced [6].

In order to get accurate solutions even when the skin depth is large, an approach to analyze the eddy current has been proposed by employing surface integral equations whose unknowns are loop electric and surface magnetic currents, I_ℓ and K_s [7]. By introducing I_ℓ, the essential electromagnetic condition that the divergence of J_s is zero is satisfied automatically. By employing a constant surface element for K_s, the condition that the sum of the magnetic charge is zero is satisfied automatically. Here, we call the method employing the surface integral equations with the loop electric and surface magnetic currents as the unknowns "improved SIEM". In this paper, we employ the improved SIEM for ECT analysis. In order to check the effectiveness and adequacy of the proposed method, we solve the benchmark models of ECT and compare the computed results with the experimental ones [8].

2. Formulation of Eddy Current for ECT

2.1 Integral Representations of Electromagnetic Fields

The integral representations of electromagnetic fields are derived from Maxwell's equation by applying Green's theorem [3]. The magnetic and electric fields, H_{op} and E_{op}, in space with a permeability and permittivity, μ_o and ε_o, and the magnetic and electric field, H_{ip} and E_{ip}, in a conductor with a permeability and conductivity, μ_i and σ, are given as follows.

$$H_{op} = H_{ep} + \int_S \left[J_s \times \nabla G_o + H_n \nabla G_o \right] dS, \tag{1}$$

$$E_{op} = E_{ep} - \int_S \left[j\omega\mu_o J_s G_o + K_s \times \nabla G_o + E_n \nabla G_o \right] dS, \tag{2}$$

$$H_{ip} = -\int_S \left[\sigma K_s G_i + J_s \times \nabla G_i + H_n \nabla G_i \right] dS, \tag{3}$$

$$E_{ip} = \int_S \left[j\omega\mu_i J_s G_i + K_s \times \nabla G_i \right] dS \tag{4}$$

where ω is the angular frequency H_{ep} and E_{ep} are the exciting magnetic and electric fields at a calculating point P_0 produced by electromagnetic sources, respectively, S is the surface of the conductor, the subscript p denotes the value at P_0 and $j = \sqrt{-1}$. Also, we define

$$J_s = n \times H_s, \quad H_n = n \cdot H_s, \quad K_s = -n \times E_s, \quad E_n = n \cdot E_s$$

with the electric and magnetic fields on the surface of the conductor E_s, H_s and the unit normal n directed from the inside to the outside of the surface,

$$G_o = \frac{\exp(-k_o r)}{4\pi r}, \qquad G_i = \frac{\exp(-k_i r)}{4\pi r}$$

with the distance r from a point on the surface to P_0 and

$$k_o = j\omega\sqrt{\mu_o \varepsilon_o}, \qquad k_i = \sqrt{j\omega\mu_i\sigma}.$$

The normal components of E_s and H_s are given also as

$$E_n = -\nabla \cdot J_s /(j\omega\varepsilon), \quad H_n = -\nabla \cdot K_s /(j\omega\mu)$$

with the permittivety ε and the permeability μ of the medium.

2.2 Surface Integral Equations for Eddy Current Analysis

The surface electric and magnetic currents have to satisfy conditions that the divergence of J_s is zero and the sum of the surface magnetic charge is zero. In order to satisfy the former condition automatically, we use a loop current I_ℓ circulating along the edge of the surface element. In order to satisfy the latter condition automatically, we use a constant surface element for K_s.

Next, we derive the integral equations suitable for I_ℓ and K_s to be determined.

Obtaining the normal component of the magnetic field on the surface of the conductor from (1) and (3) and considering the boundary condition of the continuity of the magnetic flux density B, that is $\mu_o\left(n\cdot H_{op}\right)=\mu_i\left(n\cdot H_{ip}\right)$, we get

$$n_p\cdot\int_S\left[\sigma K_s\nabla G_i + J_s\times\nabla\left(G_o+\frac{\mu_i}{\mu_o}G\right)+H_n\nabla(G_o+G_i)\right]dS=-n_p\cdot H_{ep}\cdot \tag{5}$$

Choosing P_0 on the surface of the conductor and taking the vector product on (4) with n, we obtain

$$\frac{K_{sp}}{2}-n_p\times\int_S\left[j\omega\mu_i J_s G_i + K_s\times\nabla G_i\right]dS = 0\cdot \tag{6}$$

The unknowns of (5) and (6) are the loop electric and surface magnetic currents, I_ℓ and K_s, because J_s and H_n are given as functions of I_ℓ and K_s. By dividing the surface of the conductor to be analyzed into n small elements and introducing I_ℓ and K_s, we obtain from (1)

$$J_s = 2n_p\times H_{ep}+2n_p\times\sum_{i=1}^{n}\left[\oint_C I_\ell u\times\nabla G_o d\ell+\int_{\Delta S}\frac{\nabla_s\cdot K_s}{j\omega\mu_o}\nabla G_o dS\right], \tag{7}$$

$$H_n = 2n_p\cdot H_{ep}+2n_p\cdot\sum_{i=1}^{n}\left[\oint_C I_\ell u\times\nabla G_o d\ell+\int_{\Delta S}\frac{\nabla_s\cdot K_s}{j\omega\mu_o}\nabla G_o dS\right] \tag{8}$$

where u is the direction of the loop current, C is the line integral route along the loop current and ΔS is area of the small element.

3. Eddy Current Analysis

We solve a benchmark model of ECT [8]. We analyze eddy currents of a metal plate (140x140x1.25 mm) with a crack (10x0.2x0.75 mm) as shown in Fig.1.

The ratio of the crack depth to the plate thickness is 0.6 (*ID*=60 %). Conductivity and relative permeability of the plate are 10000 S/cm and 1. An exciting coil with 140 turns is placed above the crack. Inner and outer diameters of the coil are 1.2 mm and 3.2 mm, the height of the coil is 0.8 mm. The current of the coil I_c is 1/140 A and the frequency is 300 kHz. The gap between the lower surface of the coil and the upper surface of the plate is 0.5 mm (L=0.5 mm). We call the value L "Lift-off".

In the numerical analysis, the surface of the conductor is divided into equal surface elements (0.5x0.5 mm). The circulating current I_ℓ flows along the edge of each element. The surface magnetic current K_s is supposed to be constant on each element.

In solving the surface electric and magnetic current of the crack surface, we use the conventional SIEM and introduce the crack element [7]. The matching points for solving the surface integral equations are positioned at the center of each surface element.

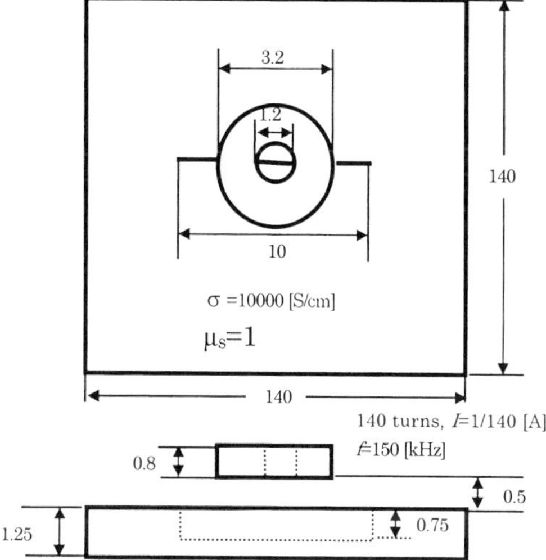

Fig.1 Benchmark model (unit: mm)

In solving the surface integral equations, the iterative sub-domain method is used [9]. The surface of the plate is divided into two regions: one is a region including the crack (near region) and the other is a region far from the crack (far region). The surface electric and magnetic currents, J_s and K_s, on the near and far regions are solved alternately until convergence. First, J_s and K_s on the near region are determined by considering those on the far region to be source currents. Next, J_s and K_s on the far region are determined by considering those on the near region to be source currents. At the first step, J_s and K_s on the far region are supposed to be zero. The rate of the convergence in the iterative procedure to obtain J_{sg} is shown in Fig.2. The convergence is checked by the equation as

$$\varepsilon = \frac{V_{k+1} - V_k}{V_{k+1}}$$

where V denotes the computed value of J_{sg} and the value with the subscript k denotes the one obtained at the k-th iteration and that with the subscript k+1 denotes the one obtained at next iteration. The maximum value of ε is in the case that the exciting coil is positioned at the center of the plate.

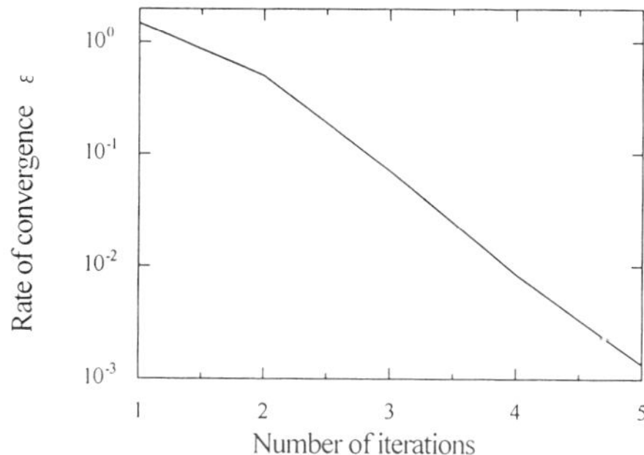

Fig. 2 Rate of convergence in the iterative procedure to obtain J_{sg}

The impedance Z_c of the coil is given by

$$Z_c = \frac{1}{I_c} \oint_c \boldsymbol{E} \cdot d\ell \tag{9}$$

where $\boldsymbol{E}$ is the electric field at the coil. Considering

$$\oint_c \int_S E_n \nabla G_o \, dS \, d\ell = 0 \,,$$

from (2), $\boldsymbol{E}$ is obtained as

$$\boldsymbol{E} = \boldsymbol{E}_{ep} - \int_S \left[j\omega\mu_o \boldsymbol{J}_s G_o + \boldsymbol{K}_s \times \nabla G_o \right] dS \tag{10}$$

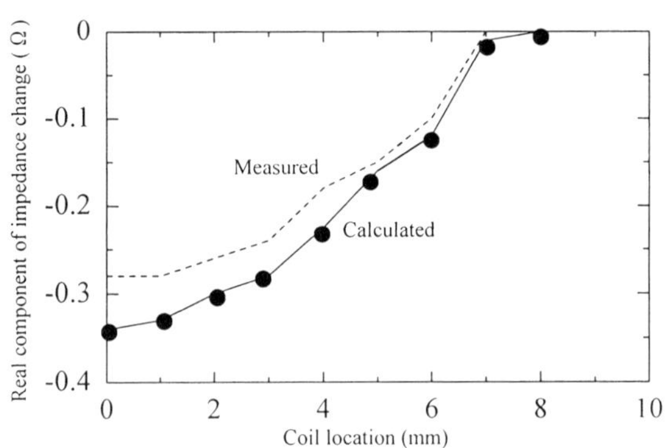

Fig. 3. Real component of impedance change
(ID=60 %, f=300 kHz, L=0.5mm)

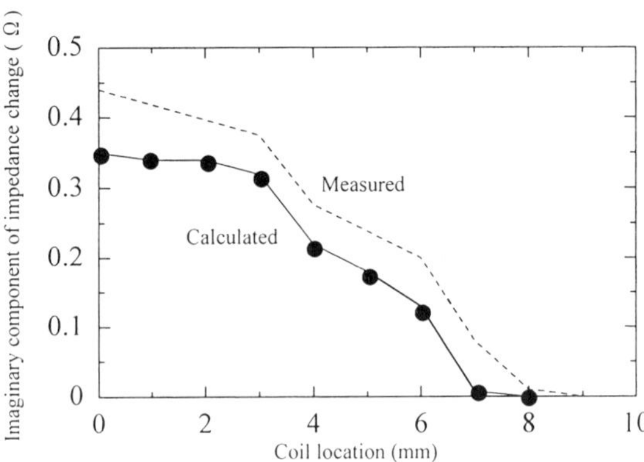

Fig. 4. Imaginary component of impedance change
(ID=60 %, f=300 kHz, L=0.5mm)

Impedance change of the exciting coil is obtained as the coil moves from the middle of the plate along the crack at every 1 mm. The real component of the impedance change along the crack from the middle of the plate is shown in Fig. 3 and the imaginary component is shown in Fig. 4. The computed results are shown by the solid line and experimental values [8] of these components are shown by the dotted line.

4. Conclusions

We have studied an approach for analyzing eddy currents induced in a conductor with a crack by employing surface integral equations. The unknowns of the surface integral equations are loop electric and surface magnetic currents. These unknowns are simple elements but these satisfy automatically the essential electromagnetic conditions that the divergence of the surface electric current is zero and the sum of the surface magnetic charge is zero.

As the width of the crack become narrower, the integral equations become ill conditioned. The crack element for formulating the eddy current at the crack is introduced. In solving the surface integral equations, the iterative sub-domain method is used. The computed results were compared with the experimental ones and we find that these results were different a little bit. We compared also with the results obtained by the conventional SIEM and we find that the results were almost the same. The unknowns of the improved SIEM are about 3/4 of those of the conventional SIEM.

Acknowledgment

The author of this paper acknowledges the members of the Integration of Eddy Current Testing Research Committee of the Japan Society of Applied Electromagnetics for their suggestive discussion about the experiment and analysis.

References

[1] R. Albanese, G. Rubinacci, and F. Villone, "An Integral Computational Model for Crack Simulation and Detection via Eddy Currents, " *Journal of Computational Physics,* Vol.152, pp.736-755, 1999.

[2] J. Bowler and S. Jenkins, "Eddy-current probe impedance due to a volumetric flaw," *Journal of Applied Physics,* Vol.70, No.3, pp.1107-1114, 1991.

[3] J. A. Stratton, *Electromagnetic Theory*, McGRAW-HILL, pp.464-466, 1941

[4] Johnson J.H. Wang, *Generalized Moment Method in Electromagnetics*, John Wiley & Sons, 1991

[5] K. Ishibashi, "Eddy Current Analysis by Boundary Element Method utilizing Impedance Boundary Condition," *IEEE Trans. Magn.*, Vol.31, No.3, pp.1500-1503, 1995

[6] K. Ishibashi, "Numerical Analysis of Eddy Current Testing by Integral Equation Method," *IEEE Trans. Magn.*, Vol.37, No.5, pp.3229-3232, 2001.

[7] K. Ishibashi, "Eddy Current Analysis by Integral Equation Method utilizing Loop Electric and Surface Magnetic Currents as Unknowns," *IEEE Trans. Magn.*, Vol.34, No.5, pp.2585-2588, 1998.

[8] T. Takagi, M. Hashimoto, H. Fukutomi, M. Kurokawa, K. Miya, H. Tsuboi, J. Tani, T. Serizawa, Y. Harada, E. Okano and R. Murakami, "Benchmark models of eddy current testing for steam generator tube," *International Journal of Applied Electromagnetics and Mechanics,* Vol.5, pp.149-162, 1994

[9] K. Ishibashi, "An Iterative Computational Method for ECT by BEM," *ELECTROMAGNETIC NONDESTRUCTIV EVALUATION*, Studies in Applied Electromagnetics and Mechanics 12, Ed. by T. Takagi, J.R. Bowler and Y. Yoshida, IOS Press, pp.31 36, 1997.

Eddy Current Testing

Electromagnetic Nondestructive Evaluation (VI)
F. Kojima et al. (Eds.)
IOS Press, 2002

Achievement of RFEC Effects in the Nuclear Fuel Rod Inspection by using Shielded Encircling Coils

Young-Kil Shin
*School of Electronic and Information Engineering, Kunsan National University,
Kunsan, Chonbuk, 573-701, Korea*

Abstract. An encircling remote field eddy current (RFEC) probe is designed to inspect the nuclear fuel rod. The exciter coil is shielded to force the electromagnetic energy to penetrate into the rod, and the sensor coil is also shielded to avoid direct influence from exciter fields. The operating frequency and effects of shielding are studied by the finite element analysis and the location for the sensor coil is decided. Numerical simulation with the designed probe shows similar signal characteristics to those of inner diameter RFEC probe. The signals show almost equal sensitivity to inner diameter and outer diameter defects, and the linear relationship between phase signal strength and defect depth is observed.

1. Introduction

The nuclear fuel rod that houses nuclear fuel pellets is the first defense wall against the catastrophic radioactive contamination. Consequently, special attention has been paid to maintain the perfect integrity of the rod and even the irradiated spent fuels have been stored underwater for a long time. Nuclear fuel rods are inspected by the helium leakage testing during the manufacturing process. However, tiny pinhole type defects are sometimes discovered while they are in service [1]. Other than the helium leakage testing, the ultrasonic and eddy current testing are frequently used [2-4].

In this paper, a new inspection method is proposed that uses the remote field eddy current (RFEC) phenomena [5-7]. The RFEC technique has been applied for the inspection of conducting tubular products such as oil and gas pipes using the inner diameter bobbin probe. This technique is characterized by the double through-wall transmission of the energy before it reaches the sensor located in the remote field region. Between the two through-wall transmission areas, *potential valley* and *phase knot* had been found by the finite element modeling [6]. In ferromagnetic pipe inspection, typical spacing between exciter and sensor coils is about 3 pipe diameters and the operating frequency is usually very low, from 10 Hz to 400 Hz. The probing signal is the electromotive force (*emf*) induced in the sensor coil and both amplitude and phase information can be used as the signal. The phase signal is in fact the phase difference between the exciter and sensor. These RFEC signals are closely related to the pipe wall condition, thickness, permeability, and conductivity and show distinctive characteristics such as 1) they are equally sensitive to inner diameter (ID) and outer diameter (OD) defects, 2) they are rather insensitive to probe wobble or lift-off, and 3) the defect signal always shows double defect indications, each occurs as the exciter and sensor coils pass a defect.

Recently, the application of this technique has been extended to the inspection of plate structure by focusing electromagnetic energy into the specimen through the use of appropriate shielding [8-11]. The use of shielding to achieve RFEC effects in the plate structure inspired the author to design a shielded encircling RFEC probe for the inspection of nuclear fuel rod. The operating frequency and effects of shielding are studied by using the finite element analysis. In this paper, the shielding for exciter to realize RFEC effects, the decision on the sensor coil location, and the effects of sensor coil shielding are presented. The simulated signals using the proposed probe have very similar characteristics to those of conventional ID RFEC signals. They show that the probe is equally sensitive to both ID and OD defects, and the phase signal strength is linearly related to the defect depth.

2. Finite Element Modeling

The governing equation for RFEC testing can be written as

$$\nabla \times \frac{1}{\mu} \nabla \times \overline{A} = \overline{J}_s - \sigma \frac{\partial \overline{A}}{\partial t} \tag{1}$$

where $\overline{A}$, $\overline{J}_s$, μ, σ are the magnetic vector potential, the source current density, permeability, and conductivity, respectively. Assuming sinusoidal steady state and perfect axis symmetry of the rod, coils, shielding structures and defects, the governing equation in the cylindrical coordinate system can be written as follows:

$$\frac{1}{\mu}\left[\frac{\partial}{\partial r}\left\{\frac{1}{r}\frac{\partial}{\partial r}(rA)\right\} + \frac{\partial}{\partial z}\left\{\frac{1}{r}\frac{\partial}{\partial z}(rA)\right\}\right] - j\omega\sigma A + J_s = 0 . \tag{2}$$

where ω is the angular frequency. By applying the Galerkin's weighted residual method to the above equation, the following elemental matrix equations are obtained.

$$([S] + j[C])\{A\} = \{Q\} \tag{3}$$

where elements of each matrix can be expressed as

$$S_{ij} = \int_{\Omega^e} \frac{1}{\mu}\left\{\left(\frac{N_i}{r} + \frac{\partial N_i}{\partial r}\right)\left(\frac{N_j}{r} + \frac{\partial N_j}{\partial r}\right) + \frac{\partial N_i}{\partial z}\frac{\partial N_j}{\partial z}\right\} 2\pi r\, dr\, dz \tag{4}$$

$$C_{ij} = \int_{\Omega^e} \frac{1}{\mu}\omega\sigma N_i N_j\, 2\pi r\, dr\, dz \tag{5}$$

$$Q_i^e = \int_{\Omega^e} \frac{1}{\mu} J_s N_i\, 2\pi r\, dr\, dz . \tag{6}$$

Here, N_i and N_j are shape functions at each node point in an element. These elemental matrix equations are then summed up to form a global matrix equation and solved for the magnetic vector potential at every node point.

Since RFEC signal is the *emf* of the sensor coil, nodal values over the sensor coil area are used for *emf* calculation as follows:

$$V_{emf} = -j\omega\oint_C \overline{A} \cdot \overline{dl} = -j\omega A_c 2\pi r_c \tag{7}$$

Where A_c and r_c are the centroidal values of magnetic vector potential and the radial distance at each element. Since *emf* is expressed in the form of a complex phasor, both amplitude and phase can be calculated.

3. Design of Exciter and Realization of RFEC Effects

Since the nuclear fuel rod is accessible only from outside, an encircling exciter coil covered with laminations of iron is designed to force the electromagnetic energy to penetrate into the rod. A zircaloy metal rod is used for modeling. Material properties and dimensions used in the finite element modeling are summarized in table 1.

To achieve RFEC effects in the nonmagnetic zircaloy fuel rod, many different operating frequencies are tested and 700 kHz is finally chosen in this work. Fig. 1 shows the logarithmic scaled equipotential plot and equiphase plot. Each plot has a unique pattern that was named as the *potential valley* and *phase knot* in the numerical modeling study of ferromagnetic pipe inspection [6]. The presence of these patterns indicates that the RFEC effects are achieved.

TABLE 1. Material properties and dimensions used in the finite element modeling.

Conductivity of the zircaloy rod	1.4×10^6 [mho/m]	OD of the shield	20.08 [mm]
Relative permeability of the rod	1	Width of the shield	5.724 [mm]
OD of the rod	9.7 [mm]	Width of exciter coil	2.544 [mm]
ID of the rod	8.43 [mm]	OD of exciter coil	12.45 [mm]
Conductivity of the shield	$1. \times 10^7$ [mho/m]	ID of exciter coil	9.91 [mm]
Relative permeability of the shield	1000	Source current density	1.24×10^6 [A/m^2]

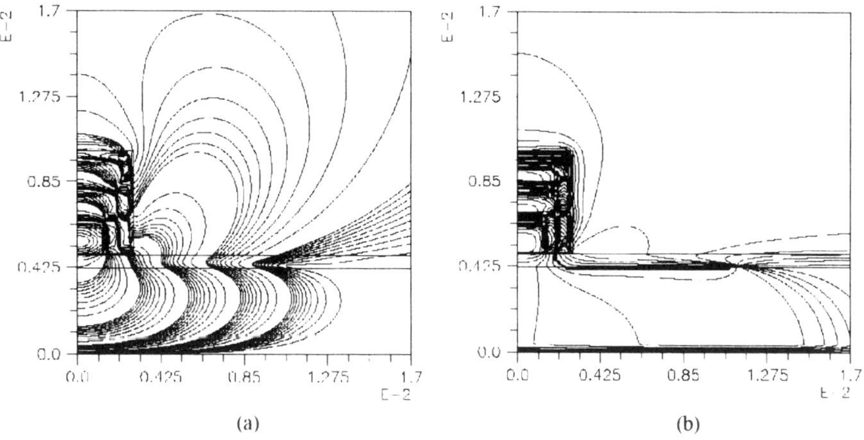

(a)   (b)

FIGURE 1. a) Log scaled equipotential plot and b) equiphase plot at the frequency of 700 kHz.

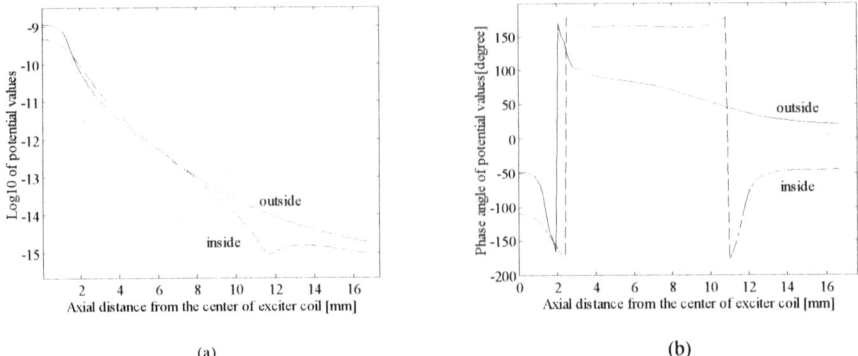

FIGURE 2. Comparison of magnetic vector potentials on the surface outside and inside the fuel rod:
a) log scaled magnitude. b) phase.

Fig. 2 reveals more explicit characteristics of encircling RFEC phenomena occurred in the vicinity of exciter. Fig. 2. (a) compares the log scaled magnitude of magnetic vector potentials on the surface outside and inside the rod. Just under the exciter coil and shield, the potential outside is higher than inside. However, in the proximity of exciter, the potential outside is lower than inside. In the region far from the exciter, the potential outside becomes higher again than inside. Fig. 2. (b), which compares the phase of magnetic vector potentials on the surface outside and inside the rod, also indicates that there exist two transition regions. From these results, it can be said that the energy forced into the rod comes out of the rod in the vicinity of exciter and goes back into the rod in the region far from the exciter.

4. Decision on Sensor Coil Location

In this work, the sensor has to be placed outside the rod. Therefore, the sensor coil needs to be placed in the region where the potential inside is higher than outside. Also, such region is where the double through-wall transmission of energy is completed. Fig. 3 shows the changes occurred in the equiphase line distribution when an OD or ID defect (0.636 mm width x 0.318 mm depth) exists in that region. Whether it is an OD defect or ID defect, changes in the equiphase line distribution direct toward the outside, where the sensor coil should be located. These changes in the plot enable us to predict that the signal would be sensitive to both OD and ID defects. From Fig. 2, proper locations for the sensor coil can be decided approximately depending on the choice of signals. Since magnitude signal would require excessive amplification in the experiment, this paper pays more attention to the phase information. Fig. 2 displays only the magnetic vector potential, not the sensor coil *emf* so that the exact location cannot be decided from the figure. In this work, the spacing between exciter and sensor coils is selected to be 9.54 mm (0.984 OD). Since exciter and sensor coils move together as one probe, the sensor location is equivalent to the exciter to sensor spacing. The size of the sensor coil used is 1.272 mm for both width and height.

Fig. 4 shows simulated phase and magnitude signals for both OD and ID defects whose depth is 50 % of the wall thickness. RFEC signals usually show double defect indications as the sensor and exciter coil pass a defect. As a result, the distance between the two indications in a signal is proportional to the spacing between exciter and sensor. In Fig. 4, the horizontal axis is the axial distance between the centers of exciter coil and defect, and it

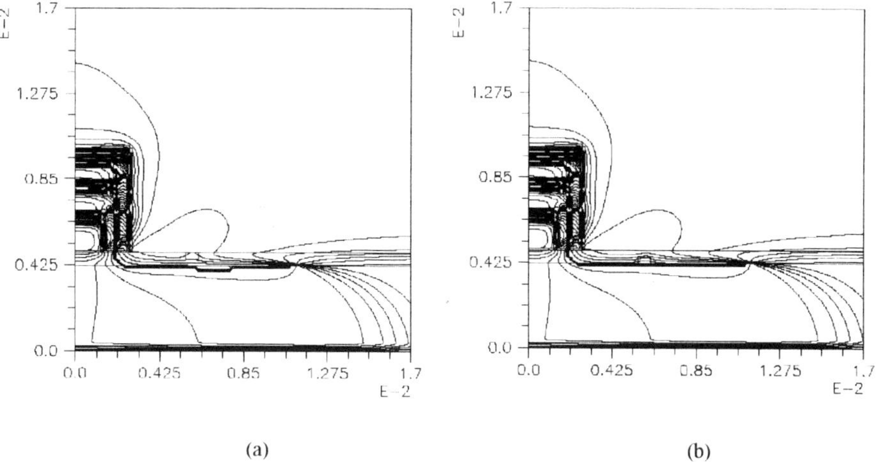

FIGURE 3. Changes in equiphase lines when a defect exists near the exciter: a) OD defect. b) ID defect.

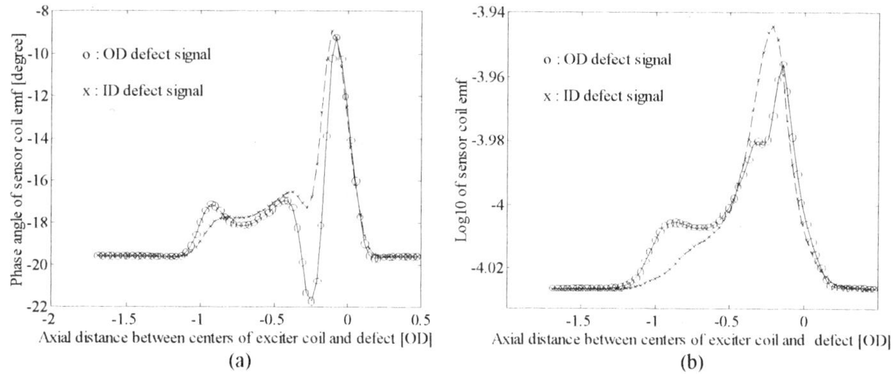

FIGURE 4. OD and ID defect signals: (a) Phase signal. (b) Log scaled magnitude signal.

is converted with respect to the OD of rod. Thus, 0 OD on the horizontal axis means that the exciter coil passes a defect. Since the sensor coil is located in front of the exciter, the first defect indication should appear at –0.984 OD on the horizontal axis. However, signals in Fig. 4 do not clearly show the defect indication at that location. OD defect signals show small indications but they are not strong as much as the indication appeared at 0 OD. Furthermore, even the indication appeared at 0 OD is affected very much by the shielding applied outside the exciter coil. When the defect passes the edge of the shield (-0.29 OD), signals change abruptly. Observing the locations of unexpected signal variations, it is evident that the external fields from the exciter directly affect the sensor signal.

5. Signal Improvement by Shielding Sensor Coil

In the conventional ID RFEC inspection, the sensor coil is located in the remote field region to avoid the direct contact with the fields from exciter and to monitor the indirect fields that have penetrated the pipe wall twice. Since the sensor coil in this work is located close to the exciter, it is necessary to prevent direct contact with the exciter fields. For this reason, the sensor coil is also shielded. The width and height of iron shielding used for the sensor coil are 2.54 mm and 2.04 mm, respectively.

The signals obtained by the shielded sensor are dramatically improved as can be seen in Fig. 5. The first indication evidently appears and it is even greater than the second indication whose strength has changed little. Also, the influence from the external fields in the second indication is removed. Fig. 5 also compares signals from both OD and ID defects. Phase signals from OD and ID defects are almost the same. Strengths of the magnitude signal are also similar. Thus, it can be said that the probe designed in this work is almost equally sensitive to OD and ID defects.

At this point, the choice of operating frequency made in this work is double-checked. Defect signals obtained from the same probe at the frequency of 100 kHz, 400 kHz, and 700 kHz are compared in Fig. 6. As the operating frequency is increased, the strength of defect indication gets higher. However, the middle portion between the two indications changes very much depending on the operating frequency.

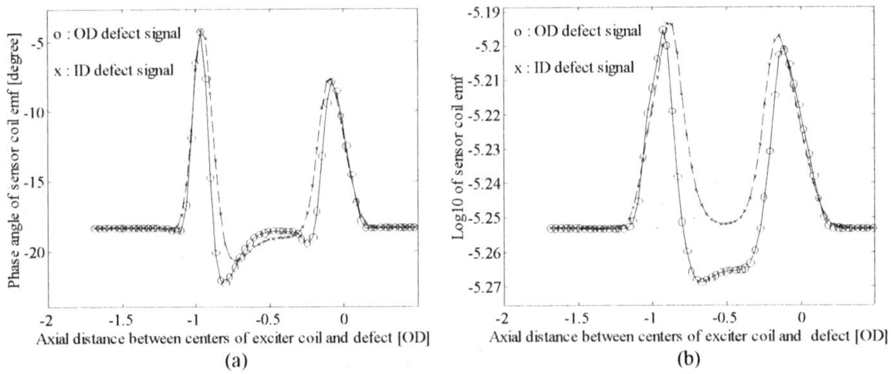

FIGURE 5. OD and ID defect signals obtained by shielded sensor; (a) Phase signal, (b) magnitude signal.

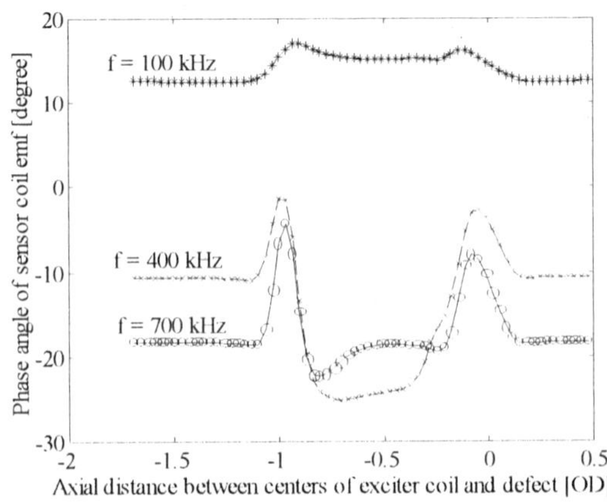

FIGURE 6. Frequency dependence of the phase signal from a 50% deep OD defect.

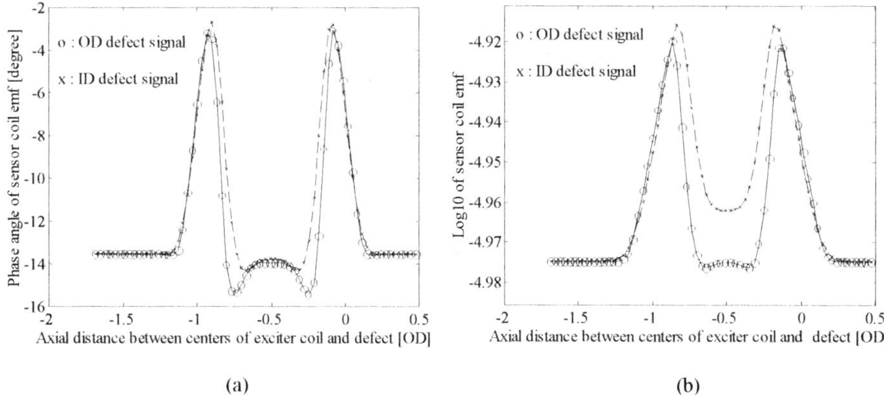

FIGURE 7. OD and ID defect signals when exciter and sensor are identical and the area between them is also shielded: (a) Phase signal, (b) Log scaled magnitude signal.

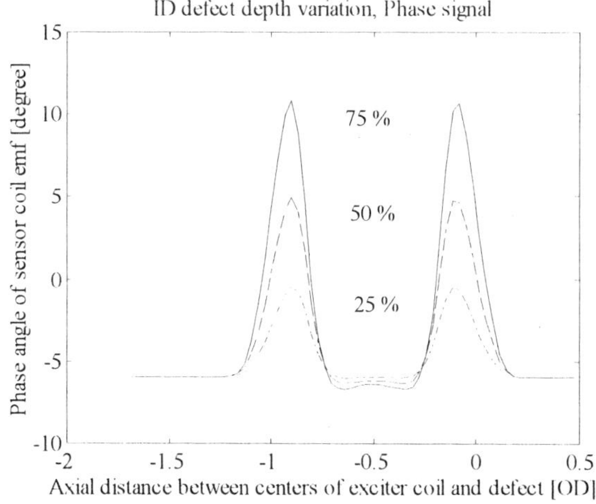

FIGURE 8. Linear relationship between the phase signal strength and the defect depth.

To improve the signal behavior in the middle portion, new shielding is added between the exciter and sensor and at this time the same size of coils and shielding structures are used. The resulting signals from the same defects are shown in Fig. 7. The two indications in both phase and magnitude signals become identical. Signal behavior between the two indications has also improved successfully, especially in the phase signal.

Another finding is that the changes in phase signal strength is linearly related to the defect depth variation. Fig. 8 shows that the phase signal strength changes to 5 °, 10 °, and 16 ° as the defect depth varies 25%, 50%, and 75%, which is approximately linear.

6. Discussion

In this work, two kinds of shielding are used. One is used for the exciter to force electromagnetic energy to penetrate into the rod and the other is used for the sensor to prevent the direct contact with the exciter fields. The results obtained in this study show that the same shielding for exciter and sensor results two identical defect indications in a signal even though their purpose of use is not the same. Also learned is that the smaller coil with lighter shielding produces a stronger signal than the larger coil with heavier shielding.

7. Conclusion

Based on numerical modeling results, this paper proposes an encircling RFEC probe for the inspection of nuclear fuel rods. The shielding of exciter coil to achieve RFEC effects in the rod is shown to be successful and the proper sensor location and effects of its shielding are discussed. The signals obtained by the proposed probe have very similar characteristics to those of conventional ID RFEC signals. They show that the probe is almost equally sensitive to OD and ID defects and the phase signal strength increases linearly as the defect depth is increased. These results show promise that the proposed probe could be used effectively in the inspection of nuclear fuel rods. Experimental verification is under way and will be reported in the near future.

Acknowledgements

This paper was accomplished with Research Fund provided by Korea Research Foundation, Support for Faculty Research Abroad. Suggestion on the sensor location by professor Yushi Sun at Iowa State University is greatly appreciated.

References

1.   D.M. Suh, K.S. Sim, W.J. Kwon, J.H. Kim, and C.H. Park, *J. KSNT* **18** (1998) 85-91.
2.   D.S. Koo, Y.K. Park, and E.K. Kim, *J. KSNT* **16** (1996) 29-33.
3.   J.D. Allinson, An Eddy Current Test Procedure for Defective End Cap Welds, Zircatec Precision Industries, Inc. Preliminary Report, 1992.
4.   J.B. Hallett, et al., *Materials Evaluation* **42** (1984) 1276-1280.
5.   T.R. Schmidt, *Materials Evaluation* **42** (1984) 225-230.
6.   W. Lord, Y.S. Sun, S.S. Udpa, and S. Nath, *IEEE Trans. Magn.* **24** (1988) 435-438.
7.   J.K. Yi, and D.L. Atherton, *Materials Evaluation* **56** (1998) 771-773.
8.   Y.S. Sun, J. Si, D. Cooley, H.C. Han, S.S. Udpa, and W. Lord, *IEEE Trans. Magn.* **32** (1996) 1589-1592.
9.   Y.S. Sun, W. Lord, L. Udpa, S. Udpa, S.K. Lua, K.H. Ng, and S. Nath, Thick-Walled Aluminum Plate Inspection using Remote Field Eddy Current Techniques, In *Review of Progress in QNDE*, D. O. Thompson and D. A. Chimenti(ed.), Plenum **16**, New York, 1997, pp. 1005-1012.
10.   Y.S. Sun, L. Udpa, S. Udpa, W. Lord, S. Nath, S.K. Lua, and K.H. Ng, *Materials Evaluation* **56** (1998) 94-97.
11.   Y.S. Sun, and T. Ouyang, Detection of Cracks in Multi-Layer Aircraft Structures with Fasteners using Remote Field Eddy Current Method, In *Proceeding of SPIE* **3994**, A.K. Mal (ed.), Newport Beach, 2000.

Behavior of the remote-field
on a tube with fan-shaped defect

Michio YASUNISHI, Michiro ISOBE and Masahiro NISHIKAWA
Supra-High Temperature Engineering Lab.,
Course of Electromagnetic Energy Engineering,
Graduate School of Engineering, Osaka University
2-1 Yamada-oka, Suita, Osaka 565-0871, Japan

Abstract. The numerical simulation for the remote-field eddy current testing has been applied to non-ferromagnetic tubes with fan-shaped defects. The results obtained from the simulation code which uses 3D finite-element calculation is good agreement with the experimental results. As scanning the small detector coil along the axis of the tube, we obtained the Lissajous figure in which the maximum magnitude appeared at just below the defect or at the side opposite the defect. However each Lissajous is inverse in the trajectory on a complex plane. To investigate these phenomena, the distribution of the magnetic flux density has been evaluated. The magnetic flux density has the maximum value below the defect when an exciter coil passes through a fan-shaped defect. Its value gradually decreases toward to the center of the inner tube, and then increases toward to the side opposite the defect. On the other hand, the phase gradually shifts from 0 degrees to 150 degrees. Therefore the magnitude and the trajectory of the Lissajous would be related to the position of the detector coil.

1. Introduction

In the in-service inspection of a nuclear system, it is indispensable to inspect the steam generator tubes. The authors have studied the Remote-Field Eddy Current Testing (RFECT) to inspect the non-ferromagnetic tubes. The RFECT is a low-frequency through-wall technique by the internal probes. It is not affected by the lift-off and detects internal and external defects with approximately equal sensitivity [1].

In our previous report [2], the azimuthal changes of the defect signals along with inner wall of the tube sample were measured by the experiment used the dual exciter RFEC probe with a small detector coil. The maximum magnitude of Lissajous appears at just below the defect or at the side opposite the defect.

In this paper, the remote-field eddy current testing simulated by means of 3D finite-element calculation has been applied to the detecting of the defect on non-ferromagnetic tubes with fan-shaped defects. We had predicted that the magnitude of the Lissajous was the maximum value at just below the defect. However, at the side opposite the defect also the magnitude of Lissajous appears the maximum value. To investigate the azimuthal dependence on the defect signal, the distribution of the magnetic flux density has been evaluated, when an exciter coil passes through a fan-shaped defect. Then the correlation between change of the magnetic field and a fan-shaped defect is important to make clear.

                    M. Yasunishi et al. / Behavior of the Remote-Field on a Tube

2. Calculation for comparison with experimental results

Fig.1 shows a schematic diagram of a RFEC probe and energy flow paths. When the current is driven on the exciter coil in a conducting tube, two components of the magnetic field are propagated. One is called as the direct-field component coupled by an exciter coil. The other is called as the indirect-field component coupled by eddy current. The indirect-field component goes through out the tube wall and return back into the inner tube wall. The propagated wave shifts 90 degrees in phase through the tube wall. Then the difference of the phase between the indirect-field component and the direct-field component is about 180 degrees. The indirect-field component dominates in the remote-field region. On the RFECT, a detector coil is placed on the remote-field region to pick up the indirect-field component.

To verify the validity of a numerical simulation code using edge-based finite elements and the magnetic vector potential method, it was applied to the case as shown Fig.1. The sample tubes are fabricated from brass. An exciter coil and a detector coil have the same dimensions.

Fig.2 (a) and Fig.2 (b) show comparisons between experimental results and calculated results on amplitude and phase of the voltage induced by the magnetic field in a tube, respectively. The induced voltages are obtained as the detector coil moved with the exciter coil fixed. In the case of no defect, the calculated results on the magnetic field correspond with the experimental results, so that the calculation on magnetic field is noted to be carried out reasonably.

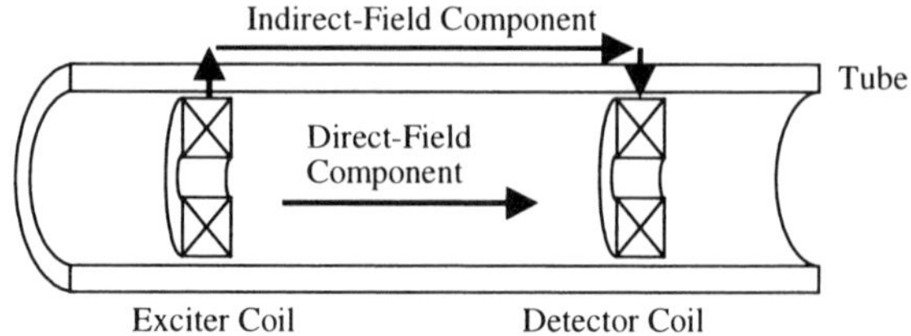

Fig.1 Schematic diagram of a RFEC probe and energy flow paths.

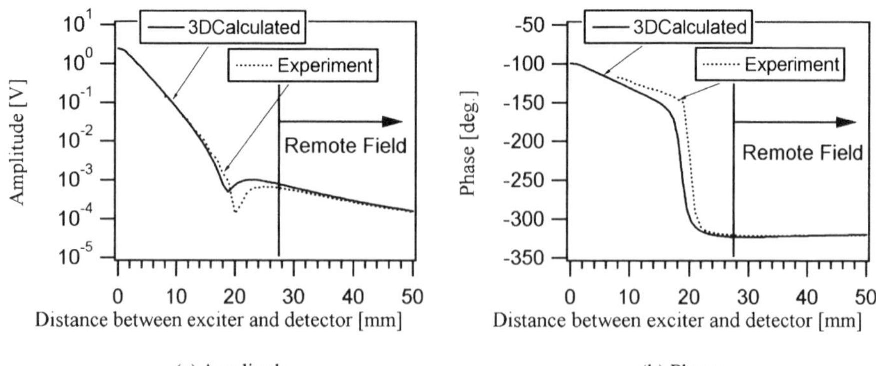

(a) Amplitude   (b) Phase

Fig.2 Comparisons between the experimental results and the calculated results

3. The simulation of defect signals on inspecting a fan-shaped defect

To identify the azimuthal dependence on the defect signal, simulations were carried out in the case of a dual exciter RFEC probe with a small detector coil moving around the inner surface of the tube. A small detector coil is set as shown in Fig.3 and Fig.4. The position **a**, the position **b** and the position **c** are the point at 0 degrees just below the defect, the point at 90 degrees and the point at 180 degrees on the side opposite the defect, respectively. The circumferential angle of the fan-shaped defect is defined as shown in Fig.5. Table 1 summarizes the dimensions, material properties and test conditions.

Fig.6 shows the Lissajous figure associated with a fan-shaped defect (width: 1.5mm, depth: 50%, circumferential angle: 90 degrees). It is observed that the magnitude of Lissajous is the minimum value at the position **b**, and the maximum value at the position **a** or the position **c**. In Fig.7, the trajectory of the Lissajous figure on the position **a** is in the direction opposite to that on the position **c**.

The behavior of calculated remote-field signals shows good agreement with experimental results [2].

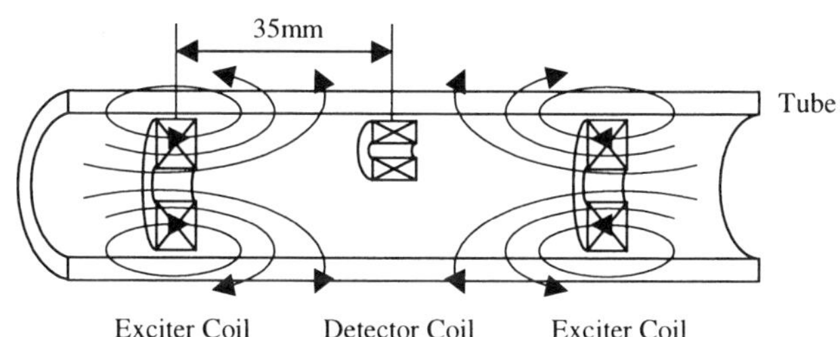

Fig.3 Schematic diagram of a dual exciter RFEC probe with a small detector coil

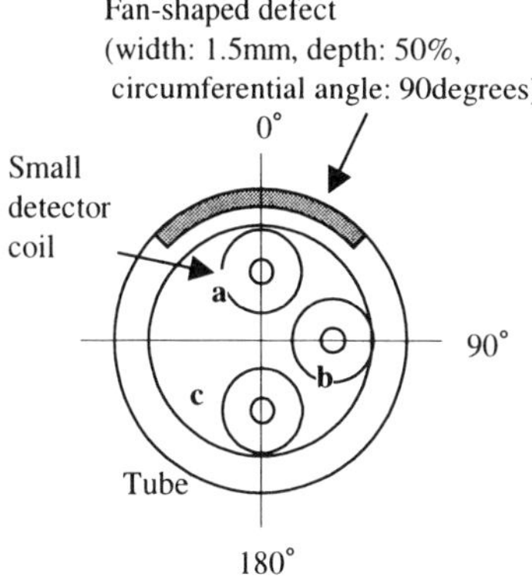

Fig.4 the set positions of a small detector coil

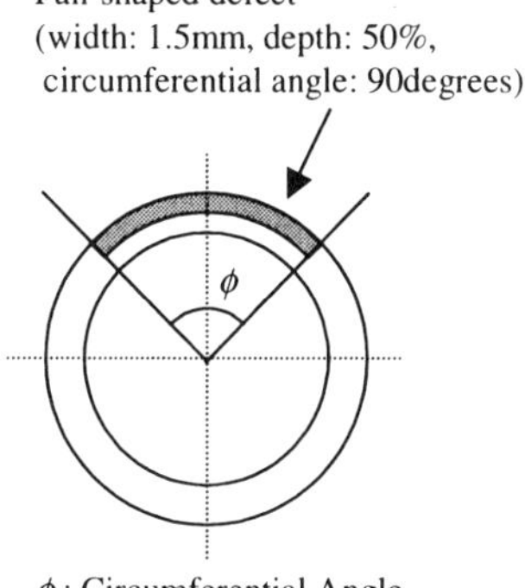

φ : Circumferential Angle

Fig.5 Definition of the circumferential angle

Table 1 Data and material properties used for the calculation

exciter coils-		
	outer diameter	19.0 mm
	inner diameter	7.5 mm
	thickness	3.0 mm
	number of turns	500
	current	57.4 mA
	frequency	3400 Hz
	interval	35 mm
a small detector coil-		
	outer diameter	7.5 mm
	inner diameter	3.0 mm
	thickness	4.0 mm
	number of turns	200
	interval	35 mm
tube-		
	outer diameter	25.4 mm
	inner diameter	19.2 mm
	thickness	3.1 mm
	material	brass
	relative permeability	1
	conductivity	1.47×10^7 S/m

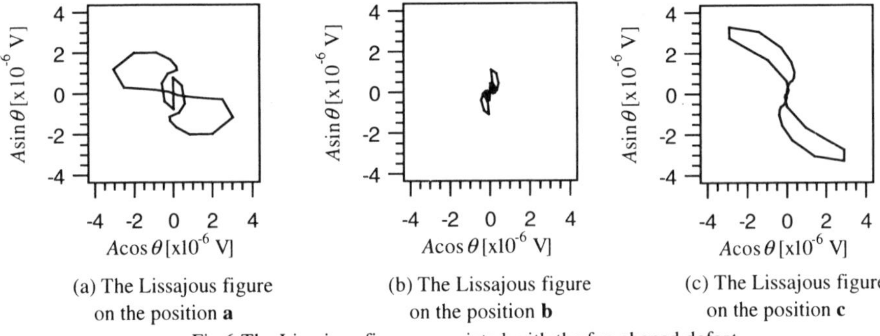

(a) The Lissajous figure on the position **a**   (b) The Lissajous figure on the position **b**   (c) The Lissajous figure on the position **c**

Fig.6 The Lissajous figure associated with the fan-shaped defect

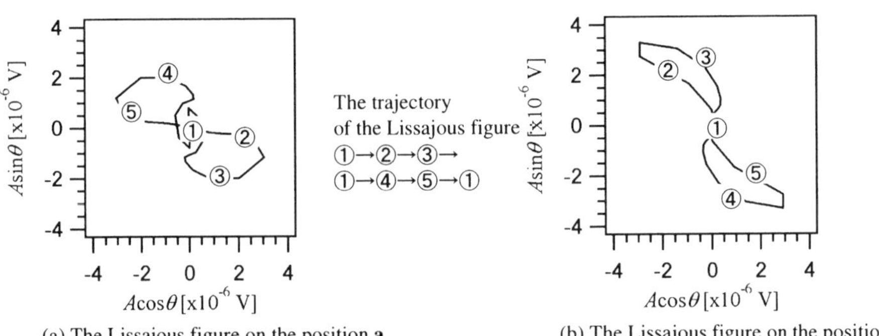

(a) The Lissajous figure on the position **a**   (b) The Lissajous figure on the position **c**

Fig.7 The trajectory on a complex plane about the Lissajous figure on the position **a** and the position **c**

4. The evaluation of the distribution of the axial magnetic flux density

From the calculated results, the phenomena of the defect signal depending on the position of the small detector coil are as follows. It is observed that the magnitude of the Lissajous is the minimum value at the position **b** (the point at 90 degrees) and the maximum value at the position **a** (just below the defect) or the position **c** (the side opposite the defect.). And the trajectory of the Lissajous figure on the position **a** is in the direction opposite to that on the position **c**. We had predicted that the magnitude of the Lissajous was the maximum value at only the position **a** because the point was just below the defect. However, on the position **c** also the magnitude of Lissajous appears the maximum value.

To explain these phenomena, the distribution of the axial magnetic flux density was evaluated, when an exciter coil passed through a fan-shaped defect. Fig.8 shows the configuration the fan-shaped defect and the exciter coils for calculation. Fig.9 shows the amplitude of the axial magnetic flux density on the small detector coil and vicinity. Fig.10 shows the phase of the axial magnetic flux density at the same state with Fig.9.

In Fig.9, the magnetic flux density has the maximum value below the defect when an exciter coil passed through a fan-shaped defect. Its value gradually decreases toward to the center of the inner tube, and then increases toward to the side opposite the defect. In Fig.10, the phase of the axial magnetic flux density shifts gradually from the position **a** toward the position **c**. The phase shows about 0 degrees on the position **a** and about 150 degrees on the position **c**.

In order to understand the behavior of the magnetic flux density at the small detector coil, the eddy current in the tube around the exciter coil is evaluated. Fig.11 shows the distribution of the eddy current density, when an exciter coil passes through a fan-shaped defect. In Fig.11, the density of the eddy current is concentrated around defect. The indirect-field component induced by the localized eddy current propagates on the outside of the tube. Then, the indirect-field component on the side of the defect more propagates than that on the side opposite the defect. Considering the law of conservation of energy, the indirect-field component on the side opposite the defect decreases as much as that increasing on the side of the defect. The indirect-field component on the side opposite the defect less propagates than that in the case of no defect. Hence, at the region of the detector coil, the indirect-field component appears much strength on the side of the defect and becomes week on the side opposite the defect in comparison with the case of no defect. Then, the distribution of the axial magnetic field is found to change gradually from the defect toward to the side opposite the defect inner tube as shown in Fig.9.

Considering the equation of skin depth, the phase shift on the side of the defect (depth: 50%) becomes half value. The phase of indirect-field component on the side of the defect appears different value from that on the side opposite the defect. At the region of the detector coil, the magnetic field consists of summation between the indirect-field component propagating from the exciter coil near the defect and that propagating from the other exciter coil. Then, the phase of the magnetic field on the side of the defect appears different value from that on the side opposite the defect. Hence, the phase of the axial magnetic flux density affected by the fan-shaped defect shifts gradually from the defect toward to the side opposite the defect inner tube as shown in Fig.10.

5. Conclusions

The remote-field eddy current testing simulated by means of 3D finite-element calculation has been applied to the detecting of the defect on non-ferromagnetic tubes. From

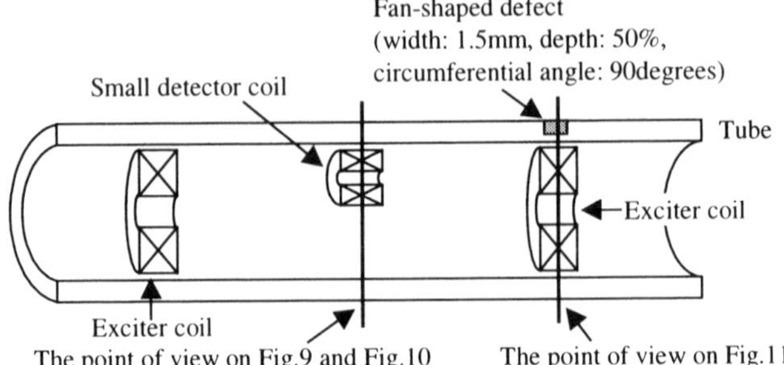

Fig.8 The configuration the fan-shaped defect and the exciter coils for calculation

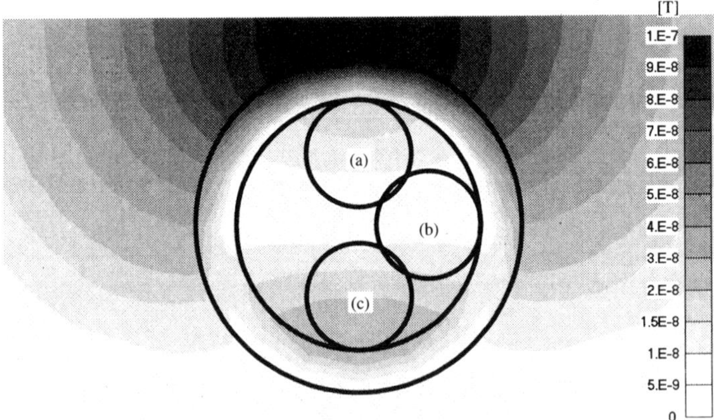

Fig.9 The distribution of the amplitude of the axial magnetic flux density

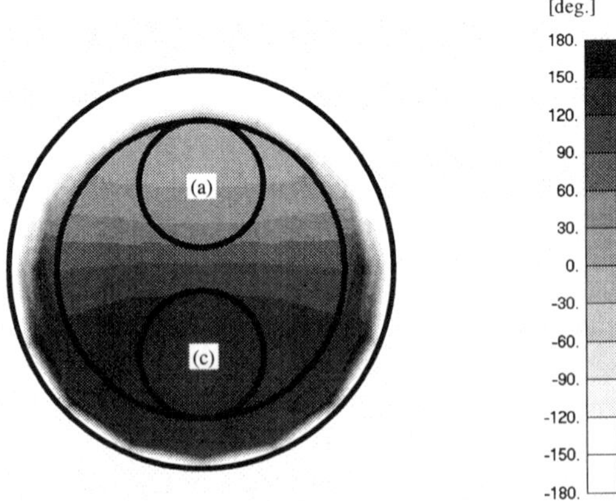

Fig.10 The distribution of the phase of the axial magnetic flux density

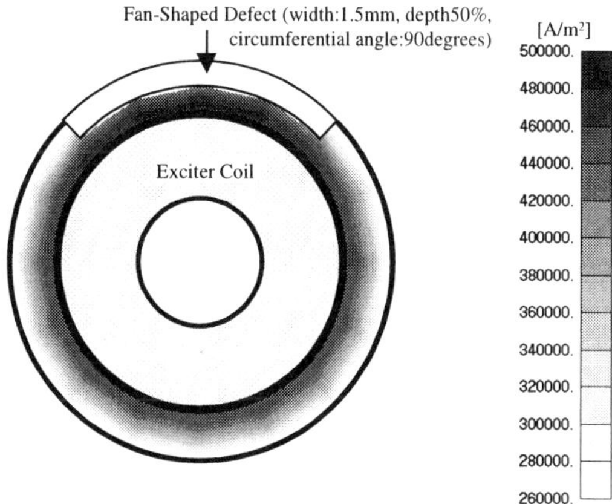

Fig.11 The distribution of the eddy-current density in the tube around the exciter coil

the calculated results, the phenomena of the defect signal depending on the position of the small detector coil are as follows. The magnitude of the Lissajous is the minimum value at the position **b** (the point at 90 degrees from the defect) and the maximum value at the position **a** (just below the defect) or the position **c** (the side opposite the defect.). And the trajectory of the Lissajous figure on the position **a** is in the direction opposite to that on the position **c**. The behavior of calculated remote-field signals shows good agreement with experimental results.

To explain these phenomena above, the distribution of the axial magnetic flux density was evaluated, when an exciter coil passed through a fan-shaped defect. The magnetic flux density has the maximum value below the defect. Its value gradually decreases toward to the center of the inner tube, and then increases toward to the side opposite the defect. And the phase of the axial magnetic flux density shifts from 0 degrees to 150 degrees gradually from the position **a** toward the position **c**.

When an exciter coil passes through a fan-shaped defect, the density of eddy current is concentrated around defect. The indirect-field component induced by the localized eddy current propagates on the outside of the tube. Then, the indirect-field component on the side of the defect more propagates than that on the side opposite the defect. Hence, the indirect-field component on the side opposite the defect less propagates than that in the case of no defect. Therefore, at the region of the detector coil, the indirect-field component appears much strength on the side of the defect and becomes week on the side opposite the defect in comparison with the case of no defect. Then, the distribution of the axial magnetic field is found to change gradually from the defect toward to the side opposite the defect inner tube.

As has been described, the change of the magnetic flux density is associated with a fan-shaped defect. If a detector coil becomes smaller and more sensitive, the more detail evaluation about the feature of a fan-shaped defect can be performed. Also the propagation of the energy flow of the electromagnetic field can be made clear.

References

1] T. R. Schmidt, The Remote Field Eddy Current Inspection Technique, *Materials Evaluation*, 42(2),pp.225-230,(1984)

[2] H. Higuchi, M. Isobe, M. Nishikawa, Defect Characterization by Using Remote Field Eddy-Current Testing, *Electromagnetic Nondestructive Evaluation, IOS Press*, pp153-160, (1997)

Electromagnetic Nondestructive Evaluation (VI)
F. Kojima et al. (Eds.)
IOS Press, 2002

Inversion of ECT Signals from Flaws with Tip Variation in Steam Generator Tubes

Sung;Jin SONG[1], Young H. KIM[1], Chang-Hwan KIM[1], Eui-Lae KIM[1], Young-Kil SHIN[2],
Hyang-Beom LEE[3], Yoon-Won PARK[4] and Chang-Jae YIM[5]
[1]School of Mechanical Engineering, Sungkyunkwan University, Suwon, Korea
[2]Faculty of Electrical Engineering, Kunsan National University, Kunsan, Korea
[3]Department of Electrical Engineering, Soongsil University, Seoul, Korea
[4]Korea Institute of Nuclear Safety, Taejon, Korea
[5]Korea Advanced Inspection Technology, Taejon, Korea

Abstract. This paper reports our recent endeavor to develop automated, systematic inversion tools by the novel combination of neural networks and finite element modelling for eddy current flaw characterization in steam generator tubes. Specially, this paper describes 1) the construction of databases with abundant flaw signals from 2-D axisymmetric flaws with tip variation using finite element models, 2) the extraction and selection of sensitive features for flaw classification and sizing, and finally 3) the inversion of ECT signals by use of two neural networks for flaw classification and sizing. In addition, this paper also presents the performance of proposed inversion tools for classification and sizing of 2-D axisymmetric flaws 1) having symmetric cross-sections with the variation in tip width, and 2) having non-symmetric cross-sections with the variation in tip deviation.

1. Introduction

Automated and systematic inversion of eddy current tesing signals is strongly desired for the reliability enhancement of the integrity evaluation of steam generator tubes in nuclear power plants. To address such an issue, we [1,2] have proposed an intelligent, systematic inversion approach by the novel combination of neural networks and finite element ECT models. The previous works [1,2] have demonstrated the excellent performance of the proposed approach by solving the flaw characterization problems of 2-D axisymmetric machined notches with symmetric cross-sections with great success.

The integrity evaluation based on the fracture mechanics concept, however, quite often requires detailed information on flaw tip geometry as well as the flaw location, depth and width. Thus, to evaluate the performance of the developed approach for addressing such a task, recently the developed approach has been applied to the classification problem of 5 types of 2-D axisymmetric flaws with various cross-sections (including I-shaped symmetric, V-shaped symmetric, I-shaped inclined, V-shaped inclined, and K-shaped) [3]. In this work [3], the discrimination of flaws according to the symmetry in their cross-sectional shape was turned out to be a rather easy task. However, the indentification of flaws in terms of their tip geometry was really difficult.

Thus, the purpose of the present work is to develop inversion tools that can predict the tip width and the tip deviation (from the central line) of 2-D axisymmetric flaws in steam generator tubes by use of the finite element ECT models and neural networks.

2. Databases constructed by finite element ECT models

The finite element ECT models used in this work can describe very carefully the geometry of 2-D axisymmetric flaws with the variations in the tip width and/or deviation, since they adopts flexible combination of quadrilateral and triangular elements. Figure 1 shows the configuration of ECT of an Inconnel 600 tube with the differential bobbin probe for FEM simulation of flaw signals to construct the databases.

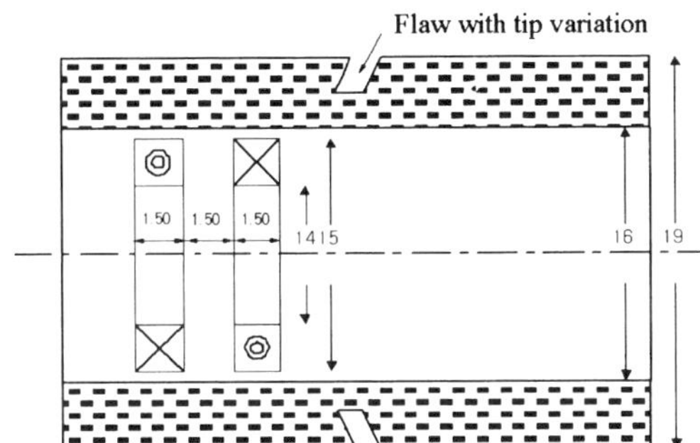

Figure 1. Parameters for the simulation of the eddy curent testing.   (unit: mm)

In this work, we have constructed two databases. Database I has 432 flaw signals simulated from 2 types of symmetric flaws with the variation in the tip width. Two flaw types include "ID" (ID flaws with different tip width) and "OD" (OD flaws with different tip width) flaws. In addition, by changing the depth and tip width of these flaws (with a fixed width at the opening) as shown in Figure 2, a total of 108 flaws were generated, and then from each flaw four signals were obtained with four different testing frequencies of 100, 200, 300 and 400 kHz

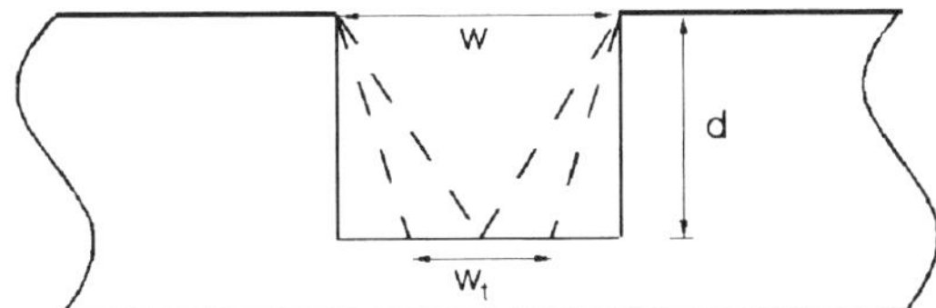

Figure 2. Schematic representation of cross-sections of flaws in the database I.

Database II has 384 flaw signals simulated from 4 types of nonsymmetric flaws with the variation in the tip deviation. Four flaw types include "I-ID" (I-shaped ID flaws with different tip deviation), "I-OD" (I-shaped OD flaws with different tip deviation), "V-ID" (V-shaped ID flaws with different tip deviation) and "V-OD" (V-shaped OD flaws with different tip deviation). Again, by changing the width (at the openning), depth and tip width of these flaws as shown in Figure 3, a total of 192 flaws were generated, and then from each flaw, two ECT signals were obtained with two different testing frequencies of 100 and 400 kHz. In the Figure 2 and 3, w, d, w_t and δ denote flaw width, flaw depth, tip width and tip deviation, respectively.

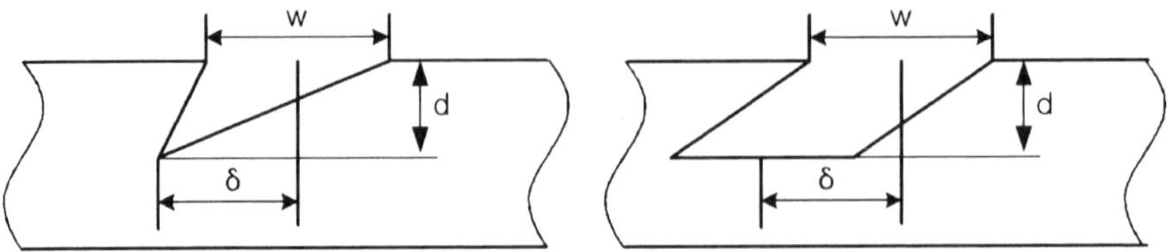

Figure 3. Schematic representation of cross-sections of flaws in the database II.

3. Feature extraction and selection

Even though it is widely recognized that features play one of the most important roles in the inversion of ECT signals, the extraction of really "good" features, however, is not an easy task. For the effective searching of sensitive features as well as their extraction, specially designed feature extraction software has been developed previously [3]. Using this software, from symmetric flaw signal in the database I, we extracted 11 features (listed in Table 1) that described the major characteristics of the impedance plane trajectory. However, from non-symmetric flaw signals in database II, we extracted 23 features (listed in Table 2) that described the major characteristics of both time domain signal and impedance plane trajectory. Figures 4 and 5 schematically present the definitions of features for symmetric and non-symmetric flaws, respectively.

Table 1. Features extracted from symmetric ECT signals.

F1. Max Resistance	F2. Max Resistance Angle	F3. Max Reactance
F4. Max Reactance Angle	F5. Max Impedance	F6. Max Impedance Angle
F7. Starting Angle	F8. Ending Angle	
F9. Turning phase angle at the point of maximum impedance of the signal		
F10. The length up to the maximum reactance point of the signal / The length from the maximum reactance point of the signal		
F11. Total length of the signal / Magnitude of the impedance at the maximum reactance point		

Table 2. Features extracted from non-symmetric ECT signals.

F1. First Peak Resistance	F2. First Peak Resistance Angle	F3. Second Peak Resistance
F4. Second Peak Resistance Angle	F5. |First Peak Resistance–Second Peak Resistance|	
F6. First Peak Reactance	F7. First Peak Reactance Angle	F8. Second Peak Reactance
F9. First Peak Reactance Angle	F10. |First Peak Reactance –Second Peak Reactance|	
F11. First Peak Impedance	F12. First Peak- Impedance Angle	F13. Second Peak Impedance
F14. Second Peak Impedance Angle	F15. Starting Angle	F16. Ending Angle
F17. First Middle Impedance	F18. Second Middle Impedance	
F19. |First Middle--Second Middle|	F20. Junction Point Magnitude	F21. Junction Point Angle
F22. Interception Vector Magnitude	F23. Interception Vector Phase Angle	

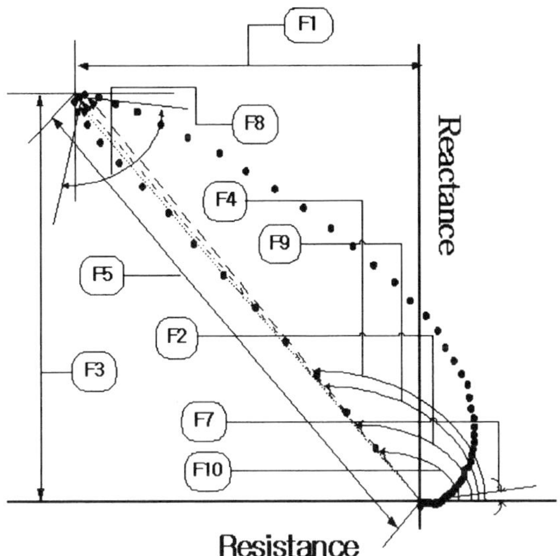

Figure 4. Definition of features for symmetric ECT signals

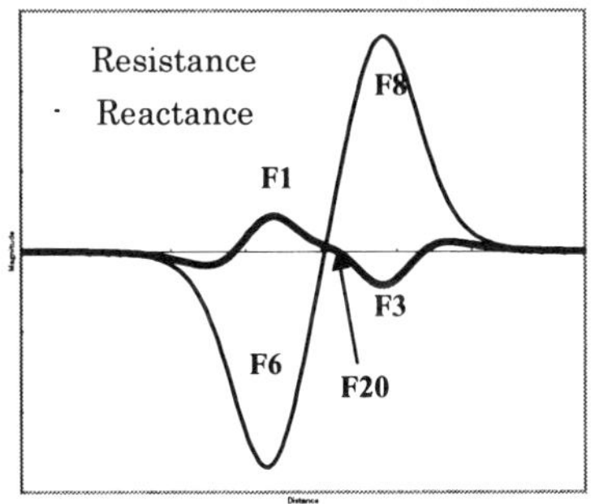

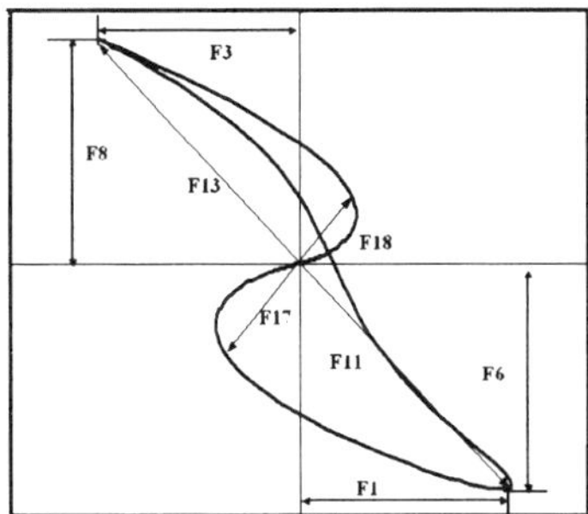

Figure 5. Definition of features for non-symmetric ECT signals .

To reduce the possible redundancy that might exist in the set of extracted features, the feature selections for classification and for sizing were performed separately. For flaw classification, two selection criteria (the single feature classification performance and the linear correlation coefficients of the specific feature to other features) were adopted. For flaw sizing, the sensitivity of individual features to the flaw size parameters (flaw depth, flaw width, tip width and tip deviation) together with the linear correlation coefficients between two different features were used as selection criteria.

Based on these criteria, feature selection has been performed for both problems. The result of the feature selection analysis will be addressed in Section 5 together with the flaw classification and sizing performances.

4. Neural networks

For the automated flaw characterization in tubes, a hybrid of two neural networks that has been used in the previous works [1,2] was used in this work. In this hybrid system, a probabilistic neural network (PNN) [4] classifier determines the flaw type, and then the back propagation neural network (BPNN) [5] estimates the size parameters (the tip width and deviation) of the flaw. Due to the space limitation, the detailed discussion on these neural networks can not be given here. However, one can refer the references [1] and [2] for the complete discussion.

5. Classification and sizing performances

5.1 For flaws with the variation in tip width

The characterization of flaws with the variation in tip width was carried out using flaw signals in the database I, with performing the flaw classification by PNN classifiers and the flaw sizing by BPNNs. For this purpose, the flaw signals in the database I were split into two parts for training and test. The training set had 240 flaw signals while the test set consisted of 192 signals. Based on the training set, feature selection for classification was carried out to choose only one kind of a feature (F11), while 3 kinds of features (F5, F9, F11) were selected for sizing.

A PNN classifier and the two BPNN size estimators were trained based on the same training set, and test samples were fed into the neural networks for the performance demonstration. The PNN showed 100% correct classification rate, since the discrimination between OD and ID flaws is a relatively easy task. Table 3 summarizes the BPNN performances for the estimation of the flaw depth and the tip width in terms of the linear correlation coefficient (r^*) between actual and estimated (by BPNNs) size parameters. BPNN showed very outstanding performance for estimating of not only flaw depth but also tip width, with r^* higher than 0.97 for all 4 cases.

Table 3. Summary of sizing performances by BPNNs for database I.

Defect type	Linear correlation coefficient (r^*)	
	Depth (d)	Tip width (W_t)
OD	0.99	0.98
ID	0.99	0.97

5.2 For flaws with the variation in tip deviation

The characterization of flaws with the variation in tip deviation was carried out using flaw signals in the database II, with performing the flaw classification by PNN classifiers and the flaw sizing by BPNNs. For this purpose, the flaw signals in the database II were split into two parts for training and test. The training set had 288 flaw signals while the test set consisted of 96 signals. Based on the training set, feature selection for classification was carried out to choose two kind of features (F15, F16) for the determination of flaw location (ID or OD) and 3 kinds of features (F15, F16, F19) for the determination of the cross-sectional shape of flaws (I-shape or V-shape), while 7 kinds of features (F1, F5, F10, F11, F19, F20, F22) were selected for sizing.

A PNN classifier and the two BPNN size estimators were trained based on the same training set, and test samples were fed into the neural networks for the performance demonstration. The PNN showed 100% correct classification rate for the determination of flaw location, while produced only 54% correct classification rate for the determination flaw shape. This implies the discrimination of OD flaws from ID ones is a relatively easy task while the identification of flaw shape is truly difficult. Table 4 summarizes the BPNN performances for the estimation of the flaw depth, the tip width and the tip deviation in terms of the linear correlation coefficient (r*) between actual and estimated (by BPNNs) size parameters. BPNN showed very outstanding performance for estimating of not only flaw depth but also tip width, with r* higher than 0.97 for all cases under consideration.

Table 4. Summary of sizing performances by BPNNs for database II.

Defect parameter	Linear correlation coefficient (r*)
Depth （d）	0.99
Width （w）	0.97
Tip deviation （δ）	0.98

6. Conclusions

In the present work, we have established automated, systematic inversion tools for eddy current testing signals for flaws with the variation either in tip width or tip deviation. For the classification and sizing of flaws with tip width variation, the proposed approach showed very outstanding performance not only for the determination of flaw types but also for the estimation of tip widths as well as flaw depths. For the classification and sizing of flaws with variation in tip deviation, the proposed approach also showed very good performances for the determination of flaw location, and for the estimation of tip deviations as well as flaw depths and widths. However, when the flaws were non-sysmmetric and inclined in their cross-sections, the determination of the cross-sectional shape of flaws was turned out to be very difiicult.

Acknowledgements

The authors are grateful for the support provided by a grant from the Korea Science and Engineering Foundation (KOSEF) and Safety and Structural Integrity Research Center at the Sungkyunkwan University.

References

1.   S. J. Song, H. J. Park, Y. K. Shin and H. B. Lee, in Rew. Prog. in Quant. NDE, Vol. 18A, eds. D. O. Thompson and D. E. Chimenti, Kluwer Academic/Plenum Publishers, New York, (1999), pp. 881-889.
2.   S. J. Song and Y. K. Shin, NDT&E International, Vol. 33 (2000), pp. 233-243.
3.   S. J. Song, C. H. Kim, Y. K. Shin, H. B. Lee, Y. W. Park and C. J. Yim, to appear in Rew. Prog. in Quant. NDE, Vol. 20, eds. D. O. Thompson and D. E. Chimenti, Kluwer Academic/Plenum Publishers, New York, (2001).
4.   D. F. Specht, Proc. of IEEE Int. Conf. on Neural Networks, Vol. 1(1988), pp. 525-532.
5.   D. E. Rumelhart, G. E. Hinton and R. J. Williams, in Parallel Distributed Processing, ed. D. E. Rumelhart and J. L. McClelland, MIT Press, Cambridge, MA, (1986), pp. 318-363.

Electromagnetic Nondestructive Evaluation (VI)
F. Kojima et al. (Eds.)
IOS Press, 2002

Development of Practical MFES System for Concrete Materials

S. Nagata[1], T. Chady[2], M. Shidouji[3] and M. Enokizono[3]
[1]Faculty of Engineering, Miyazaki University,
1-1 Gakuenkibanadai-nishi, 889-2192, Miyazaki, JAPAN
[2]Oita Industrial Research Institute,
4361-10 Takaenishi 1-chome, 870-1117, Oita, JAPAN
[3]Faculty of Engineering, Oita University,
700 Dannoharu, 870-1192, Oita, JAPAN

Abstract. An ECT system configurations for concrete material is presented. The experimental devices are designed for the practical use at the job sites, and also can be operated under high frequency. The experimental results show the validity of proposed system. Finally we discuss about suitable specifications of the practical MFES system.

Introduction

Recently, a lot of serious accidents concerned with concrete structures have happened in Japan. One of the main causes of these accidents is crack grown in low-quality concrete material, decrepitudes of concrete, fatigue or so on. In order to avoid the accidents and prolong the lifetime of structures, cracks on the concrete materials should be detected.

Many conventional NDT methods for concrete materials can be listed, AE, ultra-sonic, thermograph, microwave, and so on. However, generally speaking, these conventional methods have following common disadvantages;

    (1) Large and heavy equipments,
    (2) Need large power supplies,
    (3) Need complex data processing at laboratories after measurements.
    (4) Very expensive.

NDTs on concrete structures are supposed to be done at the job sites, that is to say, out of doors. From these disadvantageous reasons, NDTs for concrete materials are not used so often or widely nowadays.

We apply the eddy current testing (ECT) techniques to the concrete material. The ECT techniques for metallic materials such as steam generator tubes of the nuclear reactors had been rapidly developed these ten years. Some techniques are still usable for low conductivity material. Therefore we investigate the availability of ECT system for concrete materials.

We employ "multi-frequency exciting and spectrogram" (MFES) method[1][2] for ECT on concrete materials. A basic idea behind the MFES method is to use a complex signal containing selected harmonic components as an exciting signal and a spectrogram for precise crack characterization. The frequency characteristics of concrete materials are unknown, therefore, this method may be able to determine suitable frequency band for ECT.

The aim of this study is to establish the smaller and less expensive MFES system. The ECT systems developed in the laboratories tend to grow large and expensive ones. Some of the authors already presented the successful results about the NDT for concrete structures [3]. These results were measured by means of superior equipments, for example, extremely high performance A/D converter (20 Msps, 32 bit resolution). They are suitable for the fundamental studies of NDT in the laboratory, however, they are not suitable for the practical equipments at the job sites. Also it is not clear how many resolutions or how high sampling rates are sufficient. We should clarify the required and sufficient specifications such as resolutions and sampling frequencies and develop the practical MFES systems.

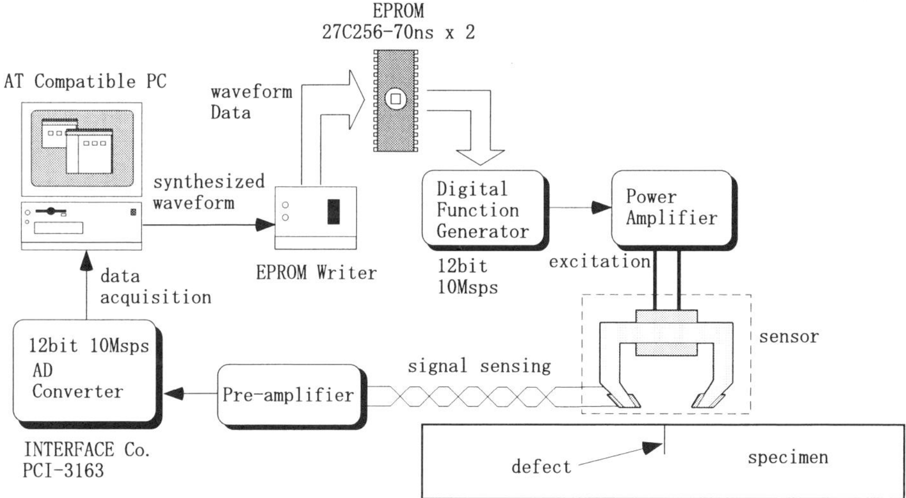

Fig. 1. Block scheme of the MFES system.

ECT system for concrete materials

Concrete mortars have low conductivity because they always contain water. It means that ECT on concrete material should be excited under high frequency due to significant output signal of crack. We estimated that high frequency operation above 10kHz up to 1MHz, therefore we designed and produced the equipments including the digital function generator, the power amplifier and pre-amplifier so that all equipments satisfied the required frequency characteristics.

A block scheme of the MFES system is shown in Fig. 1. The waveform data generated by calculation were written into two 8bit EPROMs (27C256), and set inside the function generator. The function generator supplies an exciting coil of a sensor through a power amplifier. This digital function generator includes high-speed D/A converter and it can operate at 10 mega sampling per second. It is the same sampling rate of the A/D converter (PCI-3163) due to synchronizing. For ease of digital signal processing, especially fast Fourier transform, we determined 1024 sampling numbers per a cycle at the primary frequency. Waveform were superimposed by 63 sinusoidal components having frequency from 9.8 kHz (= 10 MHz / 1024) up to 312.5 kHz each 4.9 kHz. The sensor "CON1" shown in reference [3] was also used in this system.

Figure 2 shows the details of the digital function generator. The reason why EPROMs are used for waveform memory is to achieve its sampling rate 10Msps. The access time of the waveform memories should be less than 100ns. In order to generate the waveform containing the arbitrary wave components, the waveform memories had better been the RAMs, however, system should be complicated because embeded micro computer would be needed for modifications of waveform data. Standard logic ICs (LS series TTLs) are used to perform all procedures of D/A convertion within a system cycle 100 ns.

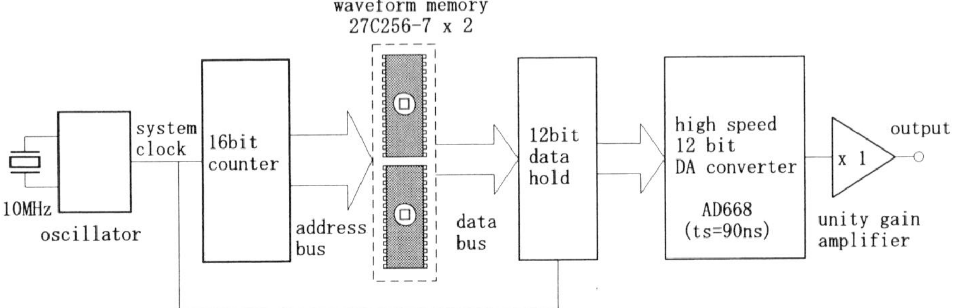

Fig. 2. Detail of the digital function generator.

Results of experiments

In order to verify usability of the proposed sensor and system, two experiments were carried out.

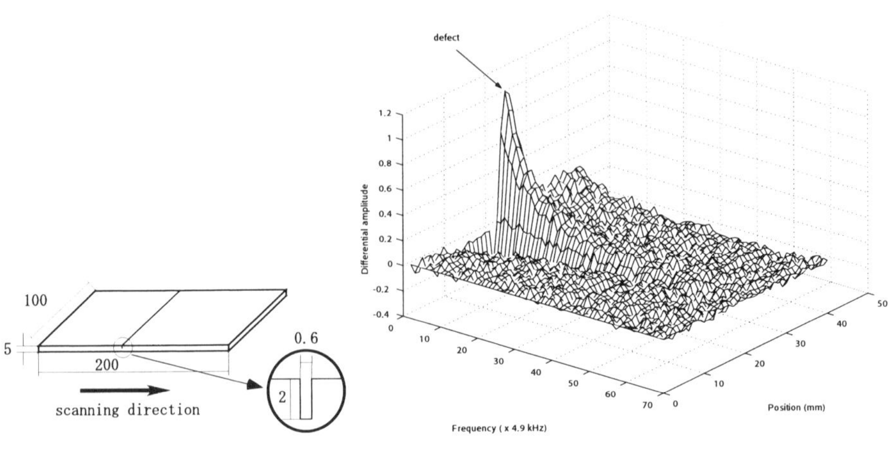

(a). Iron plate with a crack.   (b) Spectrogram result of scanning the iron plate..

Fig. 3. Present MFES system can detect a defect on the iron plate.

In the first experiment, iron plate (100x200x5mm) consisting of artificial defect were investigated. The specimen is show in Fig. 4-(a). The measurements were done by scanning the probe in X direction in steps of 1mm. The liftoff (the gap between the specimen surface and the sensor) was measured to be 0.5mm. This one is not for concrete material, but we can confirm the varidity of proposed system as a conventional NDT method.

Selected result of measurements performed on this specimens are shown in Fig. 4-(b). The defect exists at the point shown at 21mm. The defect on the iron plate is causing spectrograms with peak point for the lowest frequencies. From the result, we can say that present system is compact, simple and useful for MFES method.

Another experiment is to search for a steel bars inside concrete mortar. The specimen is show in Fig. 4-(a). Figure 4-(b) shows the result of measurement. At present, any remarkable signal of steel bar are not founded.

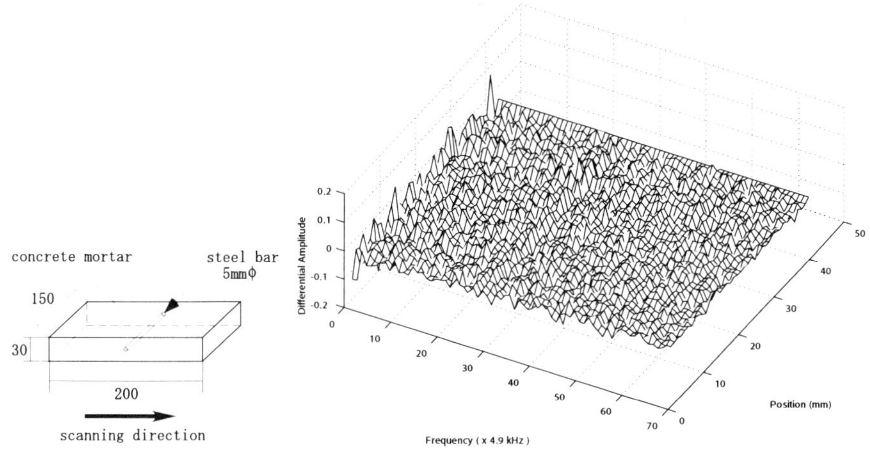

(a) Concrete block with steel bar.   (b) spectrogram result of scanning concrete block.

Fig. 4. Any remarkable signals of steel bar can not be found.

Conclusions

In this paper, we proposed a compact and inexpensive MFES system. This system contained the function generator and the A/D converter, and their resolutions are 12 bit and sampling rate 10Msps. Proposed system successfuly detected the defect on the iron plate. It can be confirmed that proposed system is valid for ECT equipment.

At present we can not find out any remarkable defect signals for concrete structure with this system. From the results of steel bar searching, fixed sampling rate and frequency components should be improved. As shown in Fig. 3-(b), lower excitation frequencies are supposed to be suitable for iron or steel detections due to serious skin effect. The sampling rate and excitation frequency components should be optimized in further studies.

Some of the authors had already confirmed the validity of MFES method on concrete defects at lower frequency components with 32bit A/D converter. More resolution of data acquisition and some digital filtering techniques are supposed to be required.

References

[1]   T. Chady, M. Enokizono, "Eddy Current Testing of Stainless Plates by Using Matrix Sensor", Proceedings of the 7th Magnetodynamics Conference, April, 1998, pp.107.

[2]   T. Chady, M. Enokizono, R. Sikora, "Crack Detection and Recognition Using an Eddy Current Differential Probe", IEEE Transactions on Magnetics, Vol. 35, No. 3, May, 1999, pp. 1849-1852.

[3]   T. Chady, M. Enokizono, K. Takeuchi, T. Kinoshita, R. Sikora, "Eddy current testing of concrete structures", ISAEM Studies in Applied Electromagnetics and Mechanics 9, in press.

Electromagnetic Nondestructive Evaluation (VI)
F. Kojima et al. (Eds.)
IOS Press, 2002

Evaluation of Crack Signals for a Superconductive Pulsed Eddy-Current Probe

J. R. Bowler[1], N. Bowler[1] and W. Podney[2]
[1]*Iowa State University, Center for Nondestructive Evaluation,*
1915 Scholl Road, Ames, IA 50011, USA
[2]*SQM Technology, Inc., PO Box 2225, La Jolla, CA 92038, USA*

Abstract. An important issue in the maintenance of aging aircraft is the detection of small cracks and corrosion embedded in the multi-layered structures of airframes. Sensitive methods of nondestructive evaluation are required to detect these deep-lying defects, particularly since they commonly occur near fasteners which themselves yield a strong signal and can mask a smaller signal from a defect nearby. A highly-sensitive measurement system has been developed by combining pulsed eddy-current technology with the unsurpassed magnetic-field sensitivity of superconductive probes. The performance of pulsed, superconductive sensors in detecting signals expected from embedded cracks has been assessed both experimentally and theoretically. The theoretical model describes the response of a superconductive probe to a crack embedded at an arbitrary depth in a layered conductor. The formulation of the field problems leads to a time-dependent integral equation which is solved by a boundary-element method and the solution used to determine probe signals. Key aspects of the theoretical model are described in this paper and the results of numerical calculations compared with experiments.

1 Probe and Pulse Excitation

A new eddy current method for detecting small subsurface cracks in aircraft uses a superconductive driver coil for exciting the electromagnetic field in the test piece and a differential pick-up loop for sensing perturbations in the field due to flaws, Figure 1. The driver consists of approximately one hundred closely spaced strips of niobium deposited on a silicon substrate. Only the central separation between strips is significant on the scale of common flaws, so in calculations of the electromagnetic field the driver can be approximated as two parallel conducting strips. The driver carries a sequence of current pulses at regular intervals with a pulse repetition rate of around 100 Hz.

The differential pick-up coil consists of a single niobium wire formed into two adjacent semi-circular loops, 6 mm diameter, wound in opposite directions. This is known as a double-D coil. The pick-up circuit couples inductively to a remote SQUID in a 5 Kelvin refrigerator. The refrigerator keeps the coils, cryogenic cable and SQUID at approximately 6 Kelvin. If the windings are perfect, mutual inductance between the source and pick-up coils vanishes. The configuration and response function of the probe is described in more detail in reference [1].

The pick-up loop signal has been calculated for a subsurface crack in a stack of aluminum plates using a boundary element scheme. As a first step, the electric field in an unflawed conductor due to the driver coil is determined. This is then used in a boundary element calculation

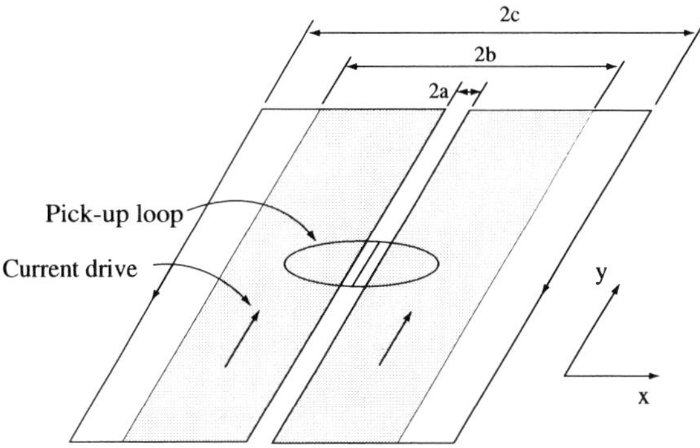

Figure 1. Superconductive probe represented as two current strips with a differential double-D pickup loop. The center of the pick-up loop is located at $x = y = 0$ and the length of the current strips (in the $\hat{y}$ direction) is $2d$.

of the discontinuity in the electric field at the crack, as represented by a current dipole layer varying in time. The differential flux through the pick-up loop is then found from the dipole density using a reciprocity theorem. Application of the theorem requires knowledge of the electric field generated by the pick-up loop in the unflawed conductor as if it were used as an excitation coil. The steps in the calculation are summarized below followed by a description of typical results.

2　Excitation Current

The current in the driver coil is a smoothed function of time known as the Moulder pulse, $M(t)$;

$$
M(t) = \begin{cases} 1 + \tanh\left(\frac{t/\tau - ncs}{s}\right), & 0 \le t \le \tau(1/4 + ncs) \\ 1 - \tanh\left(\frac{t/\tau - 1/2 - ncs}{s}\right), & \tau(1/4 + ncs) \le t \le \tau. \end{cases}
\tag{1}
$$

In (1), τ is the period of the pulse, s is a pulse sharpness parameter, $c = \mathrm{arccosh}(\sqrt{2})$ and $n = 0, 1, 2, \ldots$.

It is useful to approximate the Moulder pulse by a simple function of time that has an elementary Laplace transform allowing the inverse transformation of the field into the time domain to be carried out analytically [2]. The elemental function chosen is a linear ramp function with finite duration. A superposition of linear ramps approximates the pulse as a piecewise linear function. It is possible to approximate the current rise by a single ramp, by matching the slope of the ramp with that of $M(t)$ at half height, i.e. at $M(t) = 1$. However, a better approximation is to use an odd number of contiguous ramps whose end points lie on $M(t)$. In Figure 2, the normalized current is compared with an approximation of the current using 11 ramps.

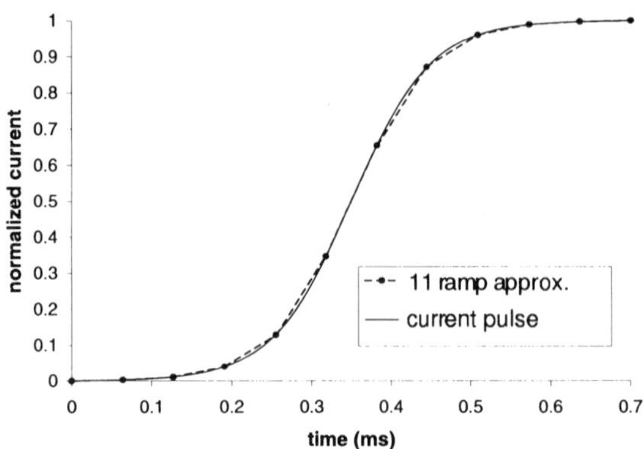

Figure 2. Comparison between the exciting current variation with time and an 11-ramp fit as used in calculations.

3   Excitation Field

For the simple case of a probe in air above a half-space conductor with conductivity σ_0, permeability μ_0 and whose surface is in the plane $z = 0$, the unperturbed electric field due to the excitation coil, $\mathbf{E}^{(0)}(\mathbf{r}, t)$, is a solution of

$$\nabla^2 \mathbf{E}^{(0)}(\mathbf{r}, t) = -\mu_0 \frac{\partial \mathbf{J}(\mathbf{r}, t)}{\partial t}, \qquad z > 0, \tag{2}$$

$$\left(\nabla^2 - \mu_0 \sigma_0 \frac{\partial}{\partial t}\right) \mathbf{E}^{(0)}(\mathbf{r}, t) = 0, \qquad z < 0, \tag{3}$$

with the auxiliary condition that the electric field has zero divergence. The unperturbed field is transverse electric (TE) with respect to the direction normal to the surface of the conductor and therefore can be written

$$\mathbf{E}^{(0)}(\mathbf{r}, t) = -\mu_0 \frac{\partial}{\partial t} \nabla \times [\hat{z}\psi'(\mathbf{r}, t)], \tag{4}$$

where $\psi'(\mathbf{r}, t)$ is a TE potential. Substituting (4) into (2) and (3) and operating with $\nabla \cdot (\hat{z}\times)$ yields

$$\nabla^2 \Psi(\mathbf{r}, t) = -j(\mathbf{r}, t), \qquad z > 0, \tag{5}$$

$$\left(\nabla^2 - \mu_0 \sigma_0 \frac{\partial}{\partial t}\right) \Psi(\mathbf{r}, t) = 0, \qquad z < 0, \tag{6}$$

where $\Psi = \nabla_z^2 \psi'$ with $\nabla_z^2 \equiv \nabla^2 - \partial^2/\partial z^2$ and $j = \nabla \cdot (\hat{z} \times \mathbf{J})$. Expressing the solution in terms of a Green function;

$$\left(\nabla^2 - \mu_0 \sigma_0 \frac{\partial}{\partial t}\right) G(\mathbf{r} - \mathbf{r}', t - t') = -\delta(\mathbf{r} - \mathbf{r}')\delta(t - t') \tag{7}$$

and

$$\Psi(\mathbf{r}, t) = \int_0^t \int_{\Omega_0} G(\mathbf{r} - \mathbf{r}', t - t')\, j(\mathbf{r}', t')\, d\mathbf{r}' dt'. \tag{8}$$

The spatial arrangement of the current drive coil lends itself to two-dimensional Fourier transform analysis. Also taking the Laplace transform of (8) with respect to time gives

$$\tilde{\psi}(u, v, z, s) = \int \tilde{g}(u, v, z - z', s)\, \tilde{j}(u, v, z', s)\, dz', \tag{9}$$

with

$$\tilde{g}(u, v, z - z', s) = \frac{1}{\kappa + \gamma} e^{\gamma z - \kappa z'}, \qquad z < 0. \tag{10}$$

In the above equations, u and v are the Fourier space variables, $\kappa^2 = u^2 + v^2$, $\gamma^2 = \kappa^2 + \mu_0 \sigma_0 s$ and for γ the root of $\kappa^2 + \mu_0 \sigma_0 s$ with positive real part is taken.

The current density in the source coil takes the form

$$\mathbf{J}(\mathbf{r}, t) = I(t)\, \mathbf{F}(x, y)\, \delta(z - z_0), \tag{11}$$

where $\mathbf{F} = \hat{x} F_x + \hat{y} F_y$ and the source coil is located in the plane $z = z_0$. F_x is odd in x and y, and F_y is even in x and y. Writing (11) in terms of j and transforming in space and time yields

$$\tilde{j}(u, v, z, s) = i(s)\, [iv\, \tilde{F}_x(u, v) - iu\, \tilde{F}_y(u, v)]\, \delta(z - z_0). \tag{12}$$

Assuming that the return current paths (e.g. at $x = \pm c$) are narrow filaments, $\tilde{F}_x(u, v)$ is given by

$$\tilde{F}_x(u, v) = -\frac{2}{u} \left\{ \frac{1}{u(a - b)} \left[\sin(ua) - \sin(ub)\right] - \cos(uc) \right\} \sin(vd), \tag{13}$$

and $\tilde{F}_y(u, v) = -(u/v)\, \tilde{F}_x(u, v)$ as required by $\nabla \cdot \mathbf{J} = 0$.

An expression for the electric field which can be evaluated numerically is obtained in the following way. Transforming (4) with respect to x, y and t gives

$$\tilde{e}^{(0)}(u, v, z, s) = -\mu_0 s \frac{iv\hat{x} - iu\hat{y}}{\kappa^2}\, \tilde{\psi}(u, v, z, s), \tag{14}$$

where $\tilde{\psi}$ is given in (9). Substituting (10) and (12) into (9) gives

$$\tilde{\psi}(u, v, z, s) = i(s)[iv\, \tilde{F}_x(u, v) - iu\, \tilde{F}_y(u, v)]\, \tilde{g}(u, v, z - z_0, s), \qquad z < 0. \tag{15}$$

Formally inverting (14), with $i(s)$ for a multiple ramp excitation given in reference [2],

$$\begin{aligned}
\tilde{\mathbf{E}}^{(0)}(u, v, z, t) = {} & -\frac{\mu_0}{\Delta\tau} \frac{(iv\hat{x} - iu\hat{y})}{\kappa^2} [iv\, \tilde{F}_x(u, v) - iu\, \tilde{F}_y(u, v)]\, e^{-\kappa z_0} \big[I_1 \chi(\kappa, z, t)\Theta(t) \\
& + \sum_{n=1}^{N-1} (I_{n+1} - I_n)\chi(\kappa, z, t - n\Delta\tau)\Theta(t - n\Delta\tau) \\
& - I_N \chi(\kappa, z, t - N\Delta\tau)\Theta(t - N\Delta\tau)\big], \qquad z \leq 0,
\end{aligned} \tag{16}$$

in which $\Theta(t)$ is the Heaviside step function and $\chi(\kappa, z, t)$ is evaluated in reference [2]. The remaining two-dimensional inverse Fourier transformation must be performed numerically, for example by means of a fast-Fourier transform routine [3].

4   Defect Detection

To calculate the response of the double-D pick-up loop to field perturbations generated by a flaw, the electric field produced by the loop in an unflawed conductor must first be calculated. Then the net flux through the loop is determined using a reciprocity relationship.

The current in the loop will be written in terms of an equivalent magnetic source, $\mathbf{M}(\mathbf{r}, t)$, related to the current by

$$\mathbf{J}(\mathbf{r}, t) = \frac{1}{\mu_0} \nabla \times \mathbf{M}(\mathbf{r}, t). \tag{17}$$

The choice of $\mathbf{M}(\mathbf{r}, t)$ is somewhat arbitrary provided that it satisfies (17). Usually a closed current filament in a plane is represented by a magnetic shell bounded by the current path. Although the shell need not necessarily be in the plane of the loop, calculations are simplified if it is confined to the plane.

If the current is transverse to the z-direction one can choose to write $\mathbf{M}(\mathbf{r}, t) = \hat{z} M'(\mathbf{r}, t)$ and hence reduce the field problem to the scalar form

$$\nabla^2 \Psi(\mathbf{r}, t) \;=\; -\frac{1}{\mu_0} M(\mathbf{r}, t), \qquad z > 0, \tag{18}$$

$$\left(\nabla^2 - \mu_0 \sigma_0 \frac{\partial}{\partial t} \right) \Psi(\mathbf{r}, t) \;=\; 0, \qquad z < 0, \tag{19}$$

where $M = \nabla_z^2 M'$. The solution of (18) and (19) may be written

$$\Psi(\mathbf{r}, t) = \frac{1}{\mu_0} \int_0^t \int_{\Omega_0} G(\mathbf{r}, \mathbf{r}', t - t') M(\mathbf{r}', t') \, d\mathbf{r}' dt'. \tag{20}$$

The potential Ψ is evaluated by taking the Laplace transform of (20) and expressing the transformed Green function as

$$G(\mathbf{r}, \mathbf{r}', s) = 4\pi \sum_{m=0}^{\infty} \epsilon_m \cos[m(\phi - \phi')] \int_0^\infty J_m(\kappa \rho) J_m(\kappa \rho') \, \widetilde{f}(\kappa, z - z', s) \, d\kappa, \quad z < 0, \tag{21}$$

which can be derived using the approach given in reference [4]. In (21), ϵ_m is the Neumann factor; $\epsilon_0 = 1$ and $\epsilon_m = 2$ ($m = 1, 2, 3, \ldots$). The function $\widetilde{f}(\kappa, z - z', s) = 2\kappa \, \widetilde{g}(\kappa, z - z', s)$ for the interior of a half-space conductor, with $\widetilde{g}$ defined in (10). For the double-D pick-up loop, define the Laplace transform of $M(\mathbf{r}, t)$ as

$$m(\mathbf{r}, s) = \begin{cases} \mu_0 \, I(s) \, \delta(z - z_0), & \rho < a \text{ and } 0 < \phi \leq \pi, \\ -\mu_0 \, I(s) \, \delta(z - z_0), & \rho < a \text{ and } -\pi < \phi \leq 0, \\ 0, & \text{otherwise.} \end{cases} \tag{22}$$

The reciprocity theorem requires that a unit step current is used to to define $m(\mathbf{r}, s)$, hence $I(s) = 1/s$. Equations (21) and (22) are substituted into the Laplace transform of (20) and the integration with respect to ϕ', ρ' and z' performed to give a summation of Bessel integrals with respect to κ. The inverse Laplace tranform can be performed analytically but the κ integrals must be computed numerically and the summation truncated at a suitable order depending on the required accuracy of the result.

This approach for modeling the double-D pick-up loop differs in detail from, but is similar to, an alternative analysis for time-harmonic excitation given in reference [5]. A truncated series of 5 terms is sufficiently accurate in most cases [5].

5 Crack in a Transient Field

The effect of the crack on the electromagnetic field is modeled as a surface distribution of current dipoles $\mathbf{p}(\mathbf{r}, t)$ [6]. Applying a dyadic form of Green's theorem and assuming that the normal component of the current density at the crack is zero leads to a Fredholm integral equation of the first kind for $\mathbf{p}(\mathbf{r}, t)$:

$$J_x^{(0)}(\mathbf{r}_0, t) = \lim_{\mathbf{r} \to \mathbf{r}_0} \left(\sigma_0 \mu_0 \hat{x} \frac{\partial}{\partial t} - \frac{\partial}{\partial x} \nabla \right) \cdot \int_0^t \int_{S_0} \mathcal{G}(\mathbf{r}, \mathbf{r}', t, t') \cdot \mathbf{p}(\mathbf{r}', t') \, dS' \, dt'. \qquad (23)$$

In (23), $\mathbf{r}_0$ denotes field points on the crack surface, $J_x^{(0)}(\mathbf{r}_0, t)$ is the normal component of the unperturbed current density at the crack and the dyadic Green function kernel is given in reference [7]. The crack region is discretized into rectangular elements and $\mathbf{p}(\mathbf{r}, t)$ is assumed piece-wise constant in space and piece-wise linear in time. Equation (23) is then solved numerically to obtain $\mathbf{p}(\mathbf{r}, t)$.

6 Numerical Results

The net flux threading the double-D pick-up loop is proportional to $\partial H_z / \partial y$ [2]. In Figure 3, the variation in values of $\partial H_z / \partial y$ with both time and position above a cracked conductor are shown. The field derivative is computed along a line above a crack 5 mm long and 1 mm high buried 4 mm below the surface of the conductor. The crack is centered at $(0, 0, -4.5)$ mm and oriented so that the unit vector $\hat{x}$ is normal to the crack surface.

The single negative peak and smaller, positive peaks are characteristic of the signal obtained as the probe is scanned along a line above the position of the crack. The signal magnitude is larger and the negative peak occurs earlier in time for a crack nearer the surface of the conductor. Thus, the position of the peak in time conveys information about the depth of the defect. In Figure 4, a comparison between theoretical and experimental data is shown. The theoretical data is shifted linearly in time to account for delays in the electronics of the measurement circuit.

The computed data shown in Figures 3 and 4 assumes excitation of the conductor by infinitely long current strips ($d \to \infty$). It is anticipated that the agreement between experimental and theoretical data will be improved by modeling the strip length more accurately.

7 Conclusion

The transient current rise, two-dimensional current source and double-D pick-up loop of a pulsed, superconductive probe have been modeled. It has been demonstrated that the superconductive, pulsed eddy-current probe has an outstanding capability for detecting deep-lying flaws in layered structures. There is good agreement between theoretical calculations and experimental data for cracks concealed in a layered metal structure, several millimeters below the surface. In the next phase of this work, the effect of a fastener adjacent to the crack will be included in the model.

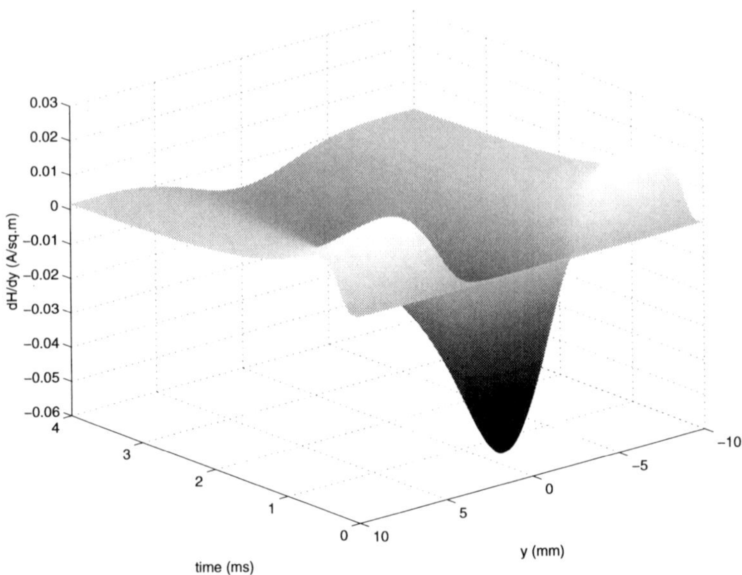

Figure 3. Calculated response from a 5 mm × 1 mm crack buried 4 mm below the conductor surface. Values of $\partial H_z/\partial y$ 2 mm above the conductor surface are shown.

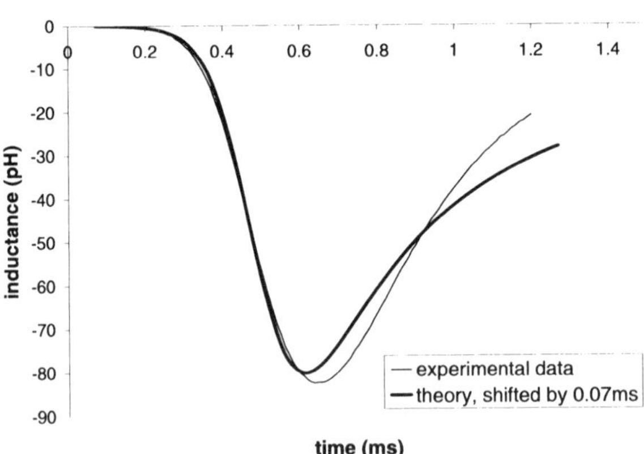

Figure 4. Experimental and theoretical data for the peak signal from a 5 mm × 1 mm crack buried 3 mm below the surface of a conductor.

8 Acknowledgement

Work sponsored by the Air Force Office of Scientific Research STTR Program, contract F49620-99-C-0056.

References

[1] W. Podney, Response Function of an Electromagnetic Microscope. In: D. O. Thompson and D. E. Chimenti (eds.), Review of Progress in QNDE, Vol. 17. Plenum, New York, 1998, pp. 1025–1031.

[2] N. Bowler, J. R. Bowler and W. Podney, Response Model of Superconductive, Pulsed Eddy-Current Probes for Detection of Deep-Lying Flaws. In: D. O. Thompson and D. E. Chimenti (eds.), Review of Progress in QNDE, Vol. 20. AIP Conference Proceedings 557, New York, 2001, pp. 941–948.

[3] W. H. Press, S. A. Teukolsky, W. T. Vetterling and B. P. Flannery, Numerical Recipes in FORTRAN: The Art of Scientific Computing, 2nd Edition. Cambridge University Press, Cambridge, 1992.

[4] P. Morse and H. Feshbach, Methods of Theoretical Physics. McGraw-Hill, New York, 1953, p. 1263.

[5] D. McA. McKirdy, S. Cochran, G. B. Donaldson and A. McNab, Theoretical Consideration of Fatigue Crack Detection and Characterisation Using SQUID Sensors. In: R. Collins et al. (eds.), Nondestructive Testing of Materials. IOS Press, Amsterdam, 1995, pp. 185-194.

[6] J. R. Bowler, Eddy-Current Interaction with an Ideal Crack. 1. The Forward Problem, J. Appl. Phys. 75 (1994) 8128–8137.

[7] J. R. Bowler, Time Domain Half-Space Dyadic Green's Functions for Eddy-Current Calculations, J. Appl. Phys. 86 (1999) 6494–6500.

Electromagnetic Nondestructive Evaluation (VI)
F. Kojima et al. (Eds.)
IOS Press, 2002

Time-Domain Finite Element Analysis of Remote Field ECT Excited by Pulse Current

Katsumi YAMAZAKI

Dept of Electrical Engineering, Chiba Institute of Technology,
2-17-1, Tsudanuma, Narashino, Chiba, 275-0016, Japan

Abstract. In this paper, we investigate the electromagnetic field in remote field eddy current testing excited by pulse current. The time-domain finite element method is applied to analyse the field and voltages of the detecting coil. It is clarified that the time-variation of the detected voltage shows the effects of magnetic fields produced by the exciting coil current and the eddy current separately. The magnetic field produced by the eddy current, which include much information of the cracks of the tube, can be detected with the appropriate separation length between the exciting and the detecting coils.

1. Introduction

The remote field eddy current testing (RFECT) is an effective method that can detect both the inner and the outer cracks of the tube [1]-[5]. In this case, the magnetic fields can be considered as the sum of two components. One is the "near field", the other is the "remote field". In order to detect the crack of the tube, the separation length between the exciting coil and detecting coil must be wide enough because the information of the crack is mainly included in the remote fields. But it is difficult to determine the appropriate separation length between the coils. In the case of time harmonic (sinusoidal) excitation, the phase shift of the detected voltage relative to the exciting coil voltage gives information about the eddy current (remote) field.

In this paper, we investigate the RFECT with pulse exciting currents. The time-domain finite element method is applied to analyse the fields and voltages of the detecting coil. In order to understand the detected voltage waveform, we introduce a new classification of the magnetic fields on the RFECT, which are "primary field" produced by the exciting coil and "secondary field" produced by the eddy current in the tube. The magnetic field can be regarded as the sum of these two components as well as the "near field" and "remote field". It is clarified that the time-variation of the detected voltage shows the effects of the primary and the secondary fields separately and that the appropriate separation length between the coils can be decided from the detected voltage waveforms.

2. Calculation method

The formulation of the time-domain finite element method for the analysis of RFECT in case of axial-symmetric arrangement is:

$$\nabla \times (\frac{1}{\mu} \nabla \times A) = J_c + J_e = J_c - \sigma \frac{A^{t+\Delta t} - A^t}{\Delta t} \tag{1}$$

where A is the magnetic vector potential, J_c is the exciting coil current, J_e is the eddy current and σ is the conductivity of the tube.

To obtain the "primary filed" and the "secondary field" separately, we introduce a procedure shown in Fig.1, which is the combination of the time-domain and the static finite element analysis. First, the total electromagnetic field is calculated by (1). Second, The primary field, which is produced by the exciting coil current, is calculated by the static analysis with the permeability of the tube and the exciting current at each time step from following equation.

$$\nabla \times (\frac{1}{\mu} \nabla \times A_1) = J_c \tag{2}$$

where A_1 is the magnetic vector potential produced by the exciting coil current.

Then the secondary field, which is produced by the eddy current of the tube, can be obtained as follows.

$$A_2 = A - A_1 \tag{3}$$

where A_2 is the magnetic vector potential produced by the eddy current.

3. Analysed Model and Result

Fig.2 shows the analysed model. The tube is made of the non-magnetic material whose conductivity is 1.5×10^7 S/m. In the finite element analysis, 13,000 triangular elements are applied to the discretization of the axisymmetric time-domain formulation. Fig.3 shows the finite element mesh.

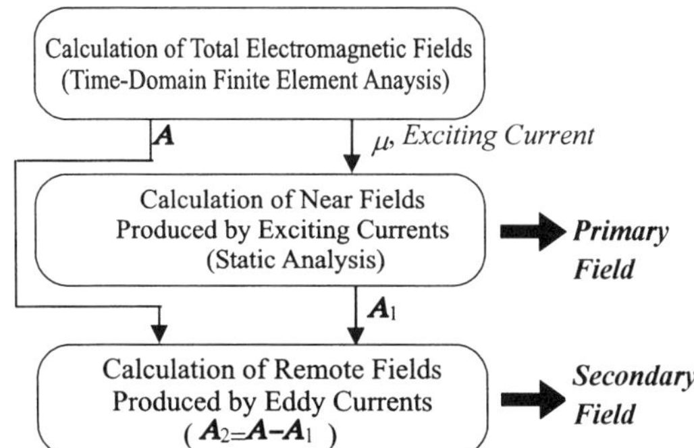

Fig.1 Procedure to separate primary field and secondary field in RFECT

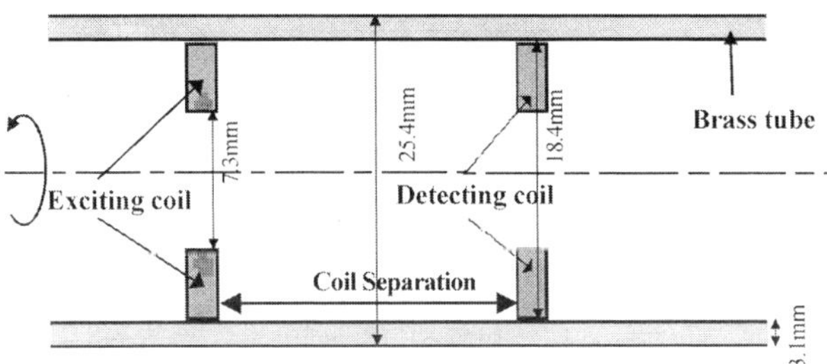

Fig.2 Analysed model

First, we examine the accuracy of the time-domain finite element analysis in the case of the sinusoidal excitation without the crack of the tube. Fig.4 shows the RMS voltage of the detecting coil due to the separation length between the coils calculated by the frequency-domain and time-domain finite element analyses. The experimental result is also shown. It indicates that the result of the time-domain analysis agrees well with the frequency-domain analysis when the number of the time step is enough. It also agrees well with the experimental result. The validity of the time-domain analysis is verified from these results.

Next, let us assume that the exciting coil current is the pulse current shown in Fig.5, which is the half sinusoidal wave of 10kHz. The pulse current flows every 1ms periodically. The calculated voltage waveforms of the detecting coil due to separation length between coils are shown in Fig.6. When the separation is short, the voltage peak that corresponds to the pulse exciting current is remarkable. This peak must be caused by the primary field produced by the exciting coil current. On the other hand, when the separation is wide, the voltage peak decrease and a delayed wave with large time-constant appears. This wave is caused by the secondary field produced by the eddy current in the tube.

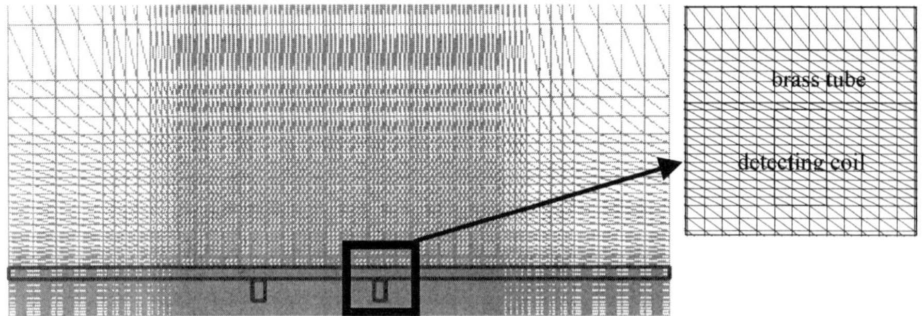

Fig.3 Finite element mesh

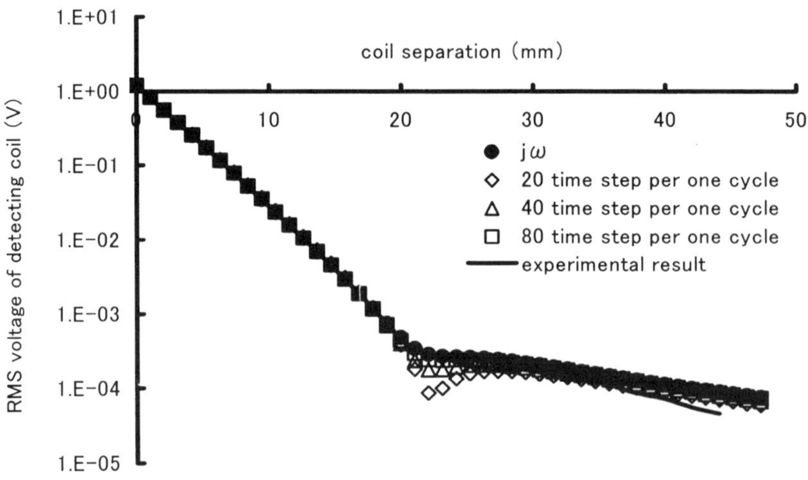

Fig.4. RMS voltage of detecting coil: Variation with coil separation

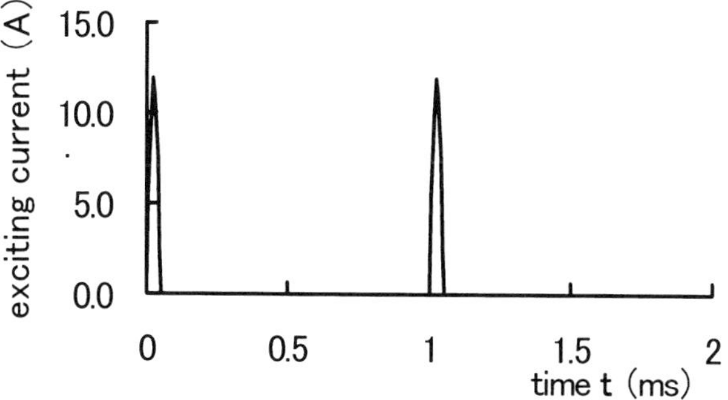

Fig.5. Pulse exciting current

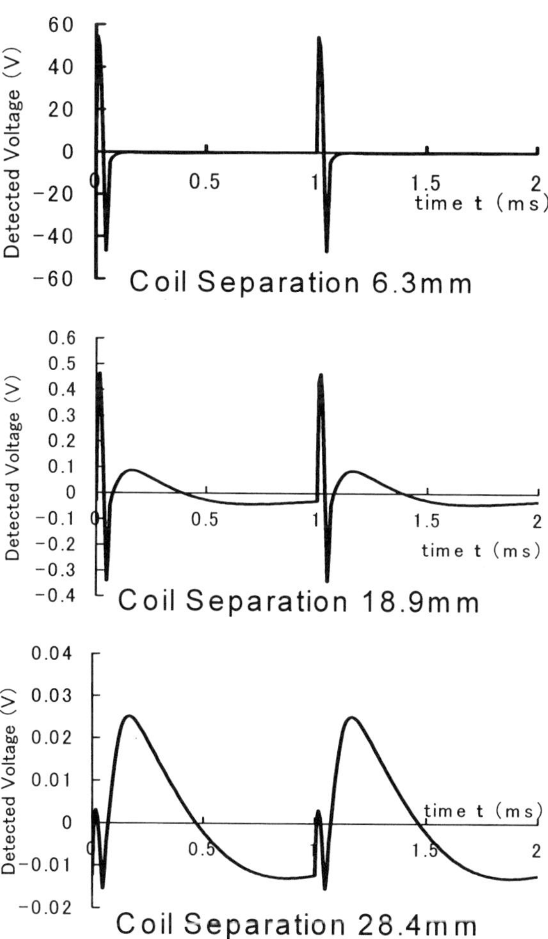

Fig.6. Waveforms of detected voltage due to separation length between coils

To verify the consideration, let us separate the total magnetic field to the primary and the secondary field by the proposed procedure shown in Fig.1. Fig.7 shows the results of the fields at three time instants. When the exciting current is increasing, the direction of the secondary field produced by the eddy current is opposite to the primary field produced by the exciting current. On the other hand, when the exciting current is decreasing, the secondary field and the primary field are in the same direction due to the Faraday's law. The eddy currents still remain and spread even when the exciting current becomes zero. In this case, the magnetic field is only the secondary field. It is clarified that the delayed wave of the detected voltage is caused only by the secondary field produced by the eddy current in the tube. It can be considered that the delayed waveform includes much information of the tube.

Fig.8 shows the RMS voltage of the detecting coil due to relative position of the crack and the coil when the coil separation is 36.8 mm. The RMS voltage of the waves shown in Fig.6 can be calculated as follows:

$$V_{RMS} = \sqrt{\frac{1}{T} \int_0^T v^2 dt} \tag{4}$$

where v is the instantaneous value of the detected voltage and T is the time-interval (1ms).

It indicates that the detected voltage changes due to the depth of the crack and that we can detect the depth of both inner and outer crack when the separation length between the coils is wide enough. On the other hand, Fig.9 shows the results when the coil separation is 3.15mm. In the case of the inner crack, the crack can be detected but the change of the voltage due to the depth of the crack is small. Furthermore, in the case of the outer crack, we cannot detect the crack except for the 100% perfect crack.

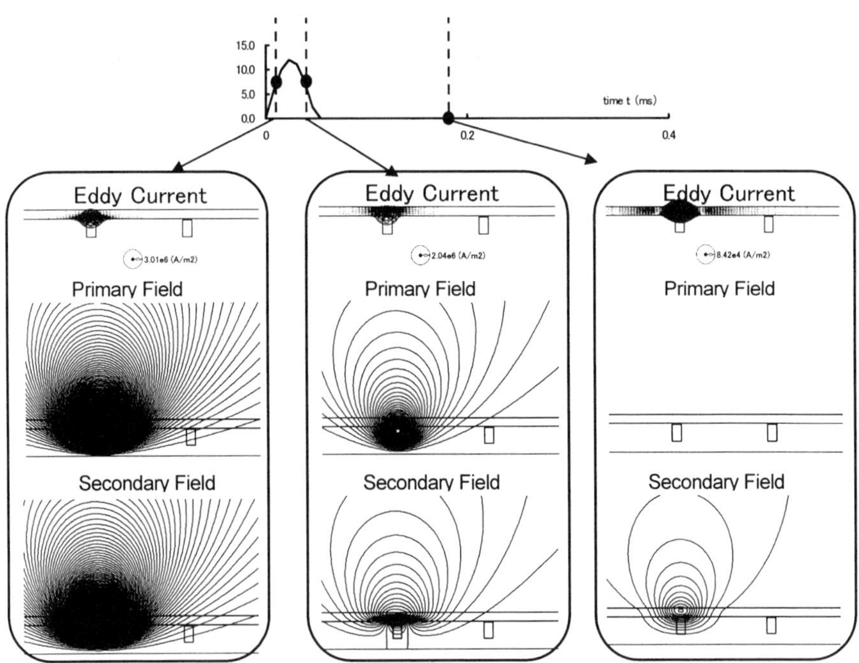

Fig.7. Separated primary and secondary field.

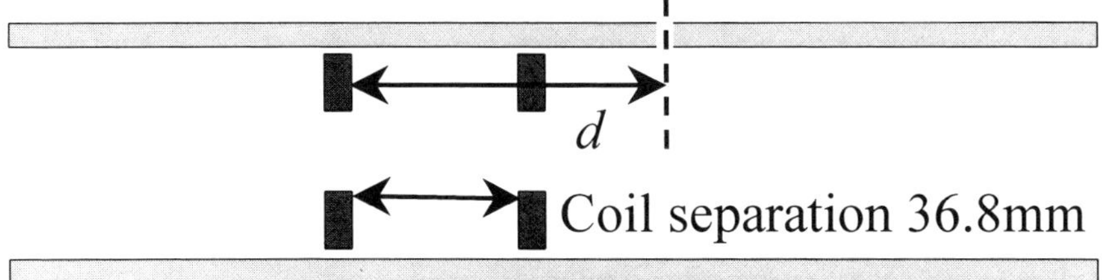

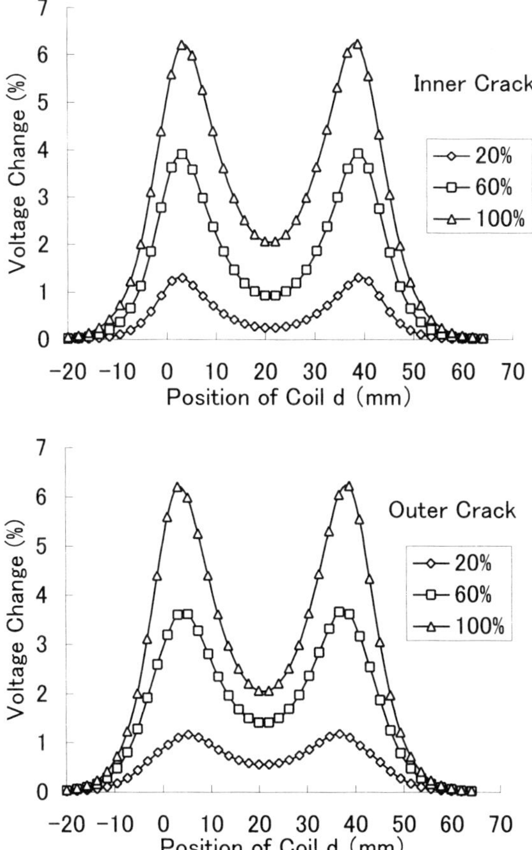

Fig.8. RMS voltages of detecting coil due to relative position of crack and coil.
(Separation length between exciting and detecting coils is 36.8mm)

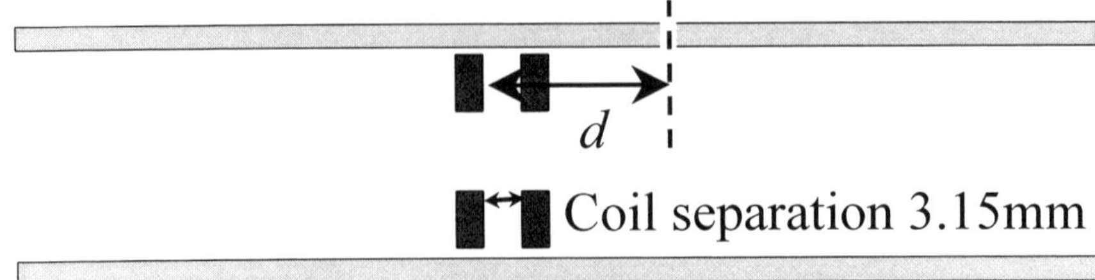

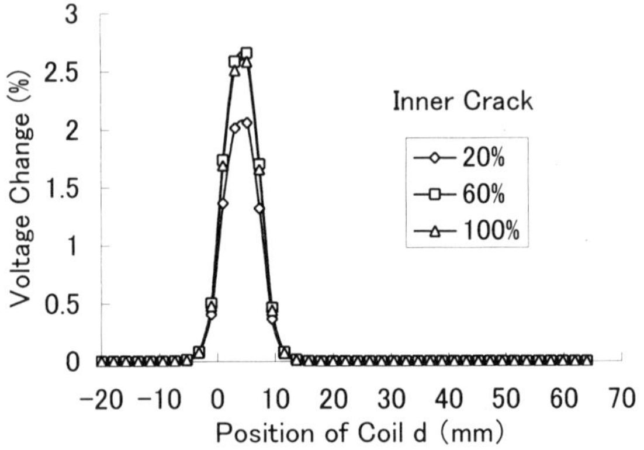

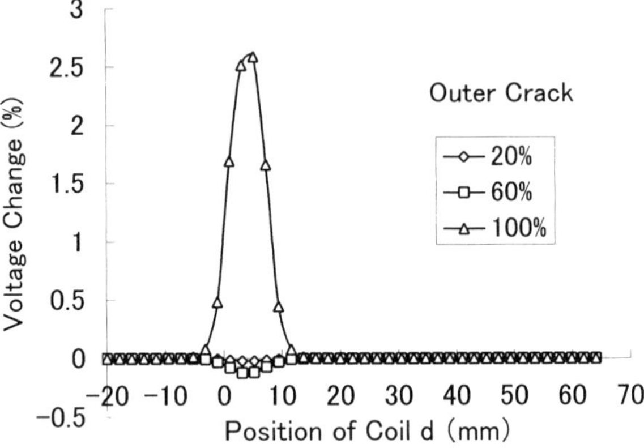

Fig.9. RMS voltages of detecting coil due to relative position of crack and coil.
(Separation length between exciting and detecting coils is 3.15mm)

4. Conclusions

The detected voltage of the remote field ECT excited by the pulse current is analyzed using time-domain finite element analysis. The primary field produced by the exciting current and the secondary field produced by the eddy current are separated and displayed by the proposed procedure. The voltage waveform of the detecting coil shows the effects of the primary field and the secondary field separately. Both the inner and the outer cracks can be detected by the pulse excited remote field ECT with appropriate separation length between exciting and detecting coils. It can be considered that the appropriate separation length can be decided from the waveforms of the detected voltage. Further work is required to understand the relation between the definition of "primary and secondary field" and "near and remote field"

References

[1] T. R. Schmidt, The Remote Field Eddy Current Inspection Technique. Materials Evaluation, 42, 2, 225, 1984.

[2] W. Load, Y. S. Sun, S. S. Udpa and S. Nath, A Finite Element Study of the Remote Field Eddy Current Phenomenon. IEEE Trans. on Magnetics, 24, 1, 435, 1988.

[3] H. Y. Lin and S. Y. Sun, Application of "Zoom-in Technique in 3D Remote Field Eddy Current Effect Computaion. IEEE Trans. on Magnetics, 26, 2, 881, 1990.

[4] M. Isobe, R. Iwata and M. Nishikawa, High Sensitive Remote Field Eddy Current Testing by Using Dual Exciting Coils, Nondestructive Testing of Materials, Studies in Applied Electromagnetics and Mechanics, 8, IOS Press, 145, 1995.

[5] H. Fukutomi, T. Takagi, J. Tani and M. Nishikawa, Three-Dimensional Finite Element Computation of a Remote Field Eddy Current Technique to Non-Magnetic Tubes, Journal of JSAEM, 6, 4, 343, 1998.

Materials Characterization

NDE of Fatigue Damage in Austenitic Stainless Steel by Measuring and Inversion of Damage-induced Magnetic Field

Zhenmao CHEN[1], Kazumi AOTO[2], and Kenzo Miya[1]
[1]*International Institute of Universality,*
SB Bldg.8F, 1-4-6 Nezu, Bunkyo-ku, Tokyo, 113-0031, Japan
[2]*Structure Safety Engineering Group, OEC/JNC,*
4002 Narita-cho, Oarai-machi, Ibaraki, 311-1393, Japan

Abstract In this paper, the possibility of applying the damage induced magnetic fields to the detection of mechanical damages in austenitic stainless steel is investigated by measuring magnetic fields induced by plastic and fatigue damages and reconstructing the magnetic charges distribution in the material. Mechanical damages are introduced to test-pieces of SUS304 stainless steel at first through a tensile and a zero tension fatigue testing. The magnetic field over the planar surface of the test-piece was measured with the use of advanced flux-gate sensors. Finally, the distribution of the magnetic charges, which is considered to be correlated to the status of mechanical damages, was reconstructed from the measured magnetic field by applying a deterministic algorithm. The reconstructed distribution of the magnetic charges agrees with that observed with the magnetic colloidal method, which gives a strong support to the inspection method using magnetic field measurement.

1. Introduction

Nowadays, for enhancing the safety of various critical structure systems such as the machinery of a nuclear power plant, new NonDestructive Evaluations (NDE) techniques are strongly expected by the related industries [1]. A typical example of such needs is an efficient way for predicting the occurrence of cracks, or in other word, to evaluate the status of damages or degradations in an aged structural component before macroscopic cracks are initiated. Recently, it has been found that magnetization appears in the vicinity of a fatigue crack even without any external magnetic field being applied to an austenitic stainless steel of SUS304 type [2],[3]. The reason of the phenomenon is considered to be that the damages occurred in the object alternate the magnetic property and even induce a magnetization without applying external magnetic field (hereafter, we call it natural magnetic field) [2]. This phenomenon means a possibility to detect the mechanical damages by predicting the distribution of magnetization in the material.

Practically, the distribution of magnetization in a material is impossible to be measured directly. To predict the magnetization in the material, a possible strategy is to measure the natural magnetic field produced by these magnetization outside the material and to reconstruct the distribution of the magnetization through an inverse analysis.

In view of these backgrounds, the correlation of mechanical damages and the natural magnetization was investigated in this work by measuring the natural magnetic flux density nearby the surface of test-pieces in which damages were introduced through a tensile or a fatigue test. Advanced film type flux-gate sensors [4] were used in the measurement [3]. The magnetic fields were measured during and after the tensile test in order to investigate the correlation of the natural magnetization and the plastic deformation. To evaluate the possibility of observing the damage status of a key structural component in a practical environment, the natural magnetic flux signal was measured in-situ during the fatigue test and its distribution was also measured but in an off-situ state after selected numbers of strain cyclic. To establish a scheme for reconstructing the distribution of the damage-induced magnetization, an approach based on the least square algorithm and the conjugate gradient method [5],[6],[7] was introduced. Concerning the ill-poseness of the inverse problem dealing with the direct reconstruction of magnetization vector, the magnetic charge was taken as the unknown to be reconstructed.

This paper is arranged in the following order. In next section, the setup, procedures and results of the experiments are described respectively. In section 3, the method for the magnetization reconstruction is presented with validations using simulated magnetic fields and measured natural magnetic fields. Some discussion and conclusions are presented in the last section.

2. Experimental measurement of damage-induced magnetic field

2.1 Details of the experiments

a) Test material and test-piece

Keeping in mind the NDE of new generation Nuclear Power Plants (NPP) such as the Fast Breeder nuclear Reactor (FBR) plant, the SUS304 stainless steel was chosen as the material to be inspected of this study. The composition of the material is shown in Table 1. Less than 1% ferrit phase exists in the virgin material. Standard button type test-pieces were fabricated for both the tensile and fatigue test. The gauge area of the test-piece is in a shape of plate of dimensions 40mm×20mm×5mm. In the central region of some test-pieces, through notches (3 mm × 0.3 mm slit) were fabricated by EDM technique for guarding the position of crack initiation. Concerning the buckling problem due to the compress loading in the zero-tension fatigue test, the thickness of the gauge region was designed as 5 mm.

Table 1 Composition of the SUS304 A3 material (Wt%)

C	Si	Mn	P	S	Ni	Cr
0.05	0.55	0.82	0.021	0.003	8.87	18.38

b) Measuring method and experimental set-ups

Fig.1 shows a flow-chart of the in-situ measuring system. 6 film type flux-gate (FG) sensors, which were newly developed by the Shimatsu Co.[4], were fixed at the test-piece with a special designed instrument for the in-situ measurement. Distributions of the natural magnetic field were measured off-situ by scanning the unloaded test-piece on a x-y stage. A Shimatsu material testing machine was used as the mechanical loading system.

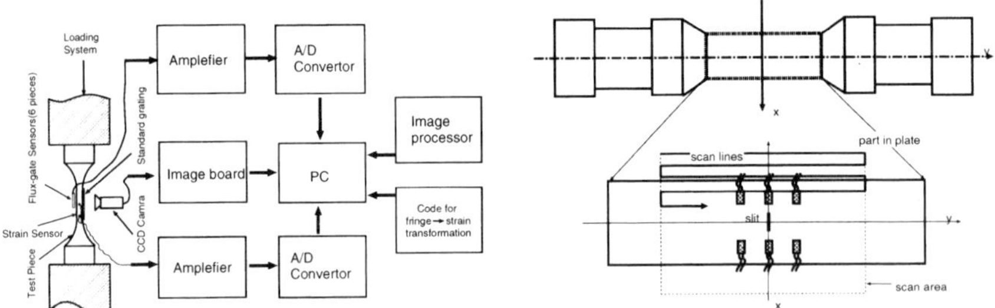

Fig.1 In-situ measurement system   Fig.2 Test-piece and arrangement of sensors

The file type FG sensors have a sensitivity over 0.001G and a spatial resolution of about 1 mm. The detected magnetic signals and the corresponding coordinate information are stored to a personal computer through a A/D converter. The test data were then processed using the PC with codes developed specially. In Fig.2, the shape of the test-piece, the arrangement of the sensors and the scanning routine of sensor are depicted. The scanning range and the pitch were chosen to be 30 mm×30 mm and 1 mm respectively.

c) Experimental procedures

The number of test-pieces used in the experiment and their testing conditions are summarized in table 1. All the test-pieces are demagnetized by using an electromagnetic demagnetizor before being mounted to the material test machine. Strain control was used for all the tensile and fatigue tests. In the case of the tension test, the magnetic flux density at the central part of the gauge area is measured in-situ. The distribution of the natural magnetic field was then measured on a x-y stage after unloading the test-piece at each ε_{max} listed in Table 1. In the case of the fatigue test, the magnetic field signal measured in-situ was picked-up at the same location of the tensile testing for each load cycle. The distributions of magnetic field are measured when the test-piece was unloaded at selected numbers of strain cycle. The cyclic loading was stopped automatically when a crack(s) was initiated.

Table 2 List of test-pieces and test conditions

	No. of test-piece	Notch	Testing Conditions
	ndt2	no slit	ε_{max} =3%/6%/9%, 0.01%/s
Tensile	ndt21	slit	ε_{max} =3%, 0.01%/s
Test	ndt14	no slit	0.2/0.5/1/2/3%, 0.01%/s
	ndt21	slit	±0.25%, 0.1%/s, N=1~940
Fatigue	ndt20	slit	±0.25%, 0.1%/s, N=1~980
Test	ndt14	no slit	±0.25%, 0.1%/s, N=1~4600
	ndt13	slit	±0.25%, 0.1%/s, N=1~64000

2.2 Experimental results

a) Results of the tensile testing

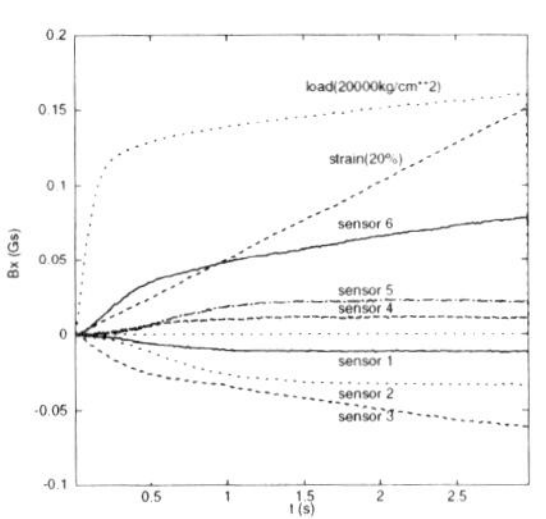
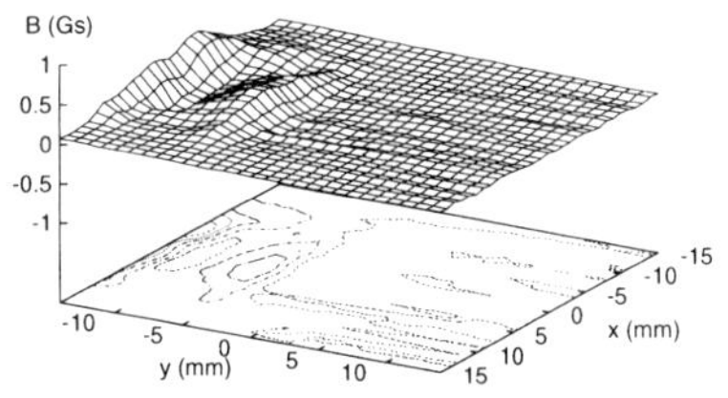

Fig.3 In-situ measurement result of NDT2　　Fig.4 Distribution of natural magnetic field
during a tensile test　　　　after tensile test (maximum strain: 3%)

Fig.3 shows an in-situ measurement results of the natural magnetic field for the test-piece NDT2 during a tensile testing. The test condition is shown in table 1 where 3%/6%/9% means that the test-piece was unloaded and measured after total strain 3%/6%/9% was reached. From the figure, it is clear that magnetization has been induced due to the plastic deformation and the magnetic field did not change much at the elastic region. In addition, the magnetization increases relatively quicker at a low plastic strain region and saturates with the increment of the plastic deformation. As shown in Fig.4, a peak value of the magnetic field appears in the corresponding distribution of natural magnetic field at the bottom side of the test-piece which may be caused by global plastic sliding (Luders band). An observation using Scanning Electronic Microscope (SEM) showed that the number of slided grains on the bottom side (where peak of natural magnetic field occurred) is about 2 times more than that on the upper side while the ferrite grains do not change much. The sliding lines lead to a natural magnetic field mainly due to two kinds of principles, i.e., anisotropy due to the sliding and martensitic transformation. Actually, Transmission Electronic Microscope (TEM) images of the samples cut from NDT2 show that martensitic transformation may happen at the cross points of the sliding lines. The results of X-ray diffraction verified the existence of martensitic phase in these regions.

A difficulty to detect the plastic damages from the natural flux density change is from the saturation. Fig.5 depicts the in-situ field change of NDT14 at residual strain of 0.08%, 0.363%, 0.945%, 2.20%, 3.78%. From the figure, we can find that the change of the magnetization becomes saturated at a high residual strain area. Fig.6 gives the measurement results for test-piece NDT21 which has a central slit (3mm,0.3mm,through) during the tensile test under conditions shown in table 1. Natural magnetic field was induced again in this sample.

b) Results of the fatigue testing

Fig.7 shows a distribution of the natural magnetic field in test-piece NDT21 after fatigue cracks were initiated at the two ends of the central slit. The peak value shows clearly the existence of concentrated magnetization. The observation of the crack tips by using a magnetic colloidal method and the observations made on a Magnetic Force Microscope (MFM) also support the results of the natural field measurement[3]. The experimental results of another test-piece NDT20 in which fatigue cracks were initiated at about the same number of strain cycle verified the phenomenon occurred in the NDT 21 though the peak value of the NDT20 is smaller than that of the NDT21. This difference may comes from the different residual strain state of the test-piece. These results clearly show that the measurement of the natural magnetic field is a possible way to identify the existence of fatigue crack.

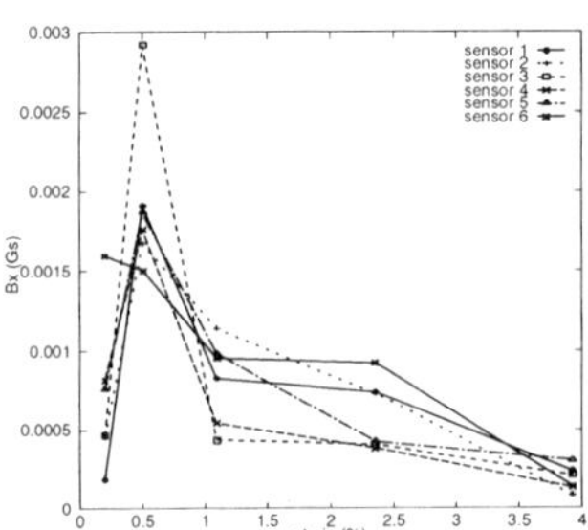

Fig.5 Saturation of natural magnetization with the increment of plastic deformation

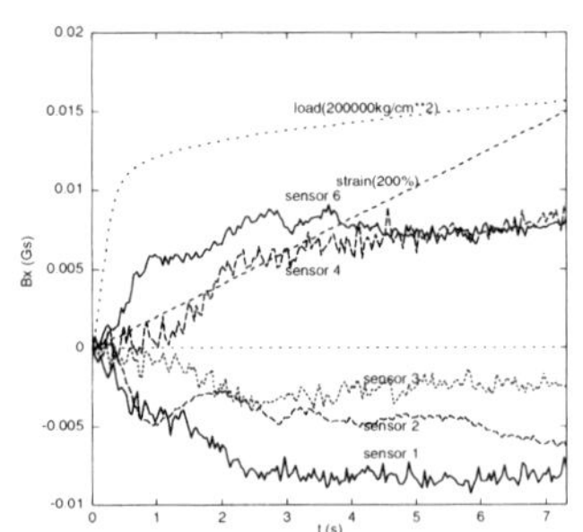

Fig.6 In-situ measuring results of NDT21

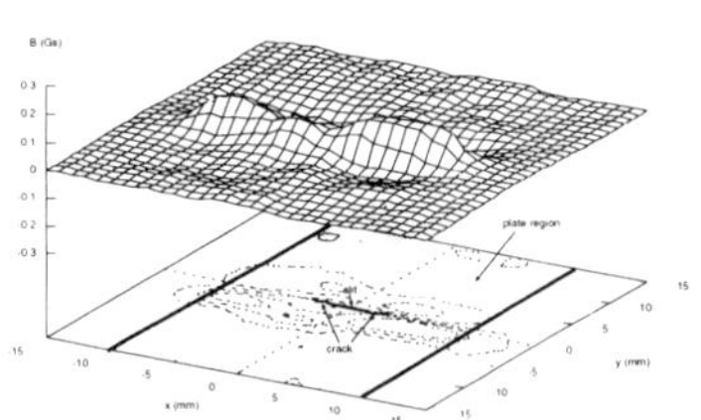

Fig.7 Distribution of natural magnetic field after crack initiated
(test-piece NDT21, residual strain: 2.8%, strain range: 0.5%, strain rate: -1)

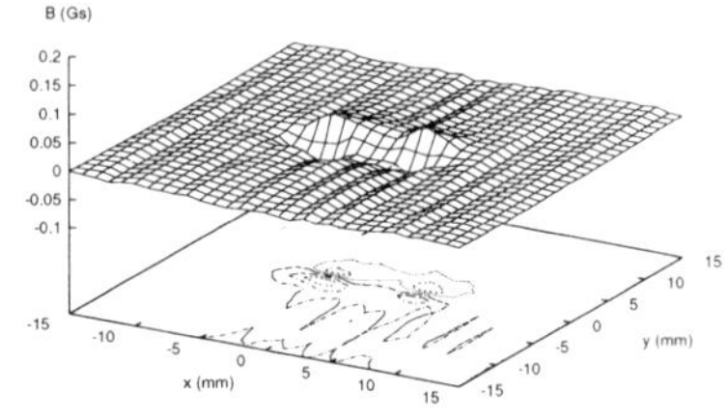

Fig.8 Distribution of natural magnetic field after crack initiated
(test-piece NDT20, residual strain: 0%, strain range: 0.5%, rate: -1)

3. Reconstruction of magnetic charges

The magnetic charge is a concept introduced referring to the definition of the electric charge. Though there is no experimental result supporting its existence, the magnetic charge still remain an important physical concept in view of its equivalence with the current dipole model in description of many physical phenomena. Concerning the bad condition number in the inversion of the magnetization vector, choosing magnetic charge as the unknown variable to be reconstructed is reasonable for the prediction of the magnetic source inside the object to be inspected. Moreover, such a treatment can significantly reduce the number of unknowns in the inverse analysis, i.e. great computer burden can be reduced.

At a field point (coordinate vector $\mathbf{r}$) outside the volume V of the magnetized object, the magnetic scalar potential induced by the magnetization at a source point (denoting as $\mathbf{r}'$) is equivalent to that induced by a related magnetic dipole. Referring to formulae for a magnetic dipole, the magnetic scalar potential due to magnetizations can be expressed as[5]

$$4\pi\mu_0\phi(\mathbf{r}) = \int_V \mathbf{m}(\mathbf{r}') \cdot \nabla' \frac{1}{|\mathbf{r} - \mathbf{r}'|} dv' =$$

$$\oint_\Omega \mathbf{m}(\mathbf{r}') \cdot \mathbf{n} \frac{1}{|\mathbf{r} - \mathbf{r}'|} ds' + \int_V \nabla' \cdot \mathbf{m}(\mathbf{r}') \frac{1}{|\mathbf{r} - \mathbf{r}'|} dv', \tag{1}$$

where, $\phi(\mathbf{r})$ is the magnetic scalar potential in air region at the field point $\mathbf{r}$, $\mathbf{m}(\mathbf{r}')$ is the magnetization vector at the source point $\mathbf{r}'$, $q(\mathbf{r}') = \nabla' \cdot \mathbf{m}(\mathbf{r}')$ is magnetic charge density at point $\mathbf{r}'$ according to its definition, and $\mathbf{m}(\mathbf{r}') \cdot \mathbf{n}$ is the surface charge density at a surface point with a normal unit vector $\mathbf{n}$. From Eq.(1), one can find that the effect of magnetization is equivalent to that caused by the volume and surface magnetic charges.

Eq.(1) can be further simplified by taking into account conditions of the practical problem. In order to reconstruct magnetic sources from the natural magnetic field of a plate test-piece subjecting to a loading of plane stress state, the surface magnetic charges at the planar surfaces can be approximated as zero considering the symmetry condition and the vanishing value of the total magnetic charge. In addition, for a problem dealing with magnetic sources near the tips of a crack initiated at the central part of a test-piece, the magnetic charges at the other surfaces can also be neglected. Under such conditions, Eq.(2), the formula for the magnetic flux density induced by the magnetic charges can be obtained by taking gradient operation at the two sides of Eq.(1) and neglecting the surface integral term, which, will be taken as the basic governing equation in the following inverse analysis.

$$\mathbf{b}(\mathbf{r}) = \frac{1}{4\pi} \int q(\mathbf{r}') \nabla \frac{1}{|\mathbf{r} - \mathbf{r}'|} dv', \tag{2}$$

where, $\mathbf{b}(\mathbf{r})$ is the leakage flux density vector at the field point $\mathbf{r}$.

3.1 Schemes for reconstruction

In the follows, we consider a problem with magnetic charges localized in a limited volume V. Upon subdividing V into $\bar{n}'$ cells and taking the charge density as a constant value in each cell, the magnetic charges in any source point $\mathbf{r}'_i$ can be expressed as

$$q(\mathbf{r}'_i) = \sum_j^{\bar{n}'} q_j \xi_j(\mathbf{r}'_i), \qquad i = 1, 2, \ldots\ldots, \bar{n}. \tag{3}$$

where, $\xi_j(\mathbf{r}'_i)$ is a box function with zero values outside the $j - th$ cell, q_j is the charge density at the $j - th$ cell, and $\bar{n}$ is the total number of the field points.

Substituting the above equation into Eq.(2), some calculations show that the magnetic flux density $\mathbf{b}(\mathbf{r}_i)$ at an arbitrary field point $\mathbf{r}_i$ outside V can be obtained from the discretized charge parameters with the following formula,

$$\mathbf{b}(\mathbf{r}_i) = \sum_j^{\bar{n}'} \frac{1}{4\pi} v_j \frac{(\mathbf{r}_i - \mathbf{r}'_j)}{|\mathbf{r}_i - \mathbf{r}'_j|^3} q_j. \tag{4}$$

where, v_j is the volume of the $j - th$ discrete cell.

Similarly, a component of the magnetic flux density vector (for instance, $\{q_x\}$, the x component) at $\bar{n}$ inspection points connects to the charge parameters by the governing system of linear equations

$$\{b_x\} = [K]\{q\}. \tag{5}$$

where, $k_{ij} = \frac{1}{4\pi} v_j (x_i - x'_j)/|\mathbf{r}_i - \mathbf{r}'_j|^3$, $i = 1, 2, ..., \bar{n}, j = 1, 2, ..., \bar{n}'$.

As in general $\bar{n}' \neq \bar{n}$, the least square method is an usual way to find a solution of such a system of equations, i.e., solving the system of equations from a point of view of a minimum square residual. The system of linear equations derived with the least square method reads

$$[K]^T \{b\} = [K]^T [K] \{q\}, \tag{6}$$

where $[K]$ is the coefficient matrix with element k_{ij}, and $\{q\}$ is the unknown vector of magnetic charges.

Matrix $[K]^T [K]$ usually has a bad condition number, i.e. the solution of Eq.(6) is not unique from a rigorous point of view. Therefore, regularization is necessary. Usually, a deterministic optimization method can gives a reasonable solution for such a problem if an appropriate initial value can be chosen somehow. Fortunately, such an initial condition is not too difficult to be selected for present problem, e.g. we at least can choose zero distribution if the charge at the boundary can be taken as vanished values. Based on these considerations, a typical algorithm of the optimization method (the steepest descent method or its accelerated version - the conjugate gradient method) is chosen to solve Eq.(6) in view of their advantage of superior convergent speed and stability.

Actually, if we let ε be the residual function, Eq.(6) can be transformed to an optimization form as,

$$Min_q \varepsilon_m (= \sum_i^{\bar{n}} w_i [b_i^{obs} - b_i(\{q\}^m)]^2), \tag{7}$$

subject to the known constraint conditions on $\{q\}$ (such as boundary conditions)[5],[6]. In Eq.(7), w_i is a weight coefficient, b_i^{obs} is an observed data of leakage magnetic flux density and $b_i(\{q\}^m)$ is a data of magnetic field from an estimated charge distribution at m-th iteration step, say, $\{b(\{q\}^m)\} = [K]\{q\}^m$.

The iteration formula for solving Eq.(7) reads

$$\{q\}^m = \{q\}^{m-1} + a_n \{f\}^m, \tag{8}$$

where, $\{f\}^m = \{\partial \varepsilon_{m-1}/\partial q_j\} + G_m \{f\}^{m-1}$ for the CG method, and $\partial \varepsilon_{m-1}/\partial q_i$ is a gradient of the objective function ε at the direction along i-th coordinate vector, which can be calculated from the coefficient matrix $[K]$ with,

$$\frac{\partial \varepsilon_{m-1}}{\partial q_j} = 2 \sum_{i=1}^{\bar{n}} k_{ij} [b_i^{obs} - b_i(\{q\}^{m-1})], j = 1, ..., \bar{n}' \tag{9}$$

3.2 Implementation and validations

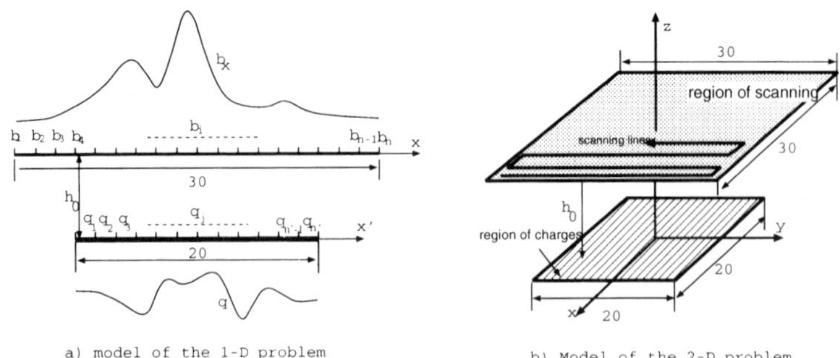

Fig.9 Numerical models for the reconstruction of the distribution of magnetic charges

The schemes described in the subsection 3.1 was implemented for 1-D and 2-D problems on a personal computer. Reconstructions of problems depicted in Fig.9 are performed to investigate the validity of the developed codes. For the 2-D problem, the charges are assumed to be located in a square region of size 20 mm × 20 mm. The field data are of either a horizontal or the normal component of the natural magnetic flux density vector which were acquired from a 2-D scanning in a 30 mm × 30 mm region located h_0 over the plane of magnetic source. These conditions are corresponding to those used in the measurement of 2-D distribution of test-piece. The natural magnetic fields from the damage-induced magnetizations of test-piece NDT21 is taken as the experimental data in the validation of the proposed method. As the magnetizations can be assumed as a constant distribution in the thickness direction, the inversion is simplified to a 2-D problem. In addition to the above measured data, several data sets of simulated magnetic field are also adopted for validating the proposed method in case of a 2-D problem. Moreover, artificial noises are also added into the input data for investigating the robustness of the proposed method. In the follows, numerical results for the least square method will be presented.

3.3 Numerical results

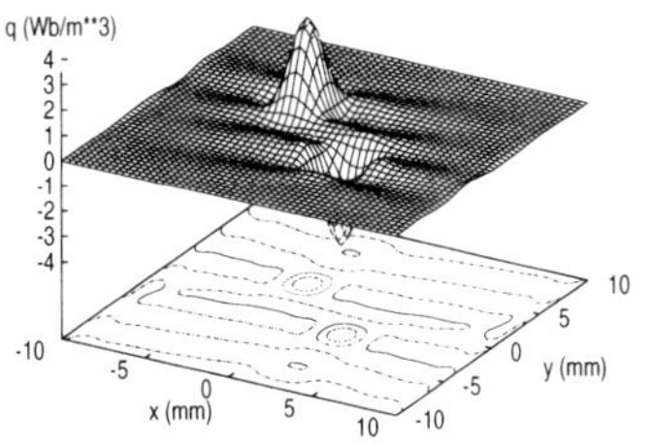

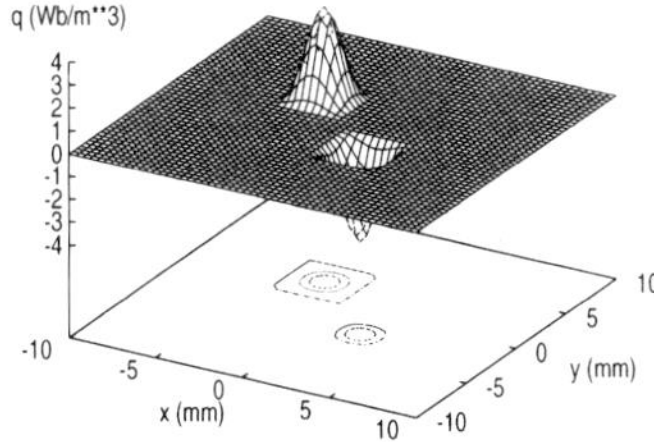

a) Reconstructed result   b) True distibution

Fig.10 Comparison of the distributions of true magnetic charges and that predicted by using the least square method

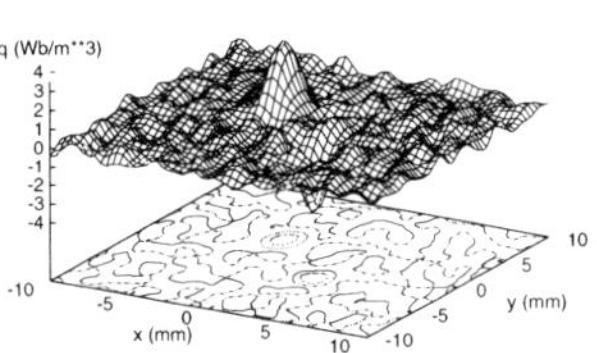

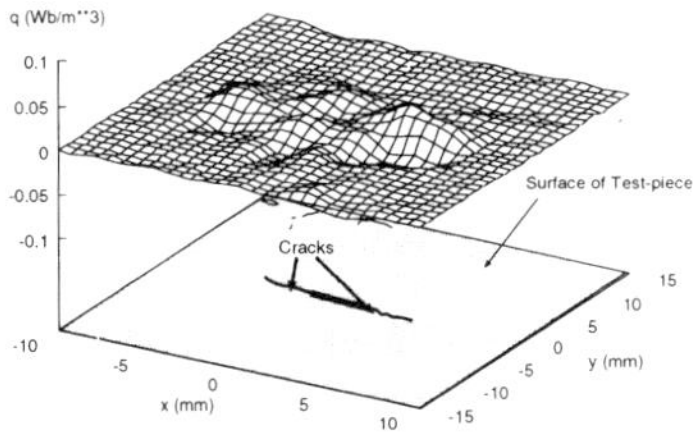

Fig.11 Reconstructed charge distribution of the least square method from a signal with 20% artificial noise

Fig.12 Distribution of the magnetic charges reconstructed from the measured natural magnetic field data

Fig.10 gives results for the two dimensional problem with the initial profile of the charges selected as zero. The figure is a comparison of the reconstructed and true distribution of the charges when noise free simulation signals were applied. The lift-off was taken as 1.0 mm and the iteration was stopped at the 100 step. The total CPU time used was about 10 minutes in a PC (Pentium II 350,). From the figure, we can find that the gradient method can give good reconstruction for noise free data. To investigate the robustness, the reconstruction was also performed with signals including artificial noise. Fig.11 is an example of reconstruction for signal with 20% white noise. Here, the percentage of noise is defined as the ratio of the maximum values of the noise and the field signal. Comparing with the true distribution shown in Fig.10(b), one can find that acceptable results are obtained again.

Fig.12 depicts a reconstructed distribution of charges in the gauge region of the test-piece NDT21. This means that the measured magnetic field data was applied in this case. The magnetic field calculated from the predicted charges shows a good agreement with the measured ones (Fig.13). Comparing the reconstructed distribution of charges with the locations of the fatigue cracks which have initiated at the two ends of the central slit, it can also be found that the reconstructed distribution of charges is reasonable if one pays attention to the position of peaks and the crack tips. The results of the magnetic colloidal experiment [3] also show qualitative agreement with the reconstructed distribution of magnetic charges. These results support the expectation that the proposed method is suitable for investigating the correlation of the magnetic and status of damage.

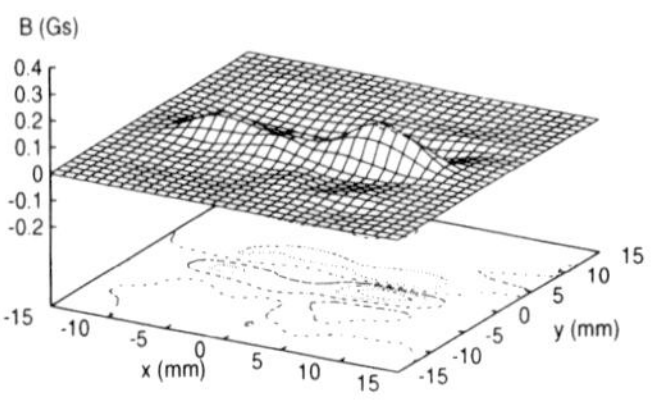

a) Reconstructed magnetic field

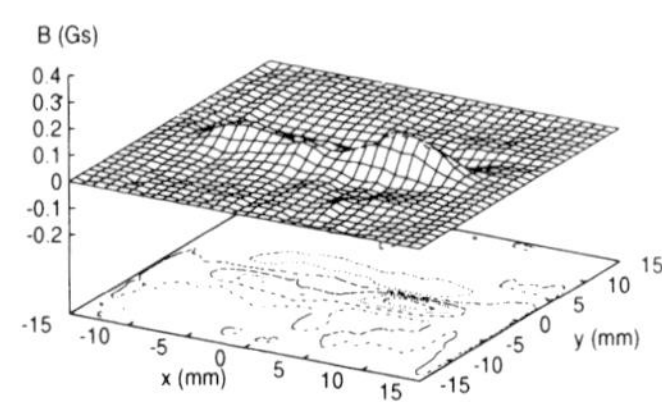

b) Mearsured magnetic field

Fig.13 Comparison of the measured magnetic field and the magnetic field from the reconstructed magnetic charges

4. Concluding remarks

Based on the experimental results described in the previous sections, the following concluding remarks are obtained:

1. The damage induced magnetization occurs for the SUS304 steel especially in the vicinity of the fatigue cracks. The natural field can be considered as the results of the plastic damages for tensile test. Detection of the residual strain from the measured natural field is difficult because of the saturation and the effect of global sliding.

2. In spite of the local minimum problem of an iteration algorithm, the gradient method has been proved feasible for magnetic charge reconstruction using a zero initial values. Moreover, it is also demonstrated that an appropriate selection of the maximum iteration number is an efficient regularization for tackling the ill-poseness of such inverse problem.

3. The reconstruction results using the measured signals of the passive leakage magnetic field near a test-piece with fatigue cracks support the validity of the present method at an experimental environment from the point of view of the reasonable charge distribution referring to the crack profiles.

References

[1] R.B.Mignogna, and et al., Passive nondestructive evaluation of Ferromagnetic materials during deformation using SQUID gradiometers, IEEE Trans. Appl. Supercond., Vol.3, No.1, p1922-1925, 1993.

[2] Report of research committee on electromagneto-fracture and its applications on nondestructive evaluation of degradation and damages, JSAEM report, JSAEM-R-9803, JSAEM-R-9902.

[3] M.Uesaka and et al.,Round-robin test for nondestructive evaluation of steel components in nuclear power plants, in Studi. Appl. Electromagn. mech., Vol.14, p39-48, 1999.

[4] Y.Yamada, and et al., Application of thin-file flux-gate sensor, New Ceramics, No.2, p15-18, 1998.

[5] S.Chikazumi, Magnetics, Kyouritsu Publisher, Lt.d, 1969.

[6] S.J.Norton, and J.Bowler, Theory of eddy current inversion, J. Appl. Phys., Vol.73, No.2, 501-512, 1993.

[7] Z.Chen and K.Miya, ECT inversion using a knowledge based forward solver, J. Nondestr. Eval., Vol.17, No.3, p167-175, 1998.

Electromagnetic Nondestructive Evaluation (VI)
F. Kojima et al. (Eds.)
IOS Press, 2002

Magnetic Nondestructive Evaluation of Ferromagnetic Steels due to Mechanical Loading

T. Sukegawa and M. Uesaka
Nuclear Engineering Research Laboratory, Graduate School of Engineering,
The University of Tokyo, 2-22, Tokai, Naka, Ibaraki, 319-1106, Japan

Abstract. Measurement of magnetic properties of low alloy steel used as a pressure vessel in nuclear reactors, which were degraded due to tensile deformation and cyclic load, was carried out. Magnetic hysteresis curves change depending on the load condition. In case of the tensile loading, there were three staged magnetic property changes. In the elastic region, magnetic properties have no change. Over the 0.2% yield stress, the residual magnetic flux density and the incremental permeability decrease while there is no change of the coercive force. When plastic deformation proceeds, the coercive force increases with the increasing residual strain while the residual magnetic flux density and the incremental permeability decrease. Those finally reach the saturation with the material work hardening. In case of fatigue degradation accumulated by the cyclic load, the same tendency observed. Based on the above results, we performed the sensitivity analysis of magnetic property changes for the mechanical load. Finally we concluded that the dB/dH at H=0 is the most sensitive among the promising magnetic parameters for early degradation due to the mechanical loads. The feasibility of application of this magnetic technique to the evaluation of degradation due to neutron mechanical and thermal loading is now under consideration.

1. Introduction

More than 50 commercial nuclear power plants are in operation and supply electricity of about 3×10^8 KWh in Japan, which is the one third of total amount of electric energy. The lifetime of Japanese nuclear rectors was declared by the government to be extended from 30-40 years to about 60 years in 1999. In the underneath of the decision, there are some important reasons based on the energy strategy. The first is the fact that many Japanese nuclear reactors have been used for over thirty years as the major supplier of electric power. Second, the construction of new nuclear energy plant is rather limited mainly due to the problem of public acceptance. Therefore, they are expected to have longer lifetime based on technical innovations and reliability study of the components. Thus, the technical development to evaluate quantitatively structural material degradation and lifetime becomes very important. One of the most serious concerns is the degradation evaluation of the reactor pressure vessel. Since that is made of low alloy steel, called A533B in Japan. We adopt the low alloy steel, A533B, as the first target in this work.

In this paper, we focus on the magnetic nondestructive evaluation to obtain the quantitative correlation between the mechanical and magnetic properties of the materials. Electromagnetic nondestructive evaluation approach is one of the most important topics in the field of nondestructive inspection and examination, since magnetic properties were greatly sensitive to micro structural changes in ferromagnetic materials through the microscopic interaction between magnetic domain walls, moments and microscopic defects. Main magnetic inspection technologies suggested recently are magnetic hysteresis measurement,

Barkhausen noise (BHN) analysis [1] and SQUID sensor [2]. All the approaches could be promising for detailed description of degraded materials. However, physical models on the correlation haven't been sufficiently established for practical poly-crystals yet. In the previous work [3,4], we performed the measurement of magnetic properties that were fatigue degraded by the cyclic load using the rod samples whose size is based on *Japanese Industrial Standard* [5]. Magnetic hysteresis curves changed depending on the loading conditions. The changes in the coercive force, residual magnetic flux density and permeability were obtained as a function of the magnitude of residual strain. In this work, we also performed the measurement of magnetic properties for low alloy steel that were degraded by the tensile and cyclic loads using the plate samples so that the surface leakage magnetic filed scanning is available. The first is to reconfirm the previous results in the correlation between the magnetic and mechanical measurements. For the second purpose, we performed the sensitivity analysis of magnetic property changes to the mechanical degradation.

2. Experimental method

The size of the plate A533B sample, the condition of heat treatment and its chemical components are described in Fig. 1.

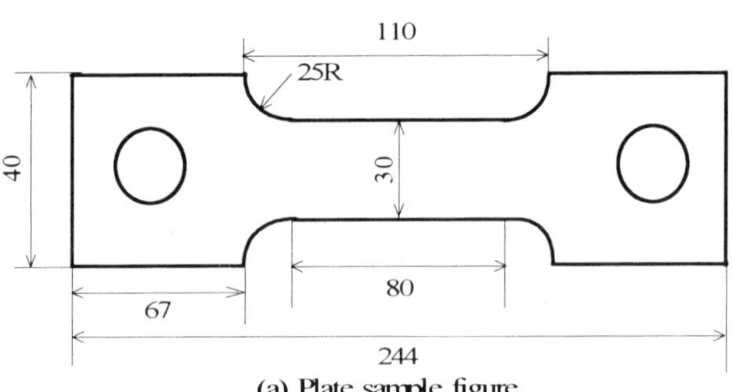

(a) Plate sample figure

Quenching : 860 ∼ 890 °C × 2hr 25min + Water Cooling

Annealing : 650 ∼ 665 °C × 2hr 19 min + Air Cooling

(b) Heat treatment

	C	Si	Mn	P	S	Ni	Mo	Cu	Al	N	As	Sn	Sb
WT%	0.18	0.14	1.53	0.004	0.002	0.66	0.56	0.03	0.010	0.0098	0.003	0.004	0.0008

(c) Chemical components

Fig. 1 Plate Sample (A533B, 2mm')

The samples were cut off from a single ingot and machined. Its thickness of 2mm' is determined so that our mechanical loading machine can make plastic deformation. . Figure 2 shows the B-H curve measurement setup. The copper wire is wound around the yoke for excitation, which produces magnetic filed, H, in the sample. To pick up the time derivation of magnetic flux density, dB/dt, and the other coil wound directly around the sample. A Hall sensor is also attached to the sample in order to measure the magnetic flux density, B. The amplitude of the applied alternating current for measurement rose quickly up to the specific level to avoid temperature rise. After the measurement was finished, the current decreased gradually so that the sample was demagnetized. Magnetic hysteresis loops were obtained by a personal computer. The frequency of the alternating current was relatively low so as to avoid the effect of eddy current.

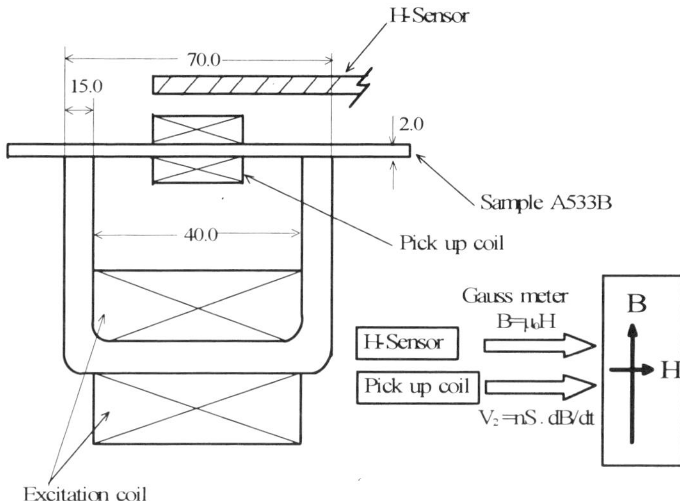

Fig. 2 B-H curve measurement setup for the plate sample

Zero-tension sinusoidal cyclic load was applied with the amplitude of 450 to 510 MPa where the yield is about 520MPa. The minimum amplitude is 30 MPa to avoid unexpected out-of- plain bending.

3. Experimental result

This load conditions for the tensile test in shown Fig.3 are described in Table 1. We investigated residual strain from 0 % to 7.5 %. We had carried out the first test with the larger plate samples (Length: 290mm. Thickness: 5mm') [3,4]. The previous loading condition is given in Table 2 as the reference.

Table 1 Loading condition for this tensile test

Sample Number	Tensile Stress (MPa)	Residual Strain (%)
A-1	620	2.1
A-2	642	3.7
A-3	0	0
A-4	534	1.2
A-5	529	0.38

#: 0.2% Strength is 517~523MPa

Table 2 Loading condition for the previous tensile test

Sample Number	Tensile Stress (MPa)	Residual Strain (%)
No.533-3	462	0
No.533-4	550	0.378
No.533-5	550	0.43
No.533-2	580	2.9
No.533-1	663	7.5

#: 0.2% Strength is 560MPa

The difference of the stress-strain relation between the current and previous tests attributes to the difference of ingot, namely manufacturing process. We calculated the strain from the elongation of the 50mm-long central longitudinal line on each samples.

Table 3 shows the loading conditions for the cyclic test. These material properties were the same as those for the tensile test in shown Fig. 1.

Table 3 Loading condition for the cyclic test

Cyclic Load (MPa)	Sample Number
510	A-26
500	A-20, A-21, A-22, A-25
480	A-19, A-27
450	A-28

#: 0.2% Strength is 517~523MPa

Figure 3 shows measured B-H curves in the current test. In the figure, the curve can be divided into the three stages. The sample A-3 stays in the elastic region. The elastic tensile deformation gives no effect on the shape of the B-H curve. The curve of A-5 corresponds to the case of just over 0.2% yield stress. Here the shape is almost unchanged and Br decreases, while the coercive force, Hc, shows no change. Especially, Luders band is observed by the leakage magnetic filed scanning by the SQUID and GaAs hall sensor [6,7,8]. The samples, A-1, A-2 and A-4, deformed in the plastic region with work hardening.

The relation of H_c and B_r as a function of the residual strain is shown in Fig. 4. Br shows the saturation around 2% residual stain, while H_c also shows

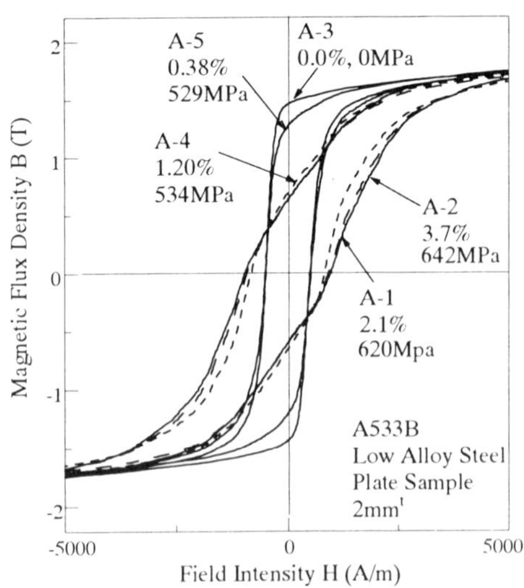

Fig. 3 Changes of magnetic hysteresis loops for tensile deformation.

the saturation around 3% residual stain as shown in the figure. Here, the sensitive check was carried out from the result of tensile tests. Table 4 shows the relative value for the c parameter. Table 5 shows the relative value for the dB/dH at H=0. Here, the c parameter is defined by Takahashi et al. by the equation $\chi = c / H^3$ [9] where χ is the susceptibility and H is the applied magnetic field. In case of 0.38% residual strain for the incremental permeability at H=0, the relative value shows almost saturation. This indicates that the dB/dH at H=0 is the most sensitive index for the early degradation due to mechanical loads. We also carried out the sensitive check for the cyclic loads. Both the relative values of the c parameter and incremental permeability can be regarded as the sensitive indices of the increase of dislocation density, namely the degradation [3,4].

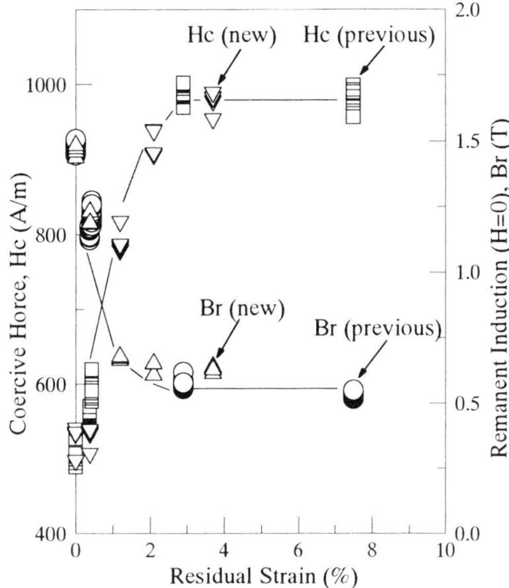

Fig. 4 The coercive force and remanent induction as a function of residual strain.

Table 4 The relative value for C parameter

σ (MPa)	Residual strain (%) (estimated by stain – stress curve)	C parameter	Relative value
0	0	1.60×10^6	1.0
462	0 (within elastic region)	1.70×10^6	1.06
550	8.8	7.10×10^6	4.44
580	11.6	1.40×10^7	8.75
663	21.6	1.50×10^7	9.40

Table 5 The relative value for the dB/dH at H=0

σ (MPa)	Residual strain (%)	dB/dH at H=0	Relative value
0	0	1.80×10^{-4}	1.0
529	0.38	4.42×10^{-4}	2.46
534	1.20	4.95×10^{-4}	2.75
620	2.10	4.85×10^{-4}	2.69
642	3.70	4.78×10^{-4}	2.65

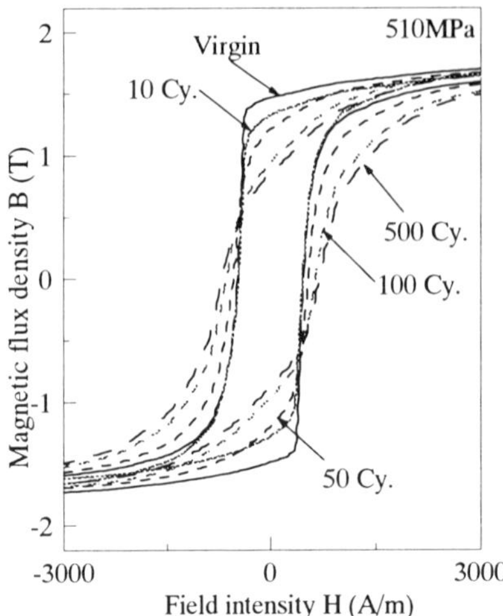

Fig. 5 Changes of magnetic hysteresis loops
for 510MPa cyclic load.

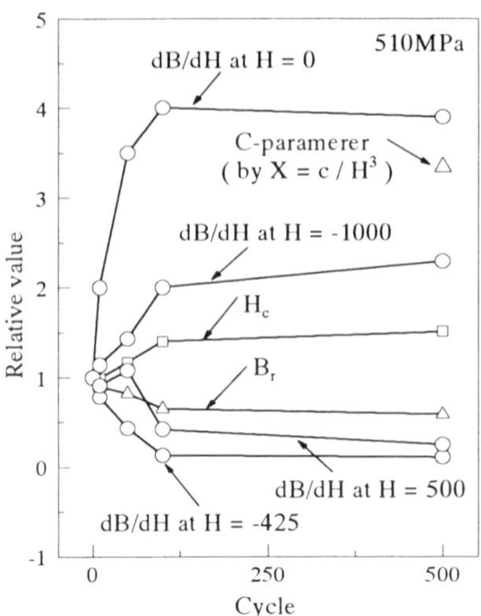

Fig. 6 The sensitive analysis for 510 MPa
cyclic load.

The change of magnetic hysistesis loops in the cyclic load is shown in Fig. 5. A-26 sample is for 510MPa that is close to the yield stress. Changes of magnetic hysteresis loops are fast with respect to the cycle. When we compare it with the result for the tensile load case as shown in Fig.3, the same tendency is observed. Namely, the shoulder of the loops declines and Br decreases for early degradation, while the total loops become oblique and H_c increases for heavy degradation. In order to analyze the reason of the similarity, we put the results of leakage flux distribution for the same sample measured by the GaAs Hall sensor. It is clear that many small Luders band are generated for the early degradation, while the large Luders bands are observed for the tensile load [7].

Thus, the declination of the shoulder and Br decrease can attribute to the generation of the Luders bands. Figure 6 shows the results of the sensitivity analysis of relative magnetic property changes for 510Mpa. Relative H_c, B_r, dB/dH at H= 0, -425, 500 and -1000 are plotted as a function of the cycle. It is clear that dB/dH at H= 0 is the most sensitive value for the early degradation even in case of the cyclic load.

We also performed the same test for the cyclic load with the lower amplitude (500MPa) as shown in Fig. 7. The results of the dB/dH and sensitivity analysis for amplitude of 500MPa are shown in Figs. 8 and 9. The sensitivity analysis clearly indicates again that the dB/dH at H=0 is the most sensitive index for the early degradation due to mechanical loads.

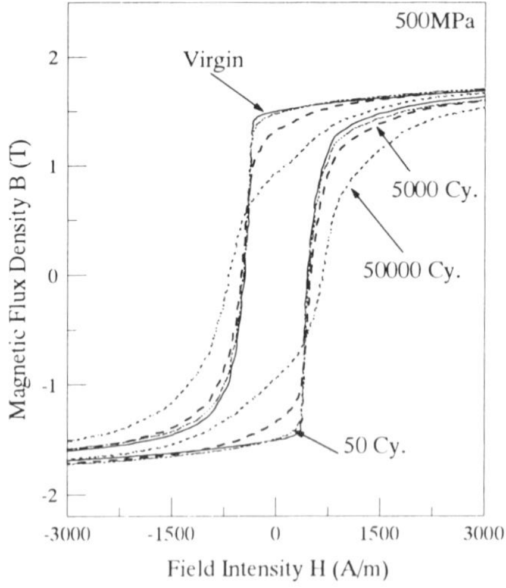

Fig. 7 Changes of magnetic hysteresis loops
for 500MPa cyclic load.

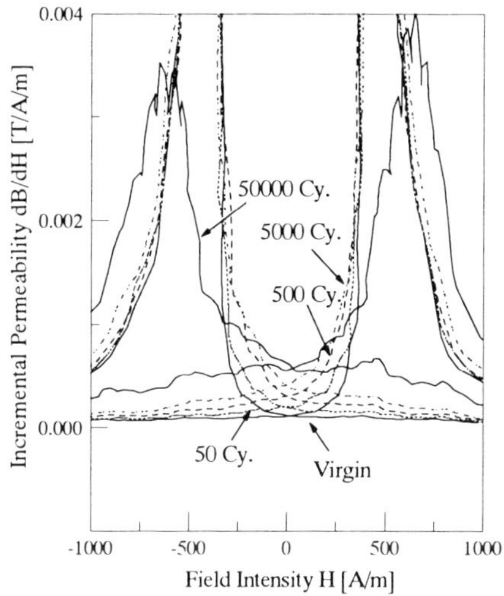

Fig. 8 Incremental permeability as a function
of applied field for 500 MPa cyclic load.

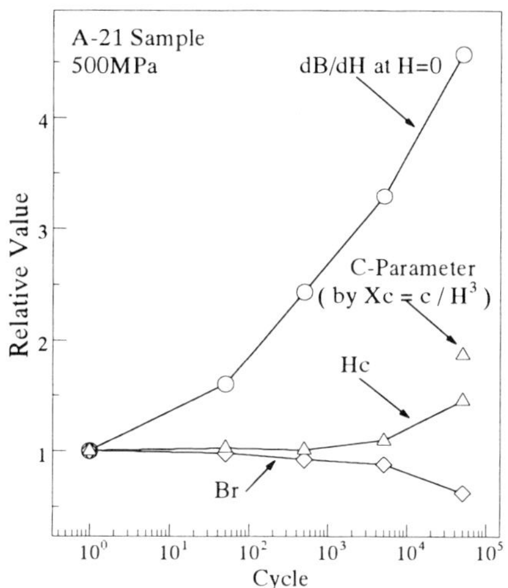

Fig. 9 Results of the sensitive analysis
for 500MPa cyclic load.

Figures 10 and 11 also show magnetic hysteresis loops and the sensitivity analysis on the magnetic properties for 480 MPa. In the figures, we don't observe any clear change of H_c and B_r, until 10^3 cycles. On the contrary, dB/dH at H=0 is increasing gradually, and turns rapidly rising from 10^4 cycles. It is clear that many small Luders bands are generated rapidly from 10^4 cycles. Plotted values of dB/dH at H=0 are the average values over 10 time measurements. These additional results also support our finding that the dB/dH at H=0 is more sensitive index than the c-parameter for the early degradation due to mechanical loads. The physical basis on the changes of the dB/dH and the c-parameter could be the same. That is, the coefficient of magnetic susceptibility, c, is a parameter that contains the synthetic information dislocation [9]. The only difference depends on the location on the B-H curve where the two parameters are evaluated.

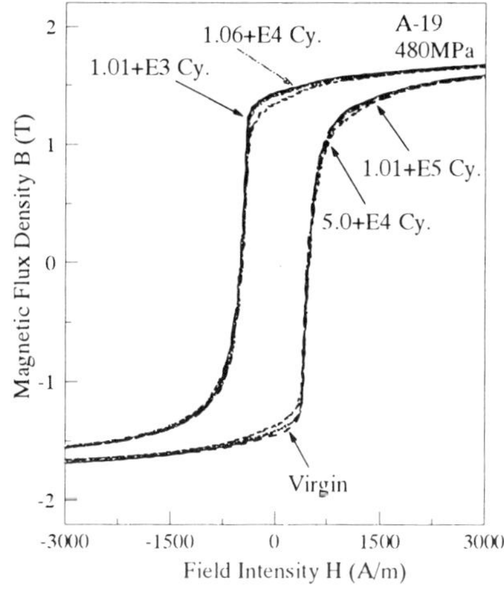

Fig. 10 Changes of hysteresis Loops
for 480MPa cyclic load.

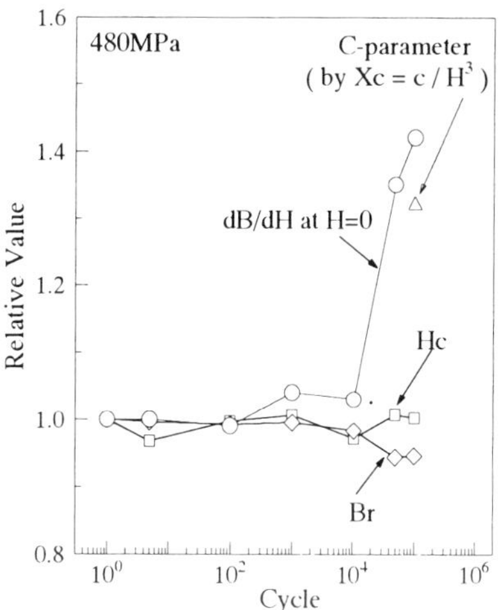

Fig. 11 Results of the sensitive analysis
for 480MPa cyclic load.

4. Future work

The next work is to evaluate changes of magnetic property due to mechanical load for stainless steels and to obtain again the relation between the magnetic and mechanical properties. In nuclear reactors, stainless steels are used in many parts of structural components such as mechanical supports. In the first step, we are going to adopt the 18-8 steel called SUS 304 in *Japanese Industrial Standard*. 18-8 stainless steel is one of nonmagnetic steels and is also austenite type.

5. Conclusion

We performed the measurement of the magnetic properties of the low alloy steel samples used as a pressure vessel of nuclear reactors. Magnetic hysteresis curves change depending on the load condition. We have got the clear correlation between the magnetic properties and the magnitude of residual strain. We also carried out the sensitivity analysis of magnetic property changes for the tensile and cyclic load. Considering only our measurement, the dB/dH at H=0 was more sensitive index above the *c*-parameter for early degradation due to the mechanical load.

References

[1] D.G. Hwang and H.C Kim: J. Phys., D: Appl. Phys.,Vol. 21, (1988), pp.1807-1813.
[2] J. Banchet, J. Jouglar, P.L. Voillermoz and H. Weinstock: IEEE Trans.on Appl. Superconductivity, Vol. 5 (1995), pp.2486-2489.
[3] A.Gilanyi, K.Morisita, T.Sukegawa, M.Uesaka and K. Miya: Fusion Engineering and Design 42, (1998), pp.485-491.
[4] K.Morisita, A.Gilanyi, T.Sukegawa, M.Uesaka and K.Miya: Journal of Nuclear Materials 258-263, (1998), pp.1946-1952.
[5] Japanese Standards Association, JIS Handbook, Iron and Steel, ISBN-542-12601-3, C3050, (1991), Z-2201, pp.121-127.
[6] M. Ueasaka T. Sukegawa, K. Miya, K.Yamada, S. Toyooka, N. Kasai, A.Chiba S. Takahashi, J. Echigoya, K. Morishita, K. Ara, N. Ebine, Y. Isoba: Proc. of ENDE' 97, Electromagnetic Nondestructive Evaluation (II), IOS Press (1998), pp.39-48.
[7] Koji Yamada, Shin-ichi Shoji and Yoshihiro Isobe: International Journal of Applied Electromagnetic and Mechanics 11, IOS Press (2000) pp.27-38.
[8] N. Kasai, S. Nakayama, Y. Hatsukade, M. Uesaka: Journal of Japan Society of Applied Electromagnetics and Mechanics, Vol. 8, (2000), pp.16-22.
[9] S. Takahashi, J. Echigoya and Z. Motoki: J. of Appl. Phys., Vol. 87(2), (2000), pp.805-813

Fatigue Evaluation of Magnetic Materials by Chaotic Attractor of Barkhausen Noise

Yuji Tsuchida, Takahiro Ando and Masato Enokizono
Faculty of Engineering, Oita University,
700 Dannoharu, Oita 870-1192, Japan

Abstract. This paper presents the fatigue evaluation of magnetic materials by a chaotic attractor of Barkhausen noise. Barkhausen noise is influenced by the dislocation density, so it can be possible to evaluate the fatigue level by measuring Barkhausen noise. It is difficult to find the change of Barkhausen noise caused by fatigue because Barkhausen noise is a very complicated signal. Therefore, we examined the chaotic behavior of Barkhausen noise to get some information on the fatigue level of magnetic materials.

1. Introduction

In recent years, non-destructive testing has been playing an important role for safety and maintenance of any structures. By the conventional non-destructive testing, the cracks of the structures can be detected by using X-ray, ultrasonic, eddy current and so on. We have been doing the research on the fatigue evaluation of magnetic materials qualitatively and quantitatively by measuring Barkhausen noise [1-4]. Barkhausen noise is a signal generated by the movement of the 180-degree magnetic domain inside the magnetic materials, and the dislocation influences this magnetic domain movement. Therefore, it can be possible to evaluate the fatigue level by measuring Barkhausen noise even before a crack appears. In the previous research, we made clear that the power of Barkhausen noise from INCONLE plates changed a little after applying the thermal fatigue [1]. It was very difficult to make clear the relationship between Barkhausen noise and the fatigue level quantitatively and qualitatively because of the complication of the signals. Barkhausen noise is a very complicated non-linear signal, so we examined the chaotic behavior of Barkhausen noise to evaluate the relationships between them because the chaos theory can treat a non-linear signal. First, Barkhausen noises from non-oriented, oriented and double-oriented silicon steel sheets were examined to reveal the basic chaotic behavior of Barkhausen noises [2, 3]. And then we made several sensors suitable for Barkhausen noise [2, 3]. In these papers, we concluded that Barkhausen noise from magnetic materials could be classified into the chaos. Moreover, we found that the constructed attractor made from Barkhausen noise changed depending on the angle of the scratch on the silicon steel sheet and described that the change of the magnetic domain can be found by the shape of the chaotic attractor [4]. In this paper, Barkhausen noises from structural steel plates, SS400 are measured and discussed the possibility of the fatigue evaluation by chaotic attractors from Barkhausen noise.

2. Chaotic behavior of Barkhausen noise

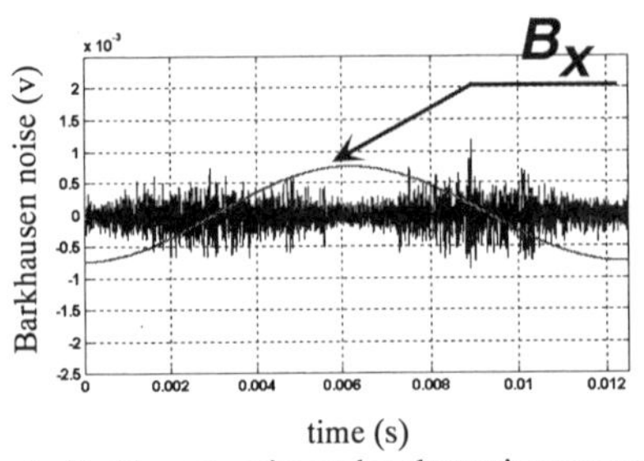

Fig. 1 Barkhausen noise under alternating magnetic field.

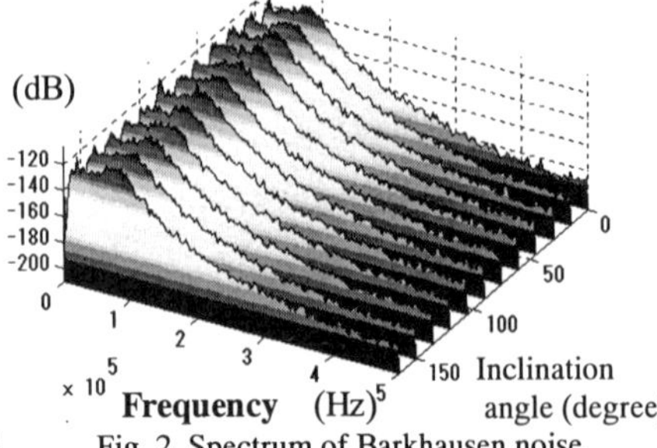

Fig. 2 Spectrum of Barkhausen noise
under alternating magnetic field.

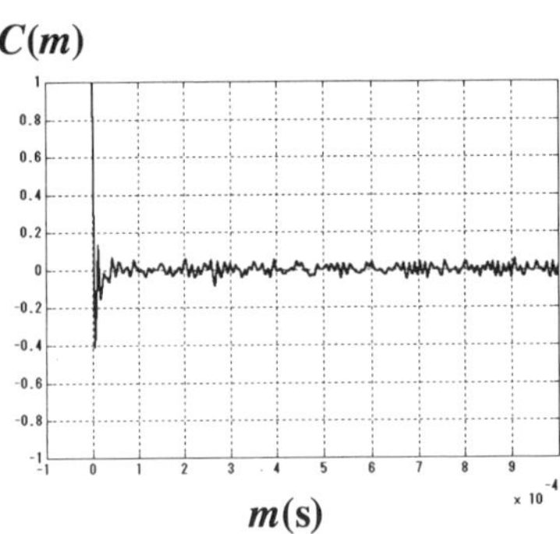

Fig. 3 Correlation function of Barkhausen noise
under rotational magnetic field.

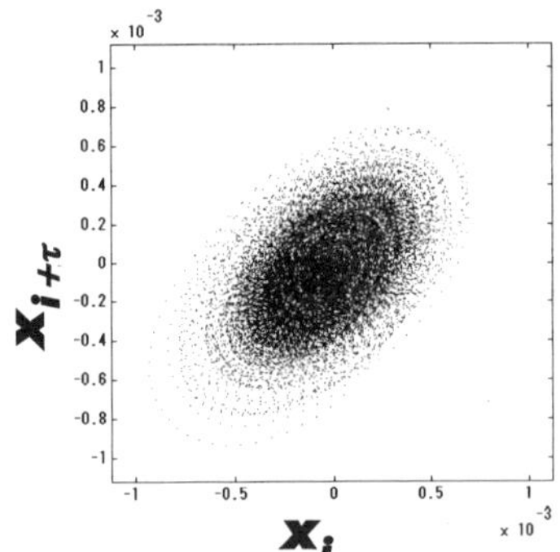

Fig. 4 Constructed attracters under
rotational magnetic field

The Barkhausen noise from the non-oriented silicon steel sheet under the alternating magnetic field is shown in Fig. 1 [1-4]. The controlled sinusoidal wave of the exciting coils is also drawn in this figure. The Barkhausen noise under the alternating magnetic field is mainly generated where the variation of the excitation wave is large as shown in Fig. 1. We examined if Barkhausen noises had a chaotic behavior by the following four points [5];

(1) The complexity of Barkhausen noise itself in the time domain. Chaotic signals are very complicated in the time domain.
(2) The existence of the specific discrete peak of the spectra after Fourier transform. Chaotic signals don't have any specific peaks in the frequency domain.
(3) The value of the correlation function. The correlation function of chaotic signals becomes zero immediately with the elapse of the time.
(4) The self-similarity of the constructed attractors from Barkhausen noises. The self-similarity of the attractor is one of the factors as a chaotic behavior.

Fig. 2 shows the spectra measured from the non-oriented silicon steel sheet under the alternating magnetic filed. It shows the relationship between the frequencies and the inclination angles of the alternating magnetic field. The spectra don't have any specific discrete peaks, so this is one factor whether Barkhausen noise can be categorized into the chaos. Fig. 3 shows the correlation function of Barkhausen noise under the rotational magnetic field. By this figure, the correlation function becomes zero immediately, so this is also one factor that Barkhausen noise can be categorized into the chaos. Fig. 4 shows the constructed attractors from Barkhausen noise under rotational magnetic field. The embedding method was used to make the attractor from time series Barkhausen noise [2, 3, 5]. It is obvious that the constructed attractor draws the self-similar elliptic orbits. This is also one of the factors for the chaos. Considering these four factors, we concluded that Barkhausen noise from magnetic materials could be classified into the chaos [2, 3].

3. Measurement system for fatigue evaluation

The shape of the steel plate for the fatigue evaluation is shown in Fig. 5. This is the shape to concentrate the fatigue on the center of the plates. Barkhausen noises are measured at the one point on the center of the plates as shown in Fig. 5. Fig. 6 shows the measurement system for the fatigue evaluation. First, the computer generates the signals to excite the specimens. The Barkhausen noise is measured by the computer from the pick-up sensor attached on the specimen through the low noise wideband preamplifier and the digital oscilloscope. The low noise wideband preamplifier has a high pass filter, so this filter eliminates the frequency components around the excitation wave from the measured signal. Fig. 7 shows the sensor to pick up the Barkhausen noises. This sensor was selected after the several sensors were examined to measure Barkhausen noise [2, 3]. Fig. 8 shows the photo of the bending stress machine. The antiplane bending stress was applied to the specimens with the amplitude, 320 MPa and 30 Hz cycles. The S-N curve of the used specimen is shown in Fig. 9. It is supposed that the used specimen might be broken around 60,000 times of applying the bending stress by this figure.

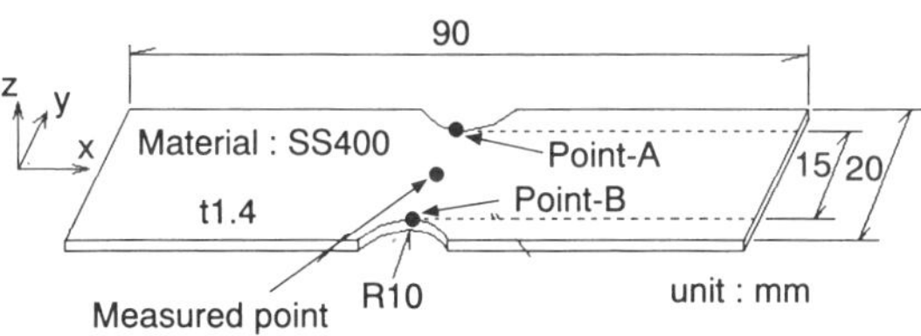

Fig. 5 Structural steel plate for bending fatigue test.

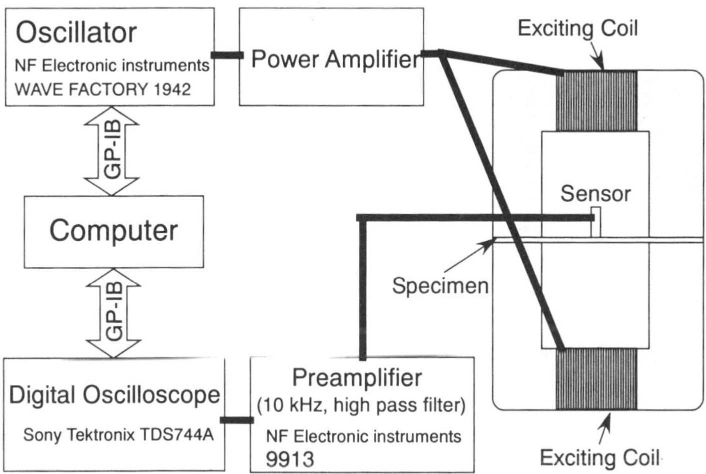

Fig. 6 Measurement system for Barkhausen noise of fatigued steel plates.

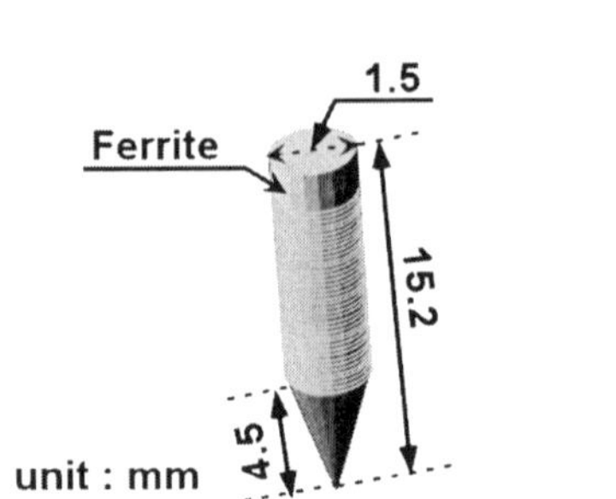

Fig. 7 Sensor to measure Barkhausen noise.

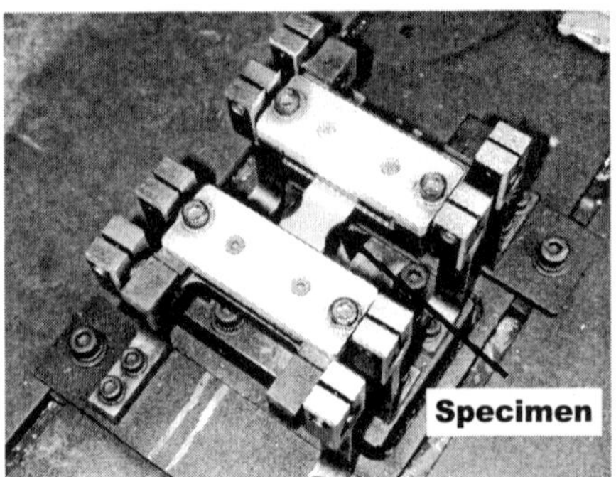

Fig. 8 Bending stress machine

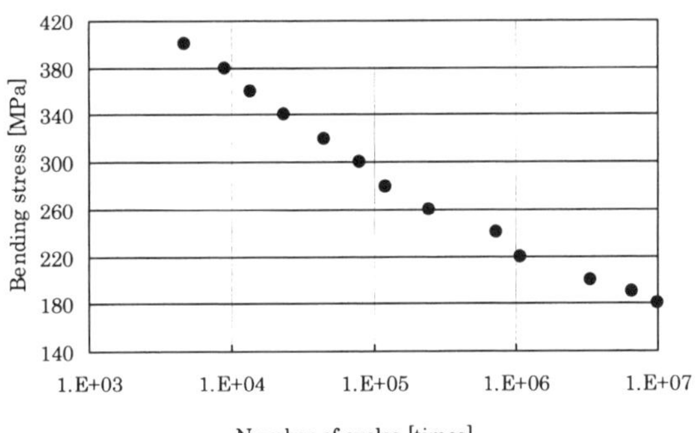

Fig. 9 S-N curve of bending stress.

4. Results and discussions

Figs. 10 show the measured Barkhausen noises from the SS400 plates in the case that the frequency of the exciting coils is 10 Hz. Fig. 10(a) shows the one before the bending fatigue stress test and Fig. 10(b) shows the one from the bending stressed plate. Here, the amplitude of the bending stress is 320 MPa and the number of the stress cycles is 20,000 times. Fig. 10(c) and Fig. 10(d) show the Barkhausen noises measured after 40,000-time and 60,000-time bending stress. From Fig. 9, the specimen was supposed to be broken around 60,000 times. Actually, some very small cracks could be recognized around Point-A and Point-B shown in Fig. 5 by using a magnifying glass. However, the visual crack could not be found around the measured point. From Fig. 10(a) to (d), it is difficult to discuss the change of the signals depending on the applied bending stress cycles. Figs. 11 show the constructed attractors from the Barkhausen noises shown in Figs 10. Here, the parameter of the time delay for the embedding method to draw the attractor is selected as $\tau=1$ [2, 3, 5]. The shapes of the attractors are changed depending on the cycle number of the applied bending stress, but it is difficult to say that each attractor has the self-similar shape. So, the chaotic behavior could not be recognized in the case that the frequency of the exciting coils is 10 Hz. Figs. 12 show the measured Barkhausen noise in the case that the frequency of the exciting coils is 100 Hz. It is also difficult to discuss the change of the signals depending on the applied bending stress by Figs. 12 (a), (b), (c) and (d). However, each attractor constructed from Figs. 12 has the self-similarity as shown in Figs. 13. These are the characteristics of the chaotic behavior.

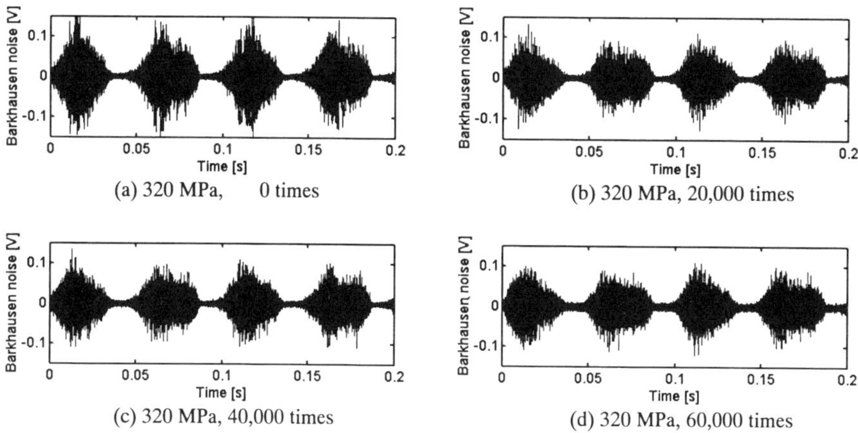

(a) 320 MPa,　0 times

(b) 320 MPa, 20,000 times

(c) 320 MPa, 40,000 times

(d) 320 MPa, 60,000 times

Fig. 10 Barkhausen noise before and after fatigue test (frequency of exciting coils : 10 Hz).

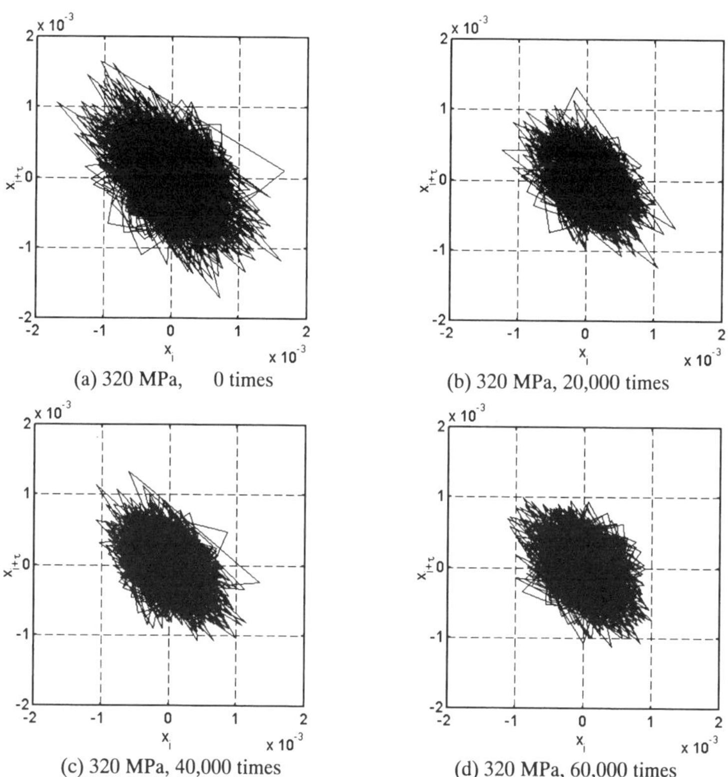

(a) 320 MPa,　0 times

(b) 320 MPa, 20,000 times

(c) 320 MPa, 40,000 times

(d) 320 MPa, 60,000 times

Fig. 11 Constructed attractors of Barkhausen noise from the stressed steel plates.
(frequency of exciting coils : 10 Hz, time delay of embedding method $\tau=1$)

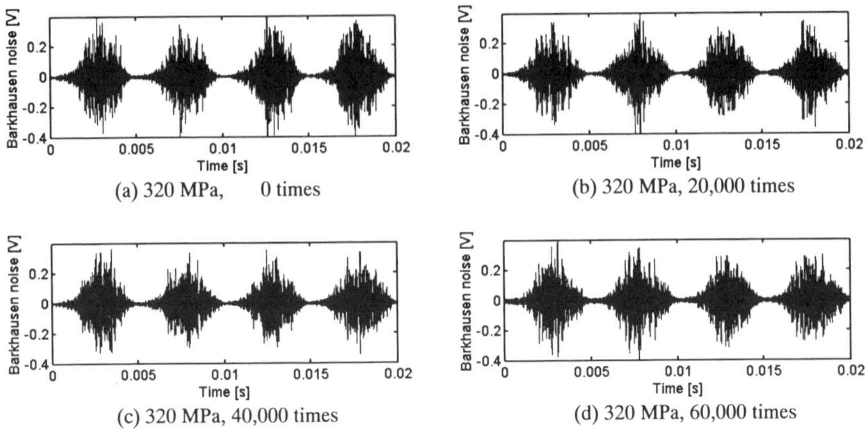

(a) 320 MPa,   0 times   (b) 320 MPa, 20,000 times

(c) 320 MPa, 40,000 times   (d) 320 MPa, 60,000 times

Fig. 12   Barkhausen noise before and after fatigue test (frequency of exciting coils : 100 Hz).

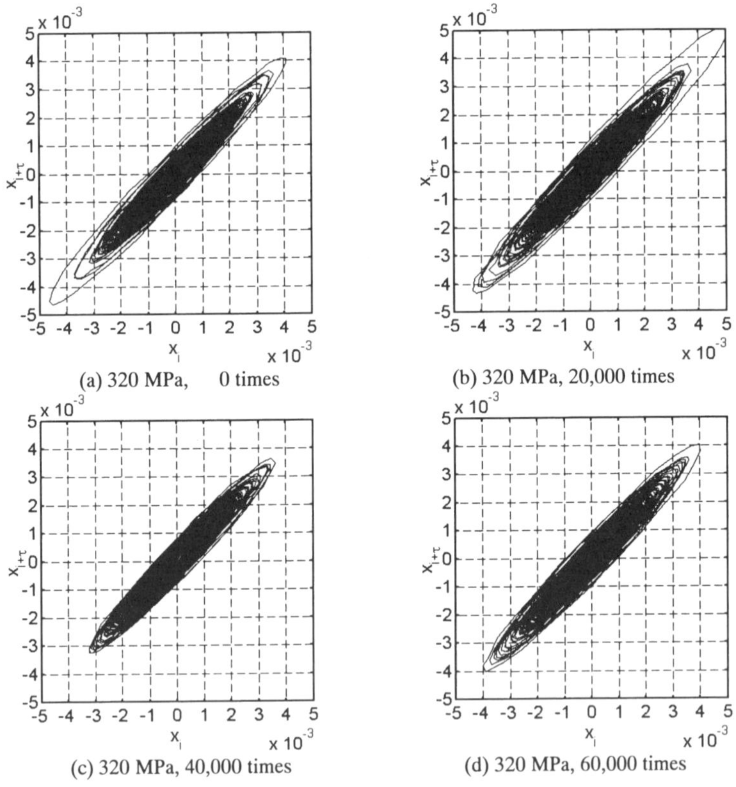

(a) 320 MPa,   0 times   (b) 320 MPa, 20,000 times

(c) 320 MPa, 40,000 times   (d) 320 MPa, 60,000 times

Fig. 13   Constructed attractors of Barkhausen noise from the stressed steel plates.
(frequency of exciting coils : 10 Hz, time delay of embedding method $\tau=1$)

Compared Figs. 10 with Figs. 12, the similar Barkhausen noises are measured on the different frequency of the exciting coils but the shapes of the constructed attractors are different as shown in Figs. 11 and Figs. 13. It is required to find the proper frequency range of the exciting to use the chaotic behavior of Barkhausen noise for the fatigue evaluation. Now, it can be said that there are some relationships between the cycle numbers of the applied bending stress and the constructed attractors by considering Figs. 13. As mentioned before, any cracks could not be found around the measured point, so the difference of the shapes of the constructed attractors could be related with the change of the magnetic domain movement due to the dislocation. The specified chaotic characteristics change should be found to use it for the fatigue evaluation before a crack appears.

5. Conclusions

We conclude the results as follows;
(1) The self-similarity of the attractors from Barkhausen noise was found in the condition that the frequency of the exciting coils is 100 Hz, even though the same characteristics could not be observed under 10 Hz. The proper frequency range should be selected to use the chaotic characteristics of Barkhausen noise for the fatigue evaluation.
(2) The shapes of the constructed attractors were different depending on the fatigue level, but the quantitative relationship between the shapes and the fatigue level could not be recognized.

So, it can be said that there are some relationships between the constructed attractors and the cycle numbers of the applied bending stress, it can be possible to obtain some information on the fatigue level from Barkhausen noise. Even though, the chaotic characteristics change should be found to use Barkhausen noise for the quantitative fatigue evaluation.

References

[1] M. Enokizono and A. Nishimizu, "Effect of thermal fatigue on Barkhausen noise," *J. of M. M. M.*, 133, pp.599-601, 1994.
[2] M. Enokizono, T. Todaka and Y. Yoshitomi, "Chaotic behavior of Barkhausen noise induced in silicon steel sheets," *IEEE Trans., on Magnetics*, Vol. 35, No. 5, pp. 5421-5423, September 1999
[3] M. Enokizono, T. Todaka and Y. Yoshitomi, "Chaotic phenomena of rotational Barkhausen noise," *J. of M. M. M.*, 215-216, pp. 43-45, 2000.
[4] Y. Tsuchida, T. Ando and M. Enokizono, "Chaotic behavior of rotational Barkhausen noises for fatigue evaluation," *JSAEM Studies in Applied Electromagnetics and Mechanics, 9*, pp. 647-648, 2001
[5] P. Berge, Y. Pomeau and Ch. Vidal, "Order within Chaos," John Wiley & Sons, 1987.

Electromagnetic Nondestructive Evaluation (VI)
F. Kojima et al. (Eds.)
IOS Press, 2002

Measurements and semi-analytical modeling of incremental permeability using eddy current coil in the presence of magnetic hysteresis

Andriy Yashan and Gerd Dobmann
Fraunhofer-Institut Nondestructive Testing IZFP,
University Building 37,
66123 Saarbrücken, Germany
dobmann@izfp.fhg.de
yashan@izfp.fhg.de

Abstract. The well known analytical solutions for standard eddy current problems have been applied to describe non-destructive testing (ndt) experiments on surface-hardened cast iron samples by measuring the incremental permeability. The agreement obtained between the measured and calculated incremental permeability profile curves allows the conclusion that measured curves can be reliably simulated and discussed in terms of the governing influencing parameters.

1. Definition and determination of the incremental permeability

According to the micro-magnetic definitions in physics we separate different types of permeability values; all of them are combined with hysteresis characteristics. The incremental permeability is a quantity which characterizes the effects of reversible micro-magnetic magnetization changes [1] and therefore provides different and independent information from an irreversible one, which, for instance, is based on magnetic Barkhausen noise [2].

Following the definition in physics, the measurement rules require a well defined magnetization of the material of interest. A magnetic field H_A is applied which – according to the hysteresis of the material and supposed the field is producing a homogeneous magnetization above the direction of H_A – defines a magnetic flux density B_A. A sinusoidal alternating magnetic field $H_\omega(t)$ is superimposed to the steady state value H_A:

$$H_\omega(t) = \Delta H \times \cos(\omega t) \tag{1}$$

The time variable is t, ω is the angular frequency and ΔH is the so-called incremental magnetic field. The value of ΔH has to be small compared with the coercivity H_c of the material and by a reasonable approach, ΔH is chosen to be $\Delta H \leq H_c/2$. The reason for this limitation is the objective to excite – by use of the superimposed $H_\omega(t)$ - only additional reversible magnetization changes, i.e., the macroscopic value B_A in the hysteresis has not to be changed. By thinking in a potential model describing the magnetization of the material in terms of a magnetic domain structure in which domains are separated by Bloch walls, the value B_A fixes the Bloch walls in potential troughs and by application of the oscillating magnetic field $H_\omega(t)$ the Bloch walls harmonically oscillate in them. The ΔH has to be so small that irreversible Bloch wall jumps, i.e. Barkhausen events, do not occur, and the walls are trapped in the troughs. The Bloch wall oscillations supply with an additional alternating magnetization ΔB which – as long as ΔH is small enough – is also purely sinusoidal without any excitation of higher harmonics. The incremental permeability value μ_Δ is defined by the equation (2):

$$\Delta B = \mu_0 \times \mu_\Delta \times \Delta H, \text{ i.e.,} \tag{2}$$

$$\mu_\Delta = \Delta B / (\mu_0 \times \Delta H), \tag{3}$$

where $\mu_0 = 4\pi \times 10^{-7}$ H/m is the permeability of the free space.

According to this definition the incremental permeability can be measured only in special devices (hystrometers) suitable for hysteresis measurements using special shaped specimen with an encircling coil for magnetic flux measurement (sphere, ellipse, cylinder), where a steady state magnetizing field H_A can be guaranteed and where the $H_\omega(t)$ is superimposed with sufficient accuracy in the same direction as H_A (Fig. 1). According to the definition it is obvious that the function $\mu_\Delta(H_A)$ is measured only at discrete values H_A. Fig. 2 documents the so determined incremental permeability function with series of 'inner loops' within the macroscopic hysteresis.

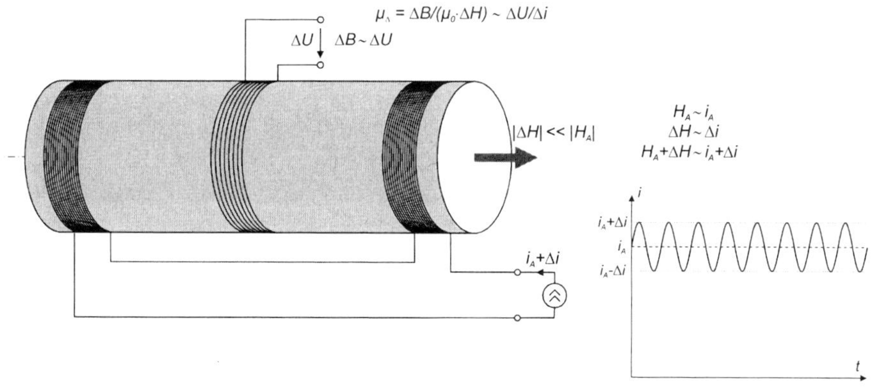

Figure 1:   Basic experimental procedure to measure the incremental permeability

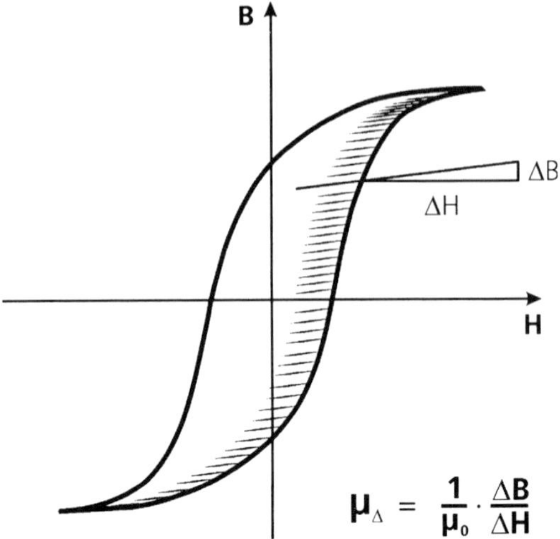

$$\mu_\Delta = \frac{1}{\mu_0} \cdot \frac{\Delta B}{\Delta H}$$

Figure 2:   Incremental permeability definition

Such a procedure is far from non-destructive testing (ndt). Ndt is normally asking for a scanning of local inspection regions of large components, and not of small specimen which can homogeneously magnetized. We have in practice a relative inhomogeneous magnetization and we have not the time to macroscopically select discrete steady state magnetic field values.

2. Using of the incremental permeability phenomena in ndt techniques

Our institute has introduced into practice a ndt-approach [3], based on eddy current testing and the use of an eddy current pick-up coil in addition to a local magnetic yoke magnetization. However, the accuracy of this approach in comparison with the exact theoretical model according to the definition neither was scientifically and basically investigated in detail nor validated by theoretical research.

By using the IZFP ndt approach the component under test is locally magnetized applying an u-shaped yoke magnetization. The magnetization is controlled by a sinusoidal electric current driving the excitation coil of the yoke, where the frequency for the magnetic oscillations is in the range between some Hz and some hundred Hz, in order to allow – by frequency adjustment – to select a certain field penetration of interest. The superimposed incremental field is produced by a pick-up eddy current coil with the advantage of the spatial resolution of that coil (Fig. 3). This, in practice, can be a pick-up air-coil or a ferrite-core-coil, but also a small u-shaped magnetic yoke transducer. The differences of these type of magnetizers - observing the definition - is their field inhomogeneity and direction compared with the macroscopic yoke magnetization. The frequency of the incremental field, i.e., the eddy current frequency has to be much higher than the hysteresis frequency; a factor of 100 is appropriate.

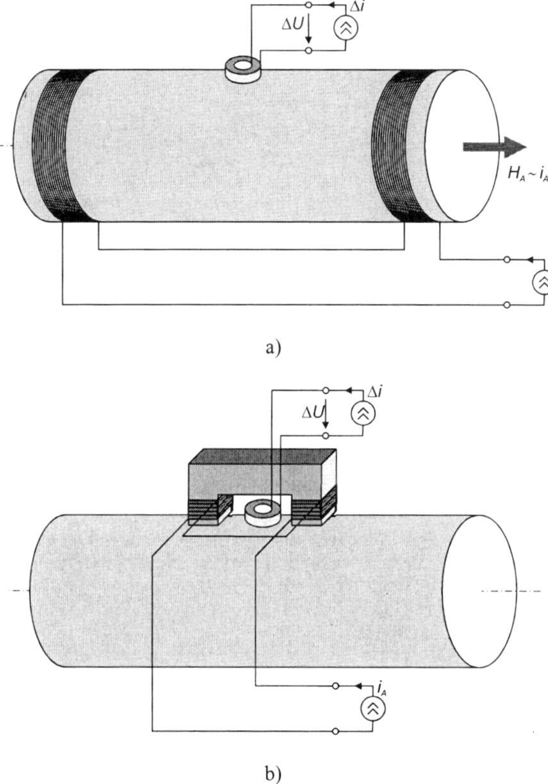

Figure 3:   Eddy current pick-up sensor arrangements to measure the incremental permeability in ndt applications with homogeneous magnetization of the specimen (a) and with one-sided assessment (b)

In the ndt approach not the inclination $\Delta B / \Delta H$ for each inner loop but the eddy current coil impedance is measured. Applying a current source as HF-generator for exciting the eddy current coil, an impressed magnetic field ΔH is assumed., i.e. an incremental field which is independent from electrical and magnetic loads. In that case, the impedance meter output is proportional to ΔB and therefore proportional to μ_Δ.

The eddy current impedance can continuously be measured during a hysteresis cycle and incremental permeability profile curves $\mu_\Delta(H)$ can be documented, as long as the magnetic field is also measured by use of a Hall-element. Fig. 4 shows such profile curves (incremental permeability versus H in a x-y-presentation, the x-axis is the H-axis) measured at laser-hardened surfaces of cast iron specimen with different hardening depths (0.5 mm and 1.0 mm). For smaller hardening depth values a typical double peak structure was observed which indicates the layered structure with the hardened martensitic surface layer covering the non-heat-affected mechanically soft core. Basic investigations document the fact that the positions of the peaks correlate with the coercivity of the hardened layer and the coercivity of the soft unhardened core. For a given eddy current frequency – in this case 750 Hz – the soft unhardened material (the two inner peaks of the observed curves) can reliably be detected only up to a certain hardening depth in between 1 mm and 2 mm. Using the ratio of the peak-amplitudes (amplitude of the soft peak / amplitude of the hard peak) a calibration can be obtained for a ndt determination of hardening depth.

hardening depth 0.5 mm   hardening depth 1.0 mm

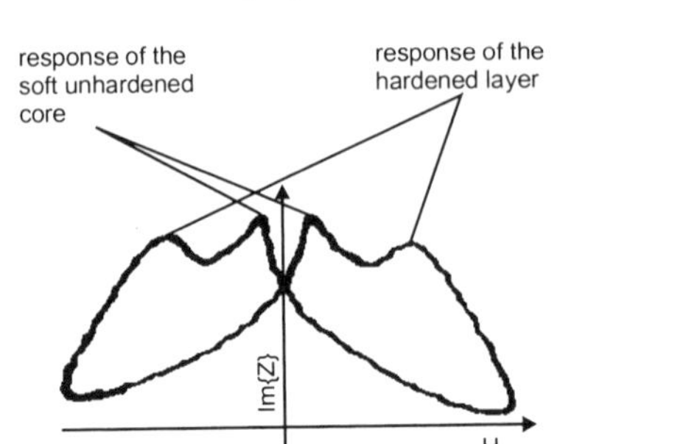

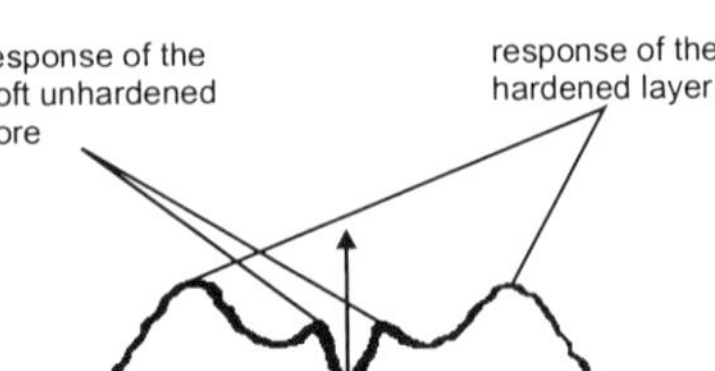

Figure 4:   Incremental permeability profile curves measured by using the experimental facility of Fig. 3, a). Measurement conditions [3]: magnetizing field magnitude $|H_A|$ = 6000 A/m, magnetizing frequency 0.8 Hz, eddy current frequency 750 Hz, eddy current amplitude ΔI = 1A (corresponds to ΔH < 250 A/m).

3. Modeling of the incremental permeability experiment by the eddy current approach

In order to understand the structure and amplitude contrast in profile curves like those shown in Figure 4, semi-analytical modeling based on Maxwell equations was performed. The solution algorithm was implemented taking into account the well-known approach of Cheng, Dodd and Deeds [4] developed many years ago, which is a valuable tool to characterize multi-layered media using eddy current testing with a pick-up coil, so far the layer properties (thickness, electrical conductivity σ, magnetic permeability μ) and the coil parameters are known.

In order to apply the solution [4], the specimen geometry was simplified from the cylinder (Fig. 3a) to the infinite half space (Fig. 5). The electrical conductivity and magnetic permeability values are assumed to be linear, isotropic and homogeneous in each layer. The chosen electrical conductivities are σ_2 = 4.0 MS/m for the soft core and σ_1 = 3.2 MS/m for the hardened layer. The used values of the magnetic permeability of both materials are functions of H_A as shown in Fig. 6, based on the measurements of the incremental permeability for homogeneous cylinders as indicated in Fig. 1 and 2. The coil inner and outer diameters are 4 mm and 9 mm respectively.

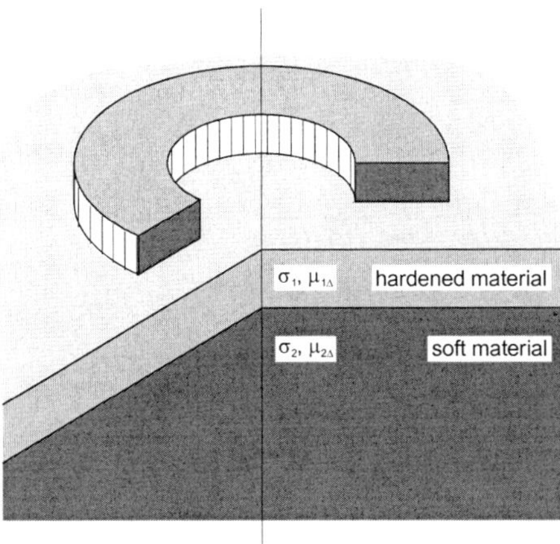

Figure 5:   Coil with rectangular cross section of the winding above the conducting and magnetic permeable half space - the simplified problem geometry for the solution of Cheng, Dodd and Deeds [4].

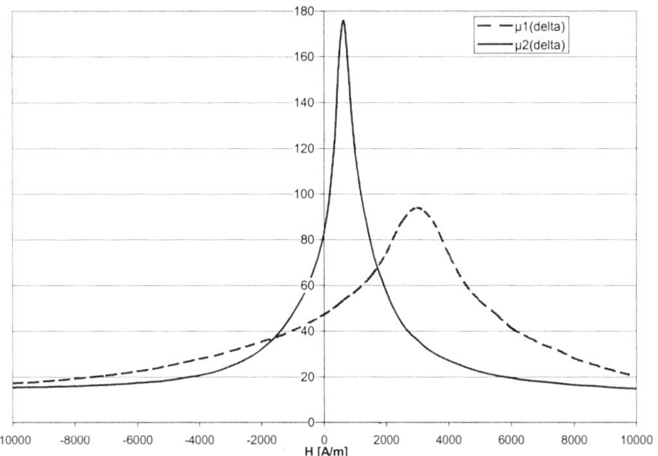

Figure 6:   The magnetic permeability curves as functions of the applied field H_A used for the calculation. $\mu1(delta)$, or $\mu_{1\Delta}$, is the measured incremental permeability of a soft steel with the coercive force $H_C = 650$ A/m. $\mu2(delta)$, or $\mu_{2\Delta}$, is the approximated curve derived from the measured incremental permeability of a hardened steel with the coercive force $H_C = 1700$ A/m by shifting along the axis H to have $H_C = 3000$ A/m.

The results of the modeling for the thickness of the hardened layer 0.5 mm, 1.0 mm, 2.0 mm and 3.0 mm are shown in Fig. 7 (right) and compared with the corresponding measurement results in Fig. 7 (left). The measured curves show the most significant projection of the complex coil impedance at 750 Hz vs. magnetizing field H which is alternating with the frequency 0.8 Hz between -6000 A/m and 6000 A/m. The calculated curves show the imaginary part of the complex coil impedance at 750 Hz vs. magnetizing field H which is alternating between -6000 A/m and 6000 A/m and affecting the incremental permeability according the procedure of Fig. 6.

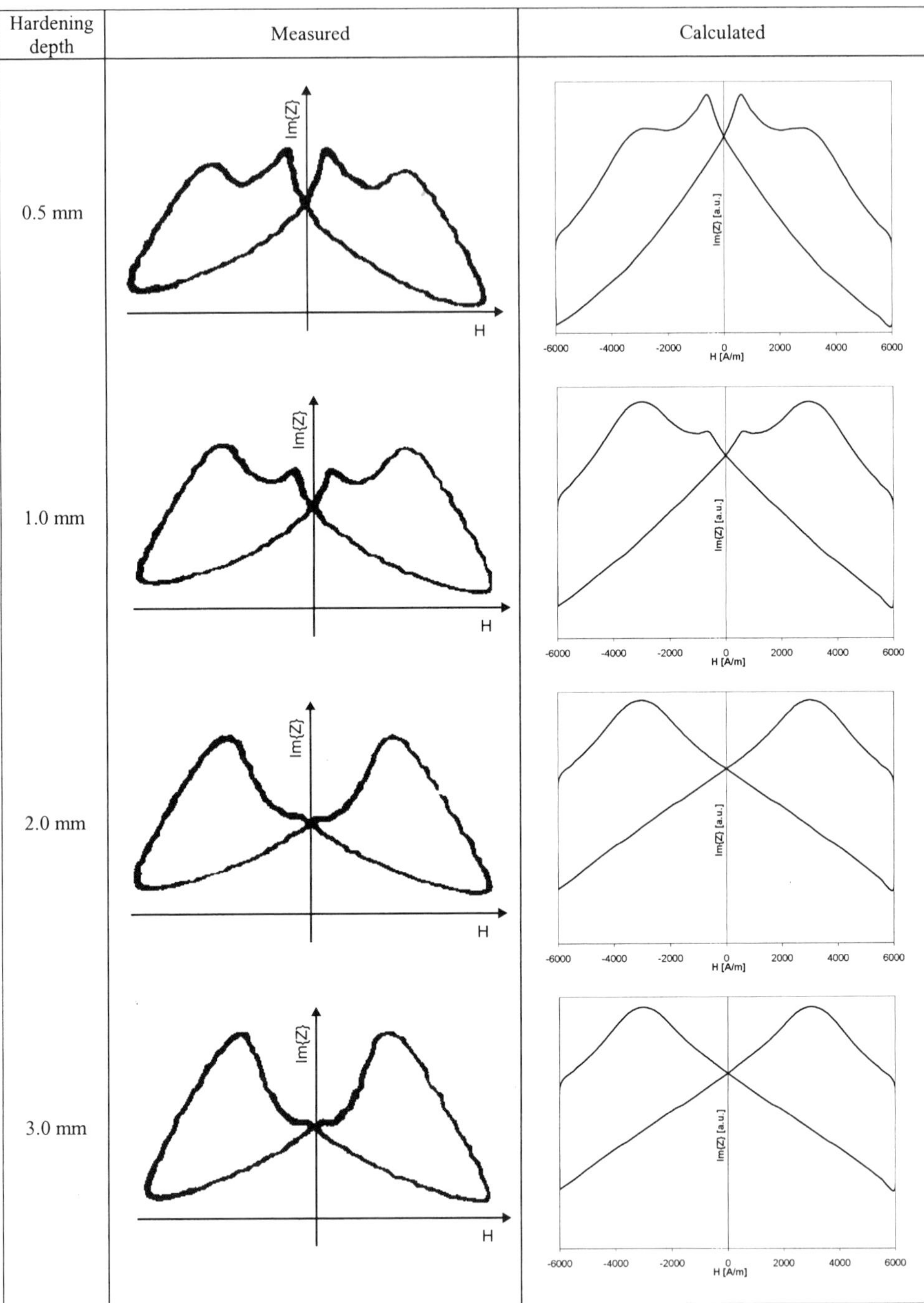

Figure 7: Incremental permeability profile curves measured (left) and calculated (right).

4. Conclusion

The chosen values of the material characteristics (electrical conductivities σ_1, σ_2 and magnetic permeability values $\mu_{1\Delta}(H_A)$, $\mu_{2\Delta}(H_A)$) result in the qualitatively good agreement between the measured and calculated incremental permeability profile curves. We can conclude that the measured impedance curves (correlating with the incremental permeability curves) can be reliably simulated using the analytical solution for the simplified eddy current approach and discussed in terms of the governing influencing parameters.

References

[1]   B.D. Cullity, "Introduction to magnetic materials", (Addison-Wesley, London, 1972)

[2]   G. Dobmann et al., Barkhausen Noise Measurements and related Measurements in Ferromagnetic Materials; in Volume 1: Topics on Nondestructive Evaluation series (B.B. Djordjevic, H. Dos Reis, editors), Sensing for Materials Characterization, Processing, and Manufacturing (G. Birnbaum, B. Auld , Volume 1 technical editors), The American Society for Nondestructive Testing, Inc., ISBN 1-57117-067-7, 1998.

[3]   G. Dobmann et al., Quantitative hardening depth measurements up to 4mm by means of micromagnetic, microstructure, multiparameter analysis (3MA), in Volume 7B Review of Progress in Quantitative NDE (D.O. Thompson, D.E. Chimenti, editors) Plenum Press, New York, 1988, p.1471-1475.

[4]   C.C. Cheng, C.V. Dodd, W.E. Deeds, General analysis of probe coils near stratified conductors, International Journal of NDT, 1971, Vol. 3, p. 109-130.

Novel E'NDE Techniques

Development of ECT Multi-Probe Detecting Axial and Circumferential Cracks using Uniform Eddy Current Excitation Coils

Mitsuo HASHIMOTO, Daigo KOSAKA, Kenichi OOSHIMA and Yasutaka NAGATA
Electronic Engineering, Polytechnic University
4-1-1 Hashimotodai, Sagamihara, Kanagawa 229-1196, Japan

Abstract. Guaranteeing the security of the steam generator tubes of the pressurized water reactor (PWR) in the nuclear power plants is a key factor in their safe exploitation. Accordingly, the eddy current testing is used for the reason of high accuracy and high speed because the test must be done in a limited period. We developed a multi probe that can detect cracks with distinction between axial and circumferential cracks. The probe is composed with two kinds of excitation coils of axial and circumferential windings and 32 pickup coils. The excitation coil of this probe is able to generate uniform eddy current distribution on a wide area compared with the detection area of the pickup coils. Therefore the probe can ignore the influence of the shape of excitation current source on detecting cracks, and numerical analysis for this probe is handled simply by devising boundary condition. From measurement results, the probe is confirmed to be able to measure direction of cracks with high sensitivity and high speed.

1. Introduction

It is one of the most important keys to make sure of the security of the steam generator tubes in the nuclear power plant of the pressurized water reactor (PWR) type. Eddy current testing (ECT) is applied for this testing [1]. Cracks in actual tubes are mostly outer axial or circumferential cracks and the crack shape is shallow and narrow. It is demanded on ECT to having abilities high accuracy and high speed because the test must be done in a limited period. To ensure the reliance of steam generator tubes, it is important to judge shape of cracks detected by the testing. We developed an ECT multi-probe that can distinguish between axial and circumferential direction cracks and also with high sensitivity and rapidity testing.

2. Development of the probe

We had developed a rotating uniform eddy current ECT probe[2] for aids of numerical analysis. The probe could distinguish between axial and circumferential direction cracks and also with high sensitivity. But detecting speed of the probe is late because of rotating type. So we developed a novel multi type ECT with object of rapidity testing. The probe has fundamentally same characteristicsthe of the developed rotating uniform eddy current probe. The probe is composed with two kinds of excitation coils of axial and circumferential windings and 32 pickup coils. The excitation coils are able to generate uniform eddy current distribution on a wide area compared with the detection area of the pickup coils. The excitation coil for detecting axial cracks is a coil of solenoidal shape. We call the coil by the name of the axial winding excitation coil. The excitation coils for detecting circumferential cracks are two coils wrapped with a fine wire in axial direction on a cylinder divided into two domains. We call the coils by the name of the circumferential winding excitation coil. The excitation coils suitable for the uniform eddy current distribution in the neighborhood of the pickup coils are designed by numerical analysis using the finite element method [1].   Fig.1 shows eddy current distributions in the tube by two kinds of excitation coils. From this figure the eddy current distribution of axial direction is existed to a wide area by range of ± 6mm for axial direction on center of the pickup coils. And one of circumferential direction is existed to a wide area by range of 22mm for circumferential direction. The two excitation coils for detecting circumferential cracks are installed in the form that is each sliding 90 degrees in the probe in order to test whole circumference. The excitation coils of this probe are the ability to generate uniform eddy current distribution in tubes on large enough area compared with the detection area of the pickup coils. Therefore the probe has the advantageous feature that specific characteristics detecting cracks is not affected by the relation between the position of the excitation coils and the cracks, and numerical analysis for this probe is handled simply by devising boundary condition.

It was confirmed that a magnetic field of perpendicular direction to the surface of the tube represents the crack shapes smartly from the numerical results. Therefore the pickup coils are installed on a parallel direction to the tube surface. The pickup coils are installed 32 coils of square type at one line in a center of the excitation coils. The pickup coils detect disturbances of eddy current distribution due to cracks. The structure and an appearance of the probe are shown in Fig.2 and 3, respectively.

These pickup coils detect distinguish between axial and circumferential direction of the cracks by changing the excitation coils. Fig.4 shows a control circuit of this testing. A computer processes controls of changing signals for the excitation coils and the pickup coils, and collections of detecting data from the pickup coils. Multiprexer circuits for changing the exciter coil and the pickup coils and amplifier circuits are attached in the probe. Ability of sampling data is 48000datas/sec of 16 bits on the measurement system. When the sampling pitch for axial direction is 1.5mm, the measuring speed becomes 375mm/sec. This is enough speed on the ECT measurement on two dimensions.

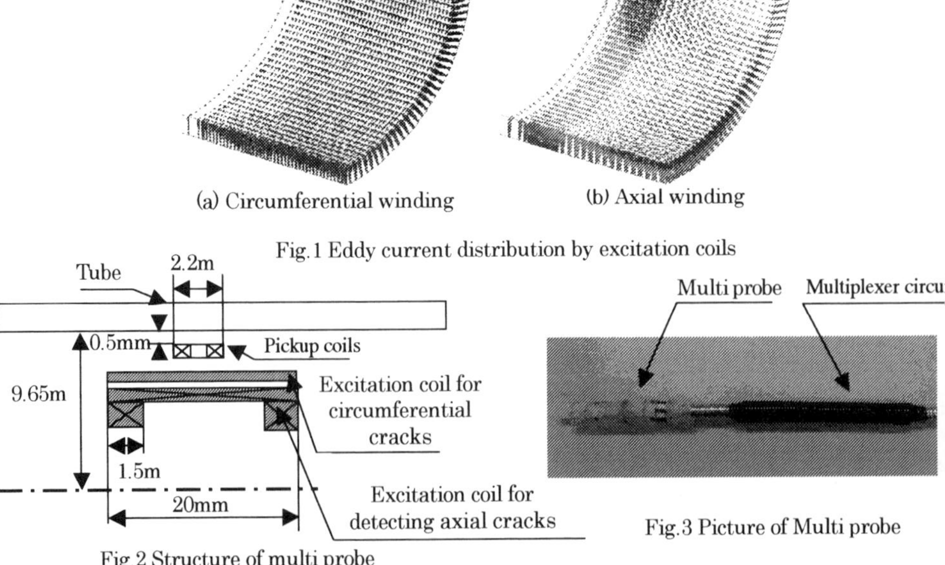

(a) Circumferential winding   (b) Axial winding

Fig.1 Eddy current distribution by excitation coils

Fig.2 Structure of multi probe

Fig.3 Picture of Multi probe

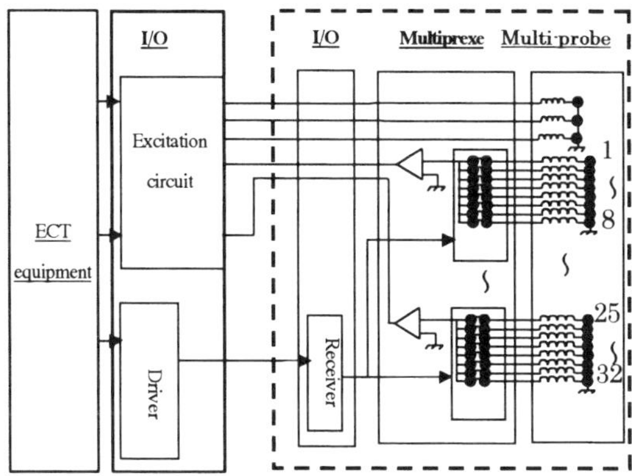

Fig.4 Control circuit

3.　Measurement results and signal processing

We performed experiment using the specimens of JSAEM model [3,4]. Fig.5 shows measurement result detected an outer axial crack of 40% depth by a mechanical stage for precision moving. In this figure two peaks of plus and minus are detected due to the existence of the crack. It is confirmed that the distance from peak to peak of the detecting signal agrees with the crack length. From measurement results, the probe is confirmed to be able to measure direction of cracks, length of cracks from peak to peak of signal and to have abilities high sensitivity.

Because the number of coils are limited, the detecting sensitivity of multi-probe is not constant by position of crack.Fig.6 shows the characteristic of detecting crack by each pickup coil. From this figure, if the crack is between pickup coils, the drop of sensitivity detected by the multi probe using 32 pickup coils is 15%. Therefore the drop of the detecting sensitivity has done underestimate for the detecting cracks.

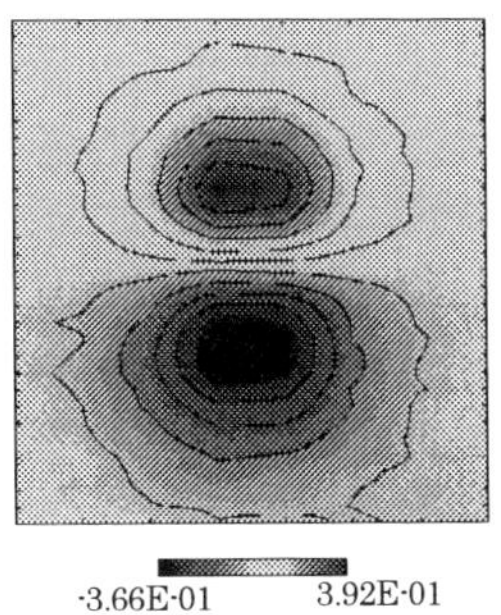

-3.66E-01　　　3.92E-01

Fig.5 Measurement result for OD 40% crack by multi probe

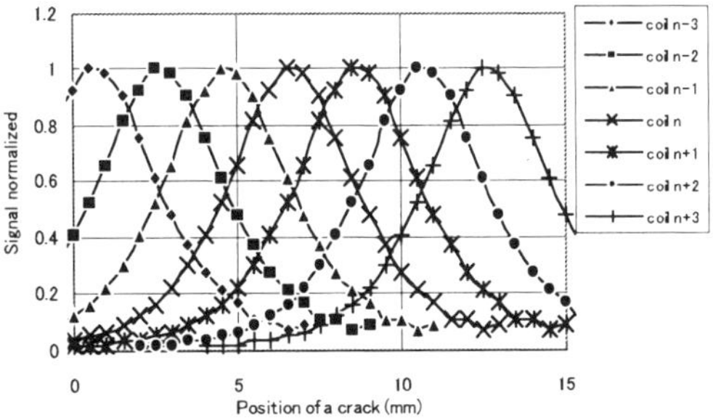

Fig.6 Detecting characteristic for 40% crack of each pick-up coils

On this account we examined an interpolate method for measurement data by the multi probe. Fig.7 shows signal strength distribution of detected result of an outer 40% axial crack for detailed detecting by the rotational uniform eddy current probe. Figure (a) shows measuring lines on measurement result of two dimensions display. Figure (b) shows signal amplitude distributions on each line shown by Figure (a). Figure (c) shows normalized signal distributions which are calculated their peak value to unify into 1 for the each line. The normalized signals are in good agreement. We

                    M. Hashimoto et al. / Development of ECT Multi-Probe

confirmed that distributions of signal amplitude on each crack of differential depths furthermore. The result is shown in Fig. 8. The distributions of the differential cracks are in good agreement, too. This means that it is an important characteristic of the uniform eddy current probe. Therefore we can handle easily the interpolation of measuring signals by 32 pickup coils using this characteristic. Figure (a), (b) and (c) of Fig.9 show a measurement result by the multi probe for 40% outer crack, an interpolation result of one and a measurement result by the rotational probe compared with interpolation result of one, respectively. We can understand that the interpolation method is useful for improvement for the inferior measurement data by the multi probe from this figure.

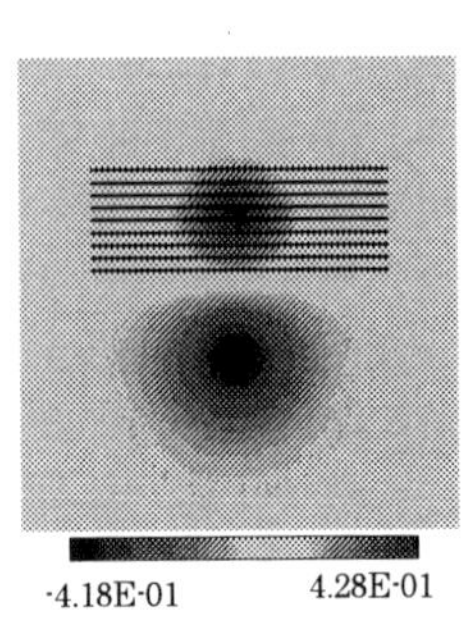

-4.18E-01   4.28E-01

(a) Measuring lines on
measurement result of two
dimensions display

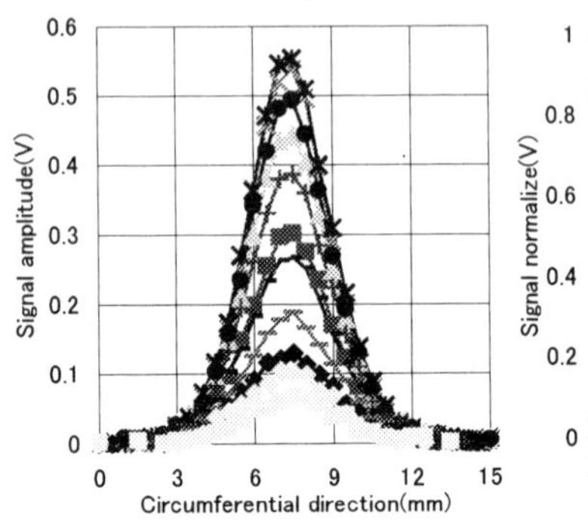

(b) Signal amplitude distributions
on the measurement lines

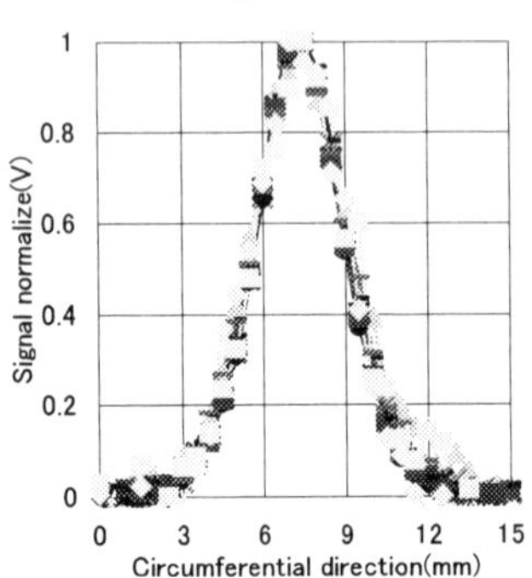

(c) Normalized signal distributions
of the each line

Fig.7 Measurement line of outer 40%crack

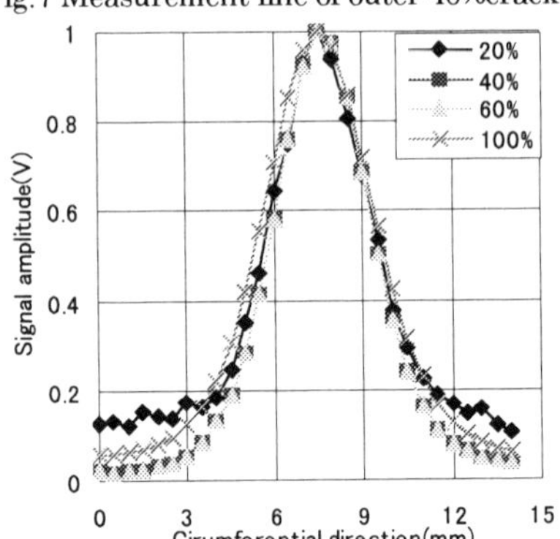

Fig.8 Signal amplitude of outer crack

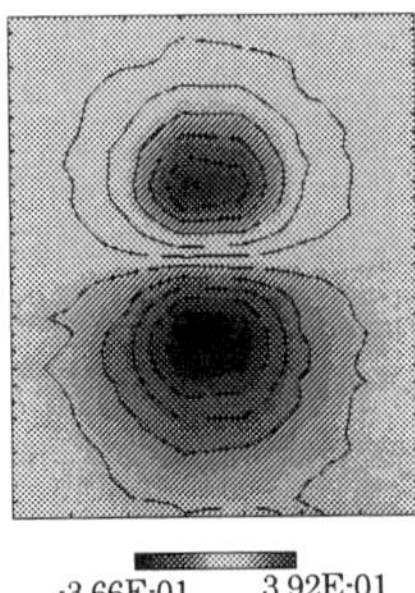

-3.66E-01   3.92E-01

(a) Measurement result using
32 multi type multi probe

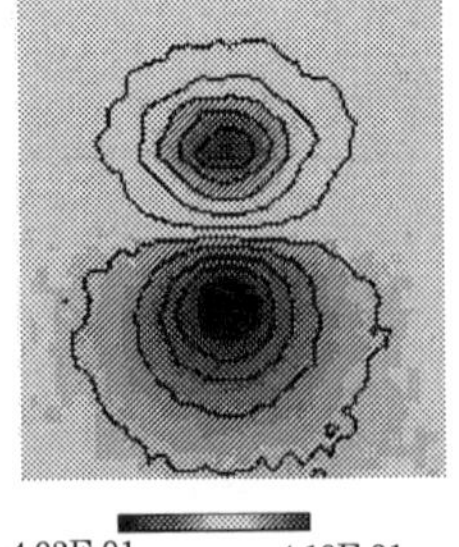

-4.03E-01   4.10E-01

(b) Interpolation result for multi probe

Fig.9 Interpolation result of the multi probe

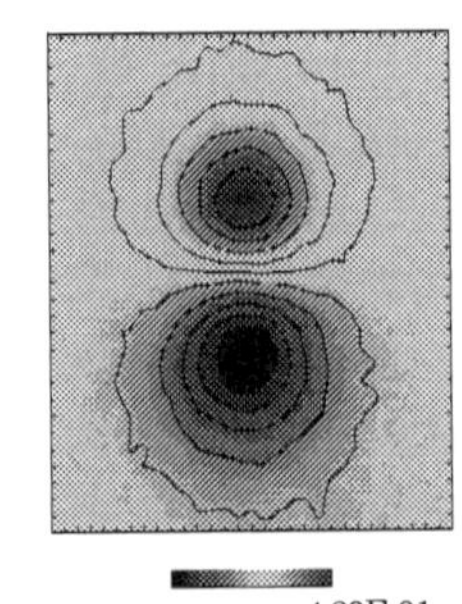

-4.18E-01   4.28E-01

(c)Measurement result using
rotating type probe

Fig.10 shows measurement results of axial and circumference cracks on high speed scanning over 300mm/sec by manual scanning. In this figure SN ratios are 8 for axial crack and 5 for circumferential crack. We could confirm that the developed probe has ability with high speed testing.

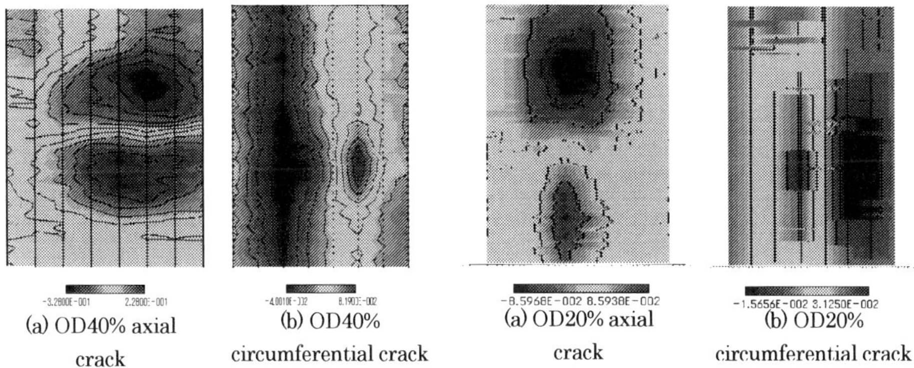

(a) OD40% axial crack   (b) OD40% circumferential crack   (a) OD20% axial crack   (b) OD20% circumferential crack

Fig. 10 Measurement result by high-speed scanning

4. Summary

We have presented a novel multi probe that is able to detect cracks with distinction between axial and circumferential directions and to generate uniformly distributed eddy current through the excitation coils in the neighborhood of pickup coils. We showed a special feature of the probe by experiments. We developed the interpolation method using the probe, and confirmed the interpolation method was useful. The results showed that the developed probe has detection capability with distinction between axial and circumferential cracks with high sensitivity and high speed testing.

References

[1] T. Takagi., M. Uesaka., and K. Miya., "Electromagnetic Non-destructive Evaluation", IOS Press, pp9-16, 1997

[2] Mitsuo Hashimoto and Daigo Kosaka: Development of Rotation ECT Probe Detecting for Axial and Circumferential Crack using Uniform Eddy Current Exciting Coils "Electromagnetic Non-destructive Evaluation(V)", IOS Press, pp.242-247, 2001

[3] T. Takagi, M. Hashimoto, H. Fukutomi, M. Kurokawa, K. Miya, H. Tsuboi, .M. Tanaka, J. Tani, T. Serizawa, Y. Harada, E. Okanoi and R. Murakami "Benchmark Models of Eddy Current testing for Steam Generator Tube: Experiment and Numerical Analysis", International Journal of Applied Electromagnet tics in Materials, 5, pp.149-162, 1994

[4] Takagi, T. and K. Miya., "ECT Round-robin Test for Steam Generator Tubes," Journal JSAEM, 8, pp.121-129, 2000

Electromagnetic Nondestructive Evaluation (VI)
F. Kojima et al. (Eds.)
IOS Press, 2002

Evaluation of the New ECT Sensor Using JSAEM Round Robin Test Samples in Presence of Fasteners Interferences

Tomasz CHADY[1], Masato ENOKIZONO[2], Ryszard SIKORA[3],
Takashi TODAKA[2], Yuji TSUCHIDA[2]

1) Technical University of Szczecin, ul. Sikorskiego 37, 70-313 Szczecin, Poland
2) Faculty of Engineering, Oita University, 700 Dannoharu, Oita 870-1192 Japan
3) Electrotechnical Institute, ul. Pożaryskiego 28, 04-703 Warszawa, Poland

Abstract. The purpose of this paper is to present new and improved version of the sensor for Inconel600 tubes testing. The sensor has a high sensitivity to shallow flaws, and good spatial resolution. Experiments with various samples and interfering structures have been carried out. Advantages of the proposed sensor and system are presented and confirmed.

1. Introduction

In this paper a new improved version of a sensor for testing tubes made of INCONEL600 is presented. Very good spatial resolution of the existing sensor PIPE-UNI2 [1] and excellent sensitivity of the sensor PIPE-MAT inspired the authors to develop a new sensor PipeUniBig1 ("PUB1"), which offers both advantages at the same time.

First, sensitivity of the sensor is evaluated using calibration tubes of the JSAEM Round Robin Test [2]. Next, the enhanced ability of the new sensor and Multi-frequency Excitation and Spectrogram system to discriminate flaw signals against interferences from fasteners is confirmed.

2. System description

In all measurements the Multi-frequency Excitation and Spectrogram (MFES) system was used. A basic idea behind the MFES method is to use a complex signal containing selected harmonic components as an exciting signal and a spectrogram to precise crack characterization. The spectrogram is a two-dimensional display of the relative amplitude of the frequency components of a signal from the search coil versus the sensor position. The system consists of function synthesizer, power amplifier, instrumentation amplifier, AD converter, X-Y-Z scanner and computer with interface boards. The detailed description of the system could be found in [3].

3. Sensor description

The sensor "PUB1" is a significantly improved version of the sensor "PIPE-UNI2" [1]. Construction of the sensor is shown in Fig. 1. The sensor contains four excitation coils and a single search coil mounted together on a core made of a high performance ferrite (MFH-9B).

The excitation coils could be connected in various ways. In a basic configuration all coils (E_1, E_2, E_3, E_4) are used and they are connected in series. Such arrangement allows us to detect all kinds of flaws and achieve the same sensitivity for axially and circumferentially oriented flaws.

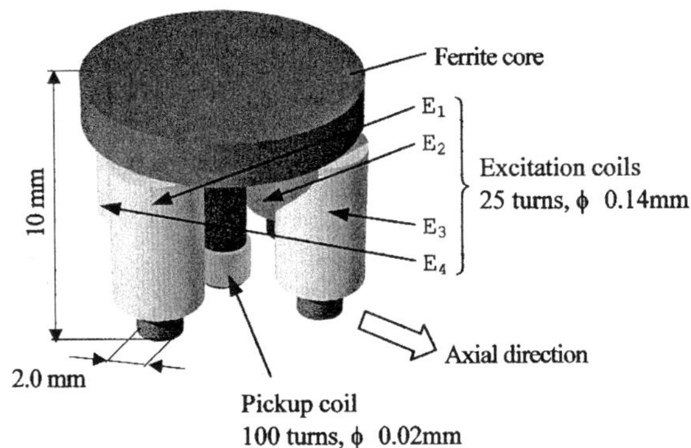

Fig. 1. View of the sensor PUB1

In case of multiple flaws having complicated structure, selective measurements are necessary. For example in order to detect only circumferential flaws, the excitation coils E_1 and E_2 are used, while E_3 and E_4 remain unconnected. In case of axial flaws, coils E_3 and E_4 are excited. With the excitation coils connected in this way, the sensor has a far higher sensitivity for certain kind of flaws, than for the others. The relevant connections have been achieved by the application of computer controlled electronic switches. In all cases the excited field is pulsating not a rotational one and the fluxes in all excitation coils are flawing in the same direction.

3. Sensor evaluation

Calibration tubes of "JSAEM round robin test" [3] were used to evaluate sensitivity and resolution of the new sensor "PUB1". The inner diameter and the wall thickness of the test pieces were 19.7 and 1.27 mm, respectively, the same as those of the steam generator tubes in nuclear power plants. The measurements were done by scanning the probe axially in steps of 0.5 mm and circumferentially in steps of 1 deg. The liftoff was measured to be 0.3 mm. The excitation signal consists of sinusoidal components having frequency from 40kHz to 220kHz (18 components).

3.1 Sensor sensitivity evaluation

The calibration tubes utilized for sensitivity evaluation consist of EDM flaws 5 mm in length and 0.2 mm in width. They were oriented in circumferential and axial directions. The relative depth of the flaws was 10%, 20%, 40%, 60% and 100%. All of them were made from the outer surface of the tubes. Results of multi-frequency measurements are shown in Fig. 2. Figure 3 and 4 shows signals obtained for the optimal frequency (frequency for which the spectrogram achieves maximum value). The single frequency measurements were utilized to calculate the signal to noise ratio: {SNR = $20\log_{10}$("p-p amplitude for crack"/"p-p amplitude for unflawed area")}. Results of calculation are summarized in Table 1. One could observe that the sensor "PUB1" has an excellent sensitivity for very shallow defects and the sensitivity is nearly exactly the same for axial and circumferential flaws.

Table 1. Results obtained for the flaws having length 5mm (specimens NEL9837, NEL9838).

	Axial flaws		Circum. flaws	
	Vpp [mV]	SNR [dB]	Vpp [mV]	SNR [dB]
No flaw	0.15	-	0.15	-
OF10	7.5	34	7.5	34
OF20	18.7	42	20.0	43
OF40	74.1	54	77.1	54
OF60	198.0	62	170	61

　　　　　T. Chady et al. / Evaluation of the New ECT Sensor

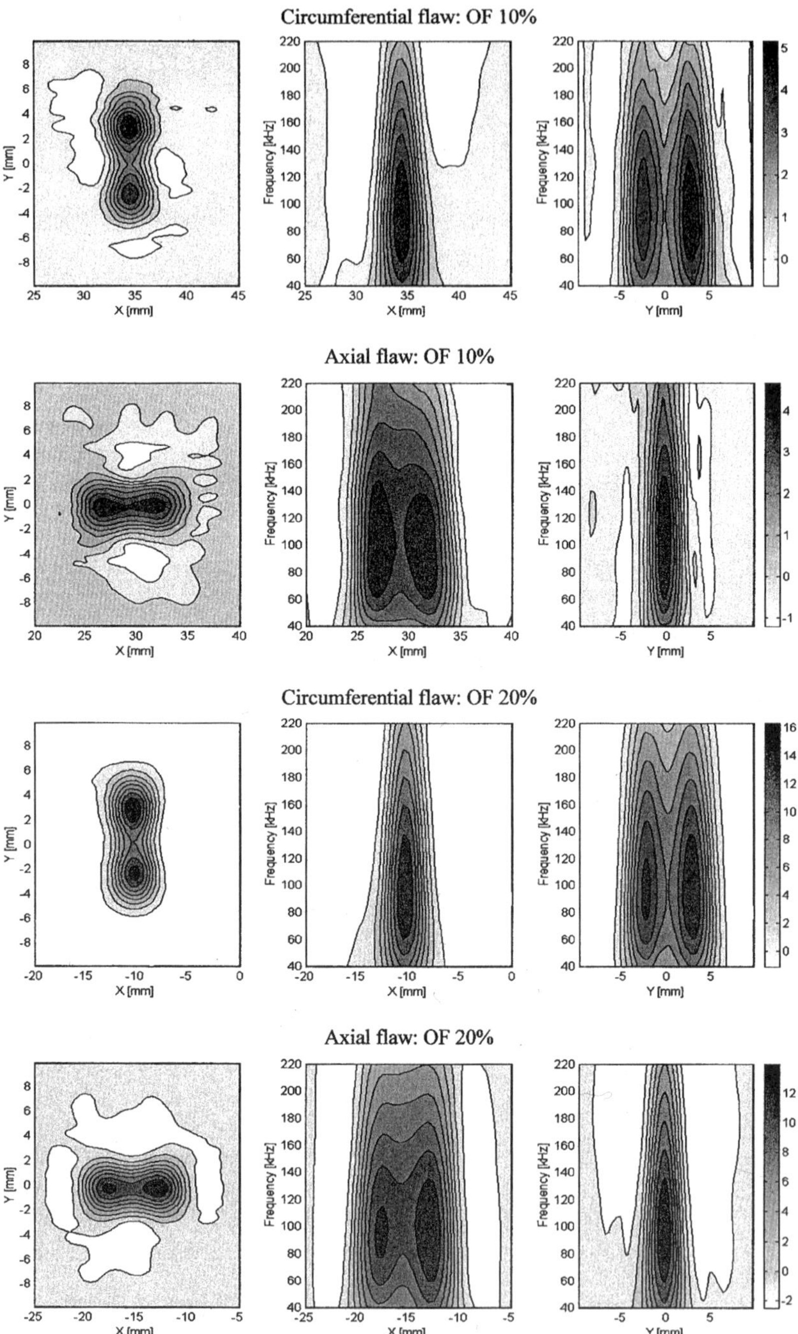

Fig. 2. Spectrograms obtained for outer flaws having different depth (axial flaws - specimen NEL9837, circumferential flaws - specimen NEL9838); flaws length 5mm.

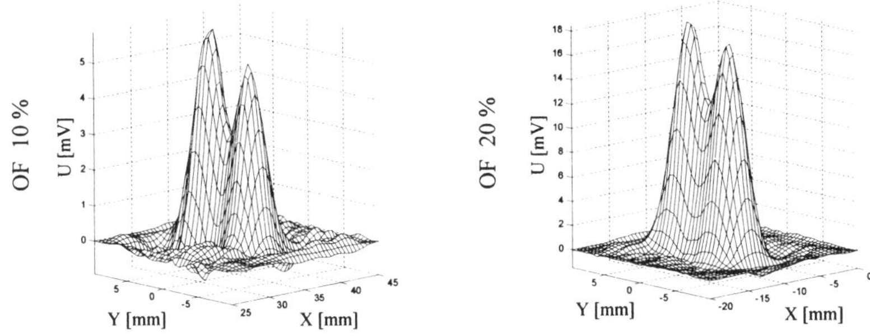

Fig. 3. Signals obtained for circumferential outer flaws having different depth (specimen NEL9838); flaws length 5mm; testing frequency for which the spectrogram achieves maximum value.

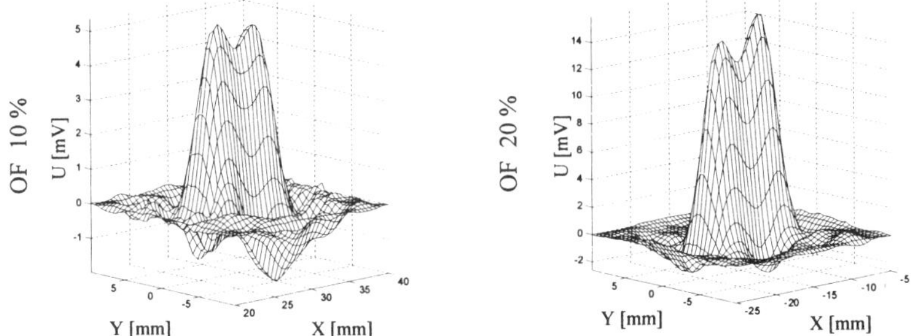

Fig. 4. Signals obtained for axial outer flaws having different depth (specimen NEL9837); flaws length 5mm; testing frequency for which the spectrogram achieves maximum value.

3.2 Sensor resolution evaluation

Masked specimen NEL98-25 [3] was used to test resolution of the sensor. This specimen contains several EDM flaws concentrated in two groups. Both groups consist of parallel flaws manufactured with very small distances (1~2mm) between each other. Such arrangement of the flaws creates possibilities to verify spatial resolution of the sensor. Results of the measurements are shown in Fig. 5. From these plots one could recognize seven flaws. This confirms that the new sensor has sufficient spatial resolution.

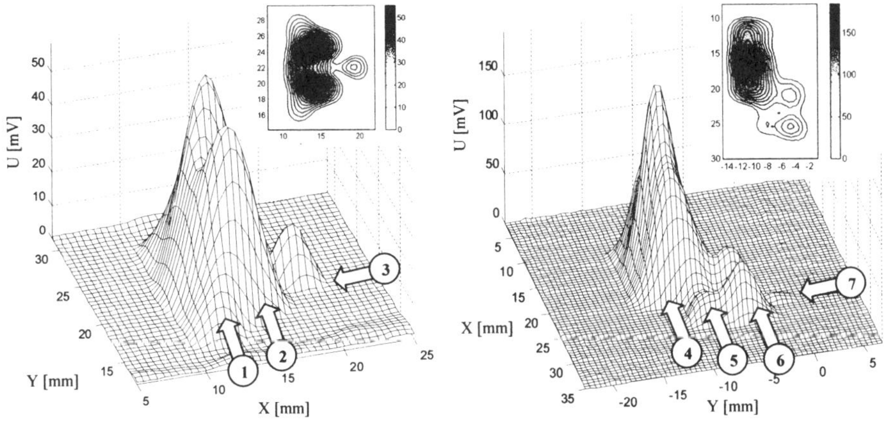

Fig. 5. Signals obtained for the masked specimen NEL9825. Two groups consist of seven flaws could be recognized without using any special inverse algorithm.

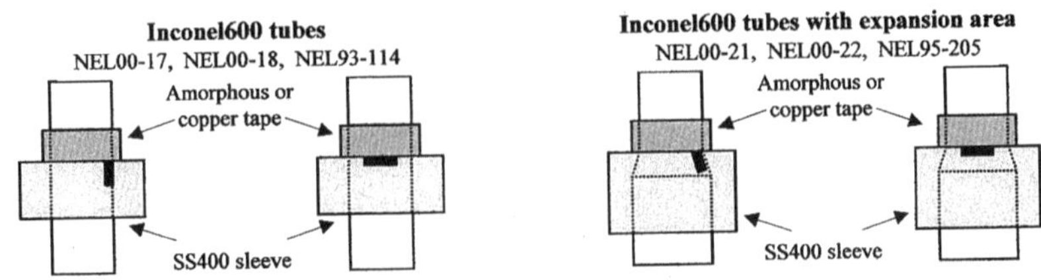

Fig. 6. Samples used to test influence of external interfering structures.

4. Flaws detection in the presence of fasteners interferences

Second set of experiments was done using tubes with fasteners to verify usability of the MFES method and new sensor for detecting flaws in the presence of interfering structures. Samples (Fig. 6) from the benchmark organized by JSAEM were used for this purpose. The set of samples consists of straight Inconnel600 tubes, tubes with expansion area, sleeves made of SS400 (inner diameter 22.5mm, outer diameter 40mm, width 24mm and 60mm), sleeve made of amorphous tape (thickness 0.03mm, width 10mm), and sleeve made of copper tape (thickness 0.03mm, width 10mm). The excitation signal and other measuring conditions were same as during tests described in Section 3. Selected results of measurements are shown in Fig. 7 ~ Fig. 9. The presented signals are usual signals obtained from MFES system [2], without using of any additional restoration techniques (only linear detrending was applied). In case of inner flaw (Fig. 7 and Fig. 8), signal obtained in presence of fasteners has slightly higher amplitude than normally. However the changes are not significant because of high frequencies used for testing. In case of outer flaws and fasteners the peak value is bigger than normal and the maximum is achieved for a lower frequency (Fig. 9 and Fig. 10). Similar tendency was observed in case of all other samples. The signal noise ratios calculated for samples with interfering structures were from 28dB (in case of outer flaws 20%) up to 43dB (in case of inner flaws 20%).

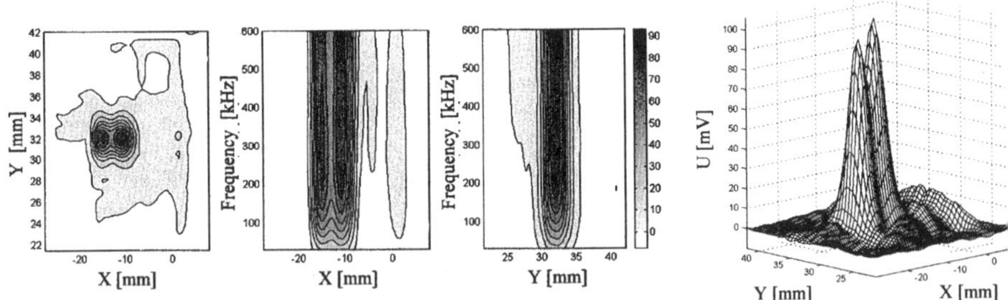

Fig. 7. Spectrograms and signal obtained for inner axial flaw
(depth 20%, length 5mm, tube with expansion area NEL00-21)

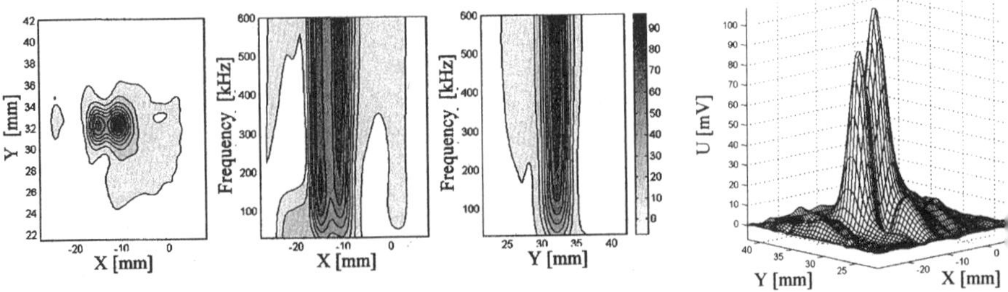

Fig. 8. Spectrograms and signal obtained for inner axial flaw (depth 20%, length 5mm, tube with expansion area NEL00-21) in presence of external interferences (steel sleeve and copper deposit)

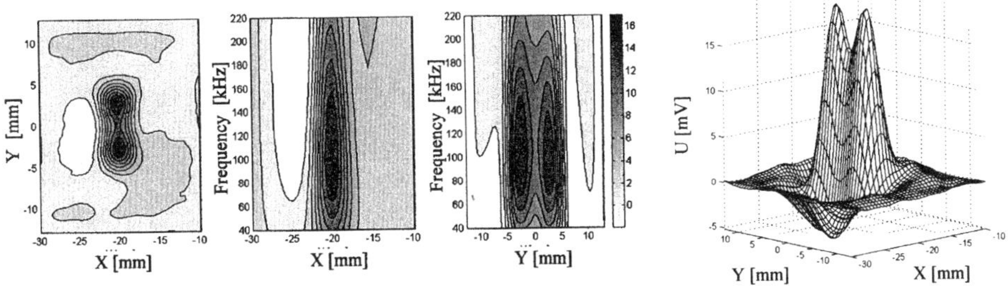

Fig. 9. Spectrograms and signal obtained for outer circumferential flaw
(depth 20%, length 5mm, tube NEL00-18)

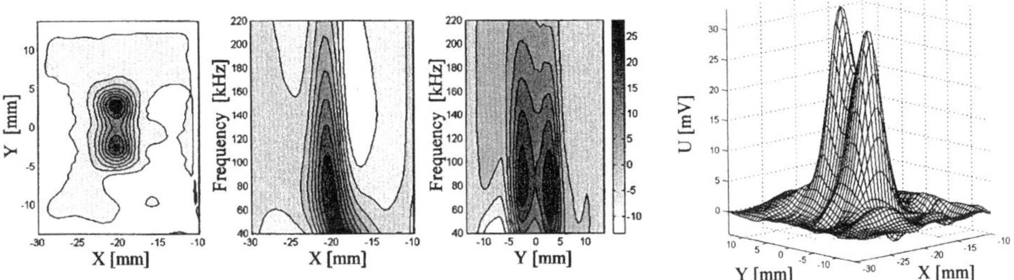

Fig. 10. Spectrograms and signal obtained for outer circumferential flaw (depth 20%, length 5mm,
tube NEL00-18) in presence of external interferences (steel sleeve and magnetic deposit)

5. Natural flaws identification

Final evaluation of the new sensor and the whole system was done using the samples
NEL93-114 (tube with unknown stress-corrosion circumferential flaw) and NEL95-205 (tube
with expansion area and unknown stress-corrosion circumferential flaw).

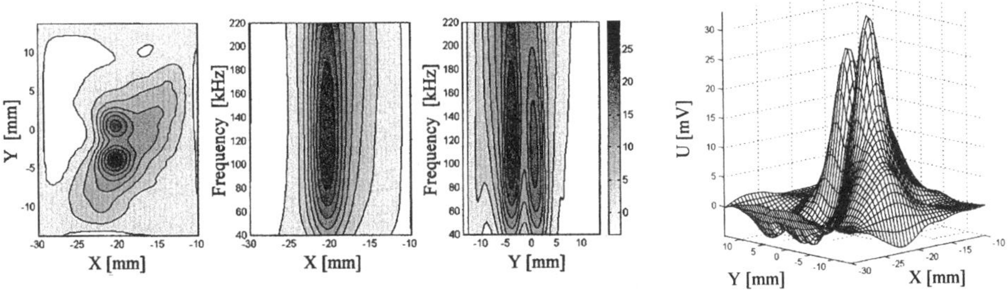

Fig. 11. Spectrograms and signal obtained for unknown stress corrosion flaw
(depth and length unknown, tube NEL93-114)

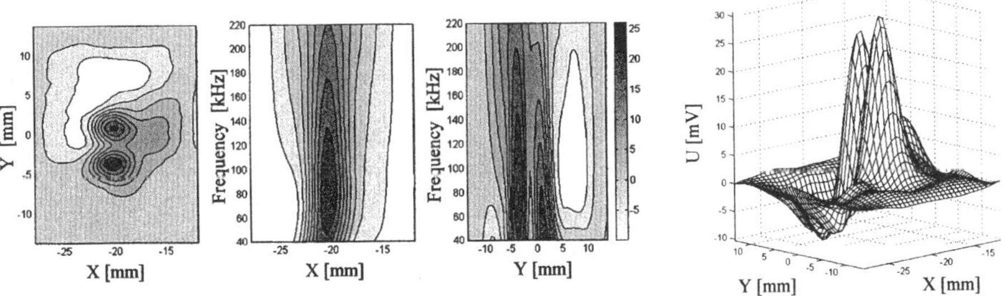

Fig. 12. Spectrograms and signal obtained for unknown stress corrosion flaw (depth and length unknown, tube
NEL93-114) in presence of external interferences (steel sleeve and magnetic deposit)

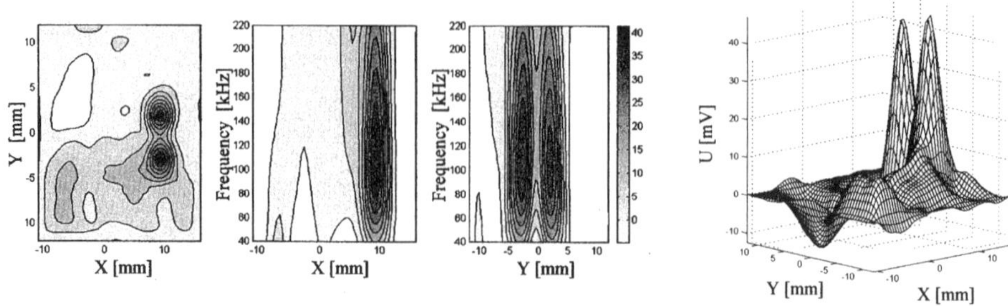

Fig. 13. Spectrograms and signal obtained for unknown stress corrosion flaw
(depth and length unknown, tube with expansion area NEL95-205)

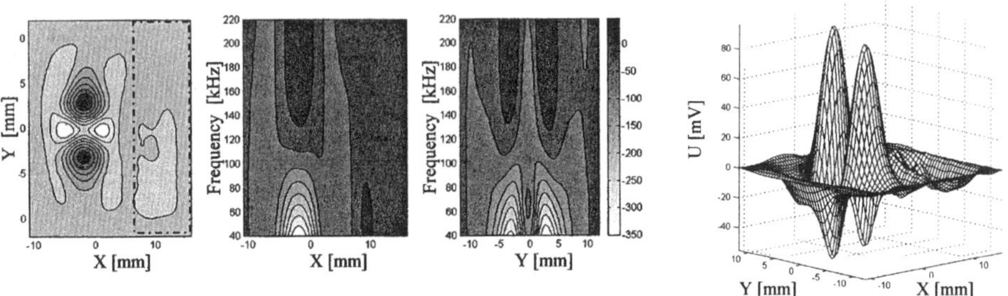

Fig. 14. Spectrograms and signal obtained for unknown stress corrosion flaw (depth and length unknown, tube
with expansion area NEL95-205) in presence of external interferences (steel sleeve and copper deposit)

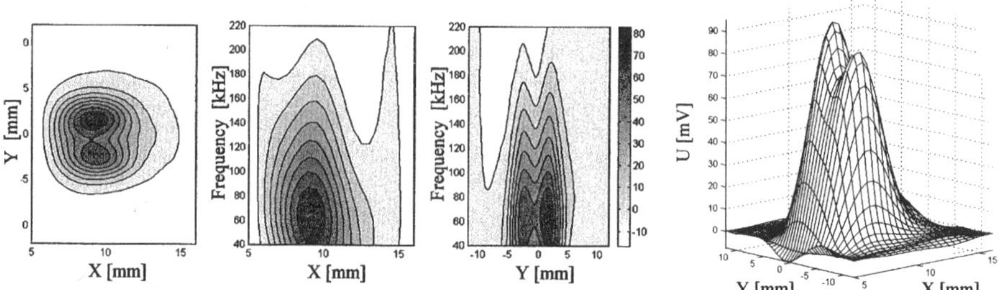

Fig. 15. Spectrograms and signal obtained for limited area around flaw
(area marked by dashed line shown in Fig. 14)

Measurements obtained for the sample NEL93-114 are shown in Fig. 11 and Fig. 12. One
could observe that also in this case the peak value of spectrogram has higher amplitude and is
obtained for a lower frequency than without fastener.

Results achieved for sample NEL95-205 are shown in Fig. 13 ~ Fig. 15. Figure 14 shows
results measured for a tube with fastener (SS400 sleeve) and copper deposit (copper tape). In
this case we can observe two peaks of the signal. The peak having highest amplitude (Fig. 14)
corresponds to "crack" in copper tape (connection of tape). The real flaw causes second peak
(Fig. 15). The "false flaw" and "real flaw" could be easily recognized by using spectrograms
(see Fig. 14 and Fig. 15).

It was found, that the relation between signal amplitude and frequency at the point where the
spectrogram achieves the maximum value, could be approximated by the following function:

$$u = \alpha \, f^2 e^{-\sqrt{\beta f}} \tag{1}$$

where: u – signal amplitude, f – frequency, α, β – approximation parameters, which depend
on a flaw depth.

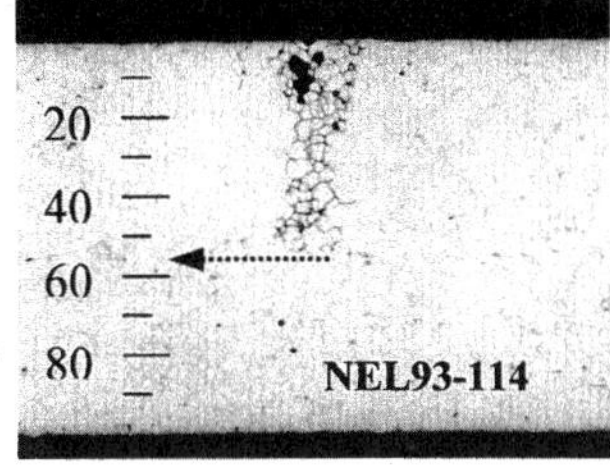

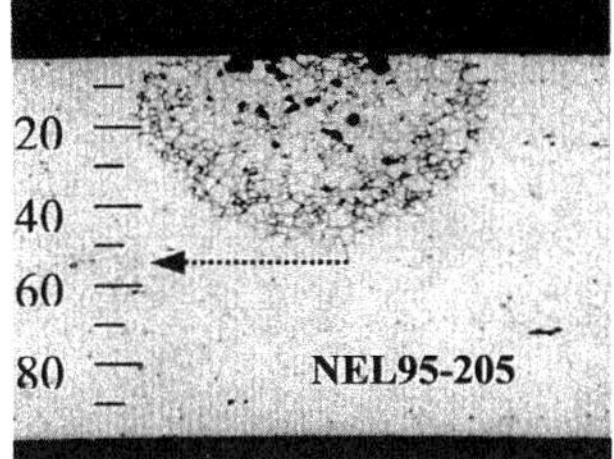

Fig. 16. Cross-section of stress corrosion flaws in sample NEL93-114 (maximum depth of flaw 54%) and in sample NEL95-205 (maximum depth of flaw 57%).

Table 2. Results of flaws' depth identification.

Specimen	Estimated depth	Real depth	Error
NEL 93-114	68%	54%	14%
NEL 95-205	59%	57%	2%

In order to identify depth of the flaws the following procedure was applied. First, parameters α, β are calculated for all of the calibration flaws. Next, the calibration curves were utilized to estimate depth of unknown flaws in sample NEL93-114 and NEL95-205. Results of estimation were compared with real data obtained a few weeks later from destructive test of the samples (Fig. 16). The estimation error was 2% and 14% respectively (Table 2).

6. Conclusions

Presented results of measurements confirmed usability of the new sensor and the MFES system. The sensor is sensitive to shallow flaws and provides high spatial resolution. MFES system allows us to distinguish real flaws from interferences caused by external structures with high probability. The algorithm for flaws' depth identification is proposed and its reliability is confirmed using samples with quasi-natural flaws (stress-corrosion flaws).

Acknowledgements

This study was supported in part by the Research Committee on Nondestructive Evaluation Technology by Eddy Current Testing of the Japan Society of Applied Electromagnetics and Mechanics through a grant from 5PWR utilities and Nuclear Engineering Ltd.

The authors would like to express thanks to Mr. Junri Shimone from Nuclear Engineering Ltd. for his valuable discussion and providing the samples and photos of natural flaws cross-sections.

Measurements used in this paper were done by the author during his stay in Oita University.

References

[1] T. Chady, M. Enokizono, T. Todaka, Y. Tsuchida, R. Sikora: "Evaluation of the JSAEM Round Robin Test Samples Using Multi-frequency Excitation and Spectrogram Method", The Sixth International Workshop on Electromagnetic Nondestructive Evaluation, August, 2000, Budapest, Hungary.
[2] T. Takagi, K. Miya: "ECT round-robin test for steam generator tubes", Journal of the Japan Society of Applied Electromagnetics and Mechanics, Vol.8, No.1 pp.121-129, 1999.
[3] T. Chady, M. Enokizono, T. Todaka, Y. Tsuchida, R. Sikora: "A Family of Matrix Type Sensors for Detection of Slight Flaws in Conducting Plates", IEEE Transactions on Magnetics, Vol. MAG-35, No.5, pp. 3655-3657, 1999.

Electromagnetic Nondestructive Evaluation (VI)
F. Kojima et al. (Eds.)
IOS Press, 2002

Non-destructive Testing by Using the Rotational Magnetic Flux Type Probe for SG Tubes

Mohachiro Oka* and Masato Enokizono**

Department of Computer and Control Engineering, Oita National College of Technology, 1666 Maki, Oita, 870-0152, Japan
**Faculty of Engineering, Oita University, 700 Dannoharu, Oita, 870-1192, Japan*

Abstract. The rotational magnetic flux type eddy current testing (RMF-ECT) probe is applied to detection of a small outer crack on a steam generator tube with sources of noise such as a pipe holder and sludge. From results of experiments, when a small outer crack exists on a boundary between a pipe holder and sludge, the output signal of this RMF-ECT probe is influenced by sources of noise. Even if the noise source exists, this probe can detect the defect according to the kind of the noise source. In this paper, crack detection characteristics of our RMF-ECT probe under the noisy environment are presented.

1. Introduction

An eddy current testing technique is very useful to detect cracks on steam generator (SG) tubes using in nuclear power plants. In order to prevent a big accident of nuclear power plants, it is very important to develop a sensitive ECT probe for small outer cracks on SG tubes [1-2]. The authors have already developed rotational magnetic flux (RMF) sensors to detect small reverse-side cracks on thick steel plates or thick stainless steel plates [3-4]. The RMF-ECT probe, which had been reported at the ENDE-Hungary is one of them [5]. It is characterized by the rotational excitation. In that paper the RMF-ECT probe was applied to detection of a small outer crack on 1.27 mm thick SG tubes without sources of noise. This RMF-ECT probe shows high detection sensitivity and high signal to noise ratio. The depth, length and width of it are 10 %, 5 mm, and 0.2 mm respectively. In this paper, the RMF-ECT probe is applied to detect a small outer crack on a SG tube with sources of noise. In this case, the output signal of this probe is occasionally influenced by sources of noise. But, this probe can detect a small outer crack depending on the condition.

2. The RMF-ECT probe

Fig. 1 shows a structural drawing of the RMF-ECT probe for a SG tube. The lift-off that is the distance between the ferrite core and a specimen is about 0.3 mm. First, B-coils are wound on connective parts of the ferrite core to measure excited maximum magnetic flux densities (B_{exmax} and B_{eymax}). Next, excitation coils are wound on them. Each of exciting coils is made of 0.16 mm diameter copper wire (60 turns). And, each of B-coils is made of 0.16 mm diameter copper wire (6 turns). The rotational magnetic field is generated in a specimen by two-phase excitation currents with 90°-phase difference. Five pickup coils, which are wound on each end of the core leg, can measure the magnetic flux density of each part near a specimen. Pickup coils on X- and Y-axis are composed of two pickup coils (See Fig. 3).

3. Measurement system

Fig. 2 shows a schematic diagram of experimental equipment. An oscillator generates two sinusoidal excitation voltages. These excitation voltages are adjusted with a computer so that B_{exmax} and B_{eymax} can be nearly equivalent to each other on a no-crack specimen. Three output voltages (e_x, e_y and e_z) and two output voltages from B-coils are digitalized by an A/D converter and input into a computer. Fig. 3 shows the wiring diagram of pickup coils. The maximum differential magnetic flux density (δB_{xmax}) on X-axis is calculated by using the output voltage (e_x) on X-axis. In the same way, δB_{ymax} is calculated from the output voltage (e_y) on Y-axis. On the other hand, B_{zmax} is calculated using e_z. If there is not a crack on a SG tube, e_x and e_y are zero. The RMF-ECT probe is moved by an X-stage every 0.5 mm automatically and rotated every 5° manually. The center of the measuring range is equivalent to the center of a specimen and an outer crack. The probe position is indicated by the center of the RMF-ECT probe. Specimens (INCONEL 600), which are SG tubes prepared by JSAEM for the round-robin test, are used in our experiments. An inner diameter and an outer diameter of SG tubes are 19.6 mm and 22.3 mm respectively. Fig. 4 shows a fixed coordinate system located on a SG tube and assembly drawing of sludge (A) and a pipe holder (B), which are sources of noise. Several kinds of a source of noise with a different shape and material are prepared.

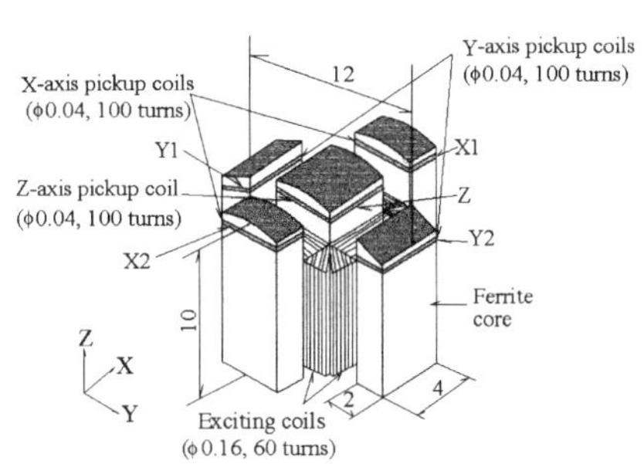

Fig. 1   The rotational magnetic flux type ECT probe.

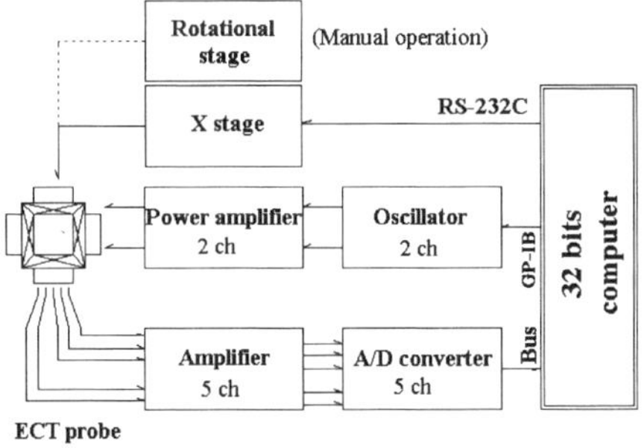

Fig. 2   The experimental equipment.

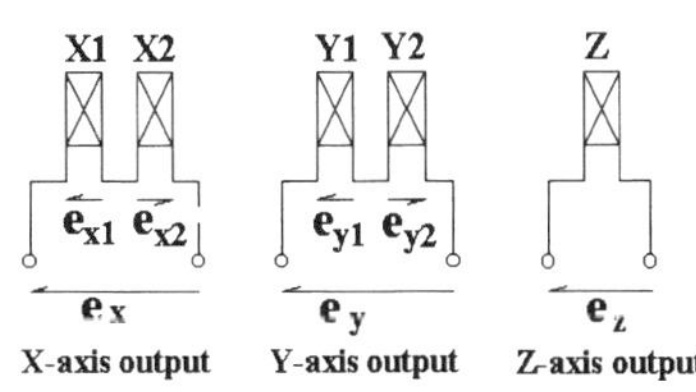

Fig. 3   The wiring diagram of pickup coils.

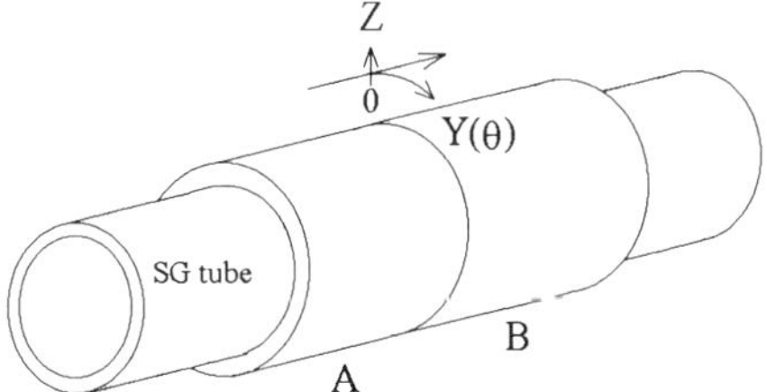

Fig. 4   Axes of a SG tube and assembly drawing of sludge and a pipe holder.

4. Experimental results and discussions

We carried out the experiment under conditions as follows: The excitation frequency is 60 kHz. Because of this the upper bound of the frequency characteristic of power amplifiers is about 100 kHz. The maximum excitation magnetic flux density (B_{emax}) was 0.005 T at connective parts of the ferrite core. Smoothing algorithm such as the method of moving averages was used for better performance of the RMF-ECT probe. We carried out the experiment as follows: At first, in order to equalize B_{exmax} and B_{eymax}, the RMF-ECT probe was put on a no-crack specimen. A computer adjusted excitation voltages so that B_{exmax} and B_{eymax} might become the same. When B_{exmax} and B_{eymax} become so, we set the RMF-ECT probe on the first measurement position, and began measurement.

A. Influence of source of noises (with a pipe holder and sludge)

Fig. 5 shows two types of a pipe holder, a support plate and sludge. Fig. 5(a), Fig. 5(b), Fig. 5(c) and Fig. 5(d) show a pipe holder with a 1 mm BEC (Broached Egg Crate) made from SUS405, a pipe holder with a round hole made from SS400, a support plate with a round hole made from SS400 and mock sludge made from 45 permalloy respectively.

Fig. 6 shows assembly drawing of mock sludge such as 45 permalloy and a pipe holder with the 1 mm BEC. Fig. 7 shows the relationship between the output signal (B_{zmax}) and the probe position. In this case, the specimen is NEL0019 (without a crack) with sources of noise such as the pipe holder and mock sludge. From this figure, it can be said that the signal (B_{zmax}) from the probe is not influenced from sludge so much though it is greatly influenced by the pipe holder (a 1 mm BEC). Fig. 8 and Fig. 9 show the relationship between the output signal (B_{zmax}) and the probe position. In the case of Fig. 8, the specimen is NEL0017 (with a 20 % OD axial crack). The specimen is NEL0018 (with a 20 % OD circumferential crack). In both cases, sources of noise are same as Fig. 6. When Fig. 8 and Fig. 9 are compared with Fig. 7, the outer crack, which is sure to exist in the circle of each figure cannot be clearly confirmed from the distribution of B_{zmax}. When source of noises is the pipe holder (a 1 mm BEC), this probe cannot detect the 20 % OD crack well.

In our experiments, higher excitation frequency than 60kHz could not be used because of that the limitation of the frequency characteristic of the power amplifier. If a higher excitation frequency is used, the skin depth gradually becomes small more than the thickness of the SG tube wall. Then, it is thought that the influence of the source of noise becomes small if the higher excitation frequency is used.

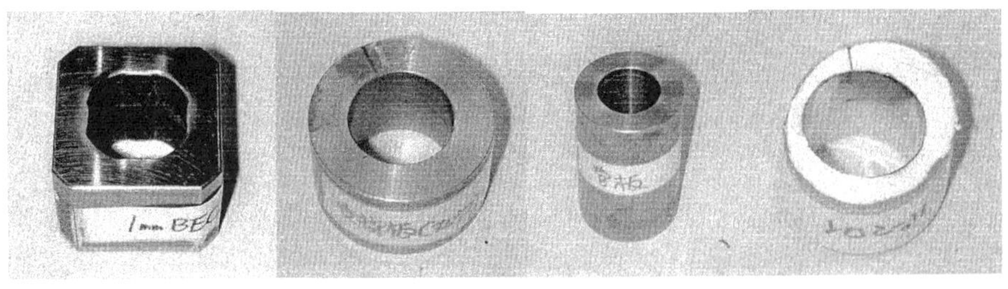

(a) Pipe holder with   (b) Pipe holder with   (c) Tube plate.   (d) 45 permalloy.
a 1 mm BEC.   a round hole.

Fig. 5 Sources of noise used in our experiments.

The influence caused
by permalloy is small.

The influence caused by
the pipe holder is large.

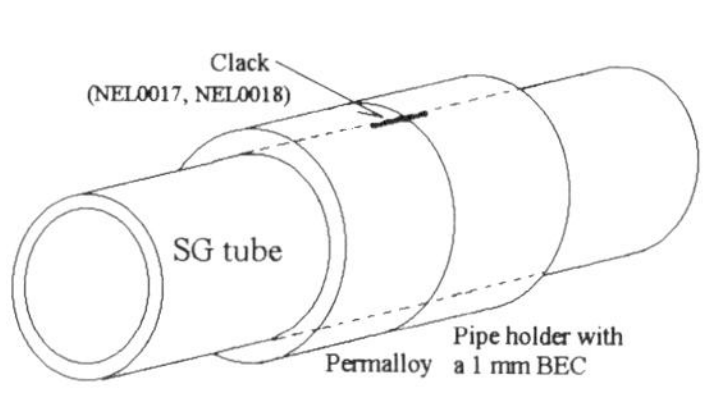

Fig. 6   Assembly drawing of a pipe holder
with a 1 mm BEC and permalloy.

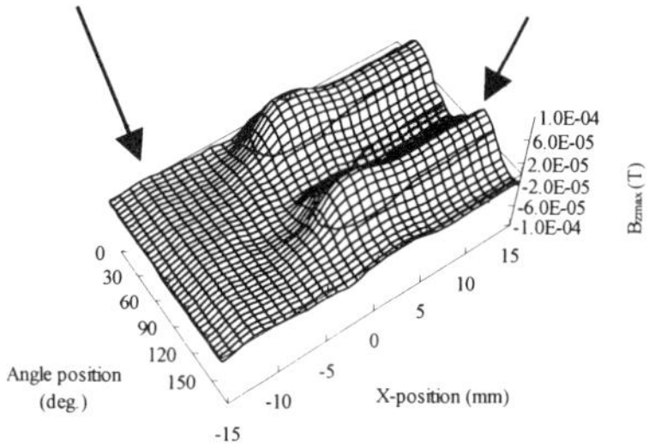

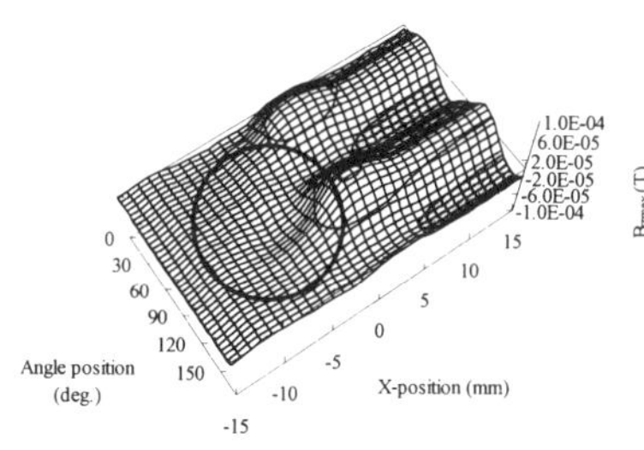

(No crack, NEL0019)
Fig. 7 B_{zmax} vs. probe position.

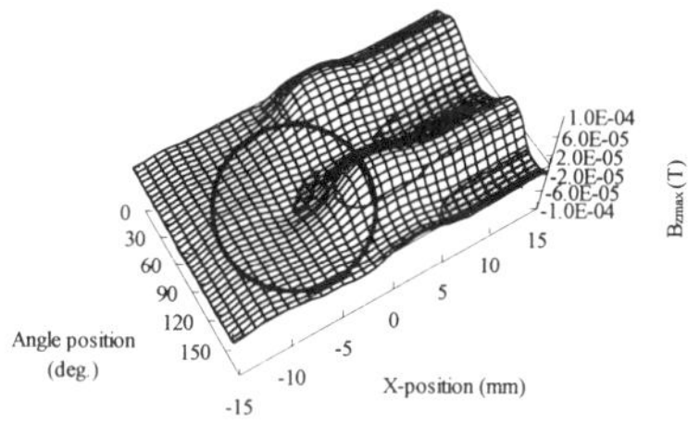

(Axial crack, OD 20 %, NEL0017)
Fig. 8 B_{zmax} vs. probe position.

(Circumferential crack, OD 20 %, NEL0018)
Fig. 9 B_{zmax} vs. probe position.

B. Crack detection characteristics and Signal-to noise ratio (without sources of noise)

Fig. 10 shows the relationship between B_{zmax} and the probe position in the case of the specimen with a 20 % outer axial crack (NEL0017). Fig. 11 shows the same relation in the case of the specimen with a 20 % outer circumferential crack (NEL0018). From Fig. 10 and Fig. 11, this RMF-ECT probe clearly detected a small outer crack. The values of S/N are 15.6 dB (NEL0017) and 9.0 dB (NEL0018) respectively. The definition of signal-to-noise ratio followed the method of JSAEM.

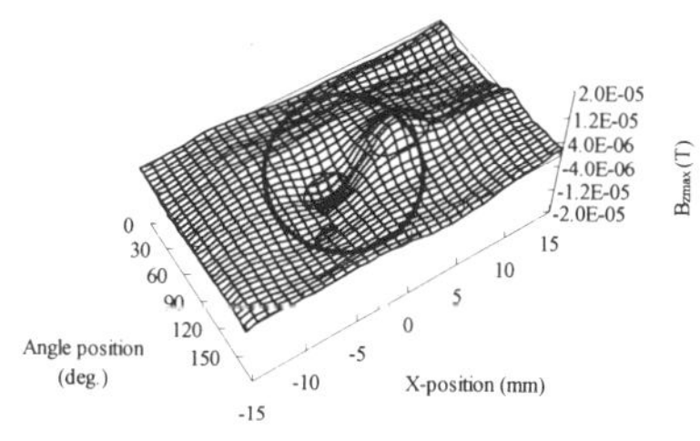

(Axial crack, OD 20 %, NEL0017)
Fig. 10 B_{zmax} vs. probe position.

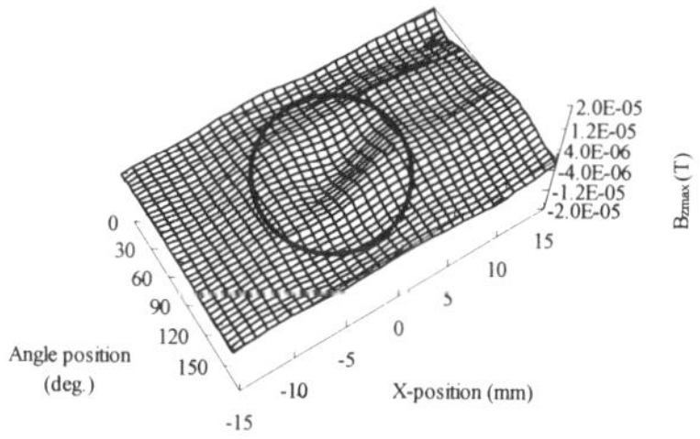

(Circumferential crack, OD 20 %, NEL0018)
Fig. 11 B_{zmax} vs. probe position.

C. Crack detection characteristics (with a tube holder or a tube plate)

Fig. 12 shows assembly drawing of a pipe holder with a round hole or a tube plate. Fig. 13 shows the relationship between B_{zmax} and the probe position in the case of the specimen without a crack (NEL0019). In this figure, arrows point a signal from the source of noise. This part is corresponding to the edge of the pipe holder with a round hole. When Fig. 14 and Fig. 15 are compared with Fig. 13, the outer crack, which is sure to exist in the circle of each figure, can be clearly confirmed from the distribution of B_{zmax}.

In Fig. 16, the arrow points a signal from the source of noise such as the tube plate with a round hole. This part is corresponding to the edge of the tube plate with a round hole. From these figures (Fig. 14, Fig. 15, Fig. 16 and Fig. 17), it can be said that this RMF-ECT probe can clearly detect a small outer crack when the outer crack do not exist near the edge of the source of noise And, this probe can obviously distinguish a direction of a small outer crack.

D. Crack detection characteristics (with a pipe holder and sludge)

Fig. 18 shows assembly drawing of a pipe holder with a round hole and mock sludge such as 45 permalloy. From this figure, it can be said that the signal from the probe is not influenced from a pipe holder and sludge so much. The influence of sources of noise appears only in the boundary of both. The relationship between B_{zmax} and the probe position is shown in Fig. 20 and Fig. 21. In these figures, a small outer crack is detected. But, this probe cannot distinguish a direction a small outer crack.

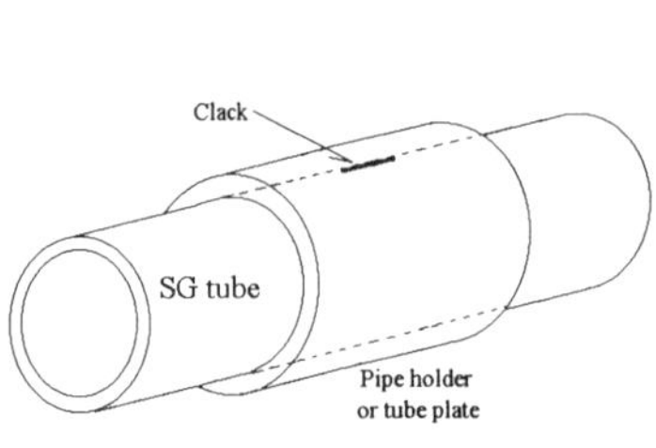

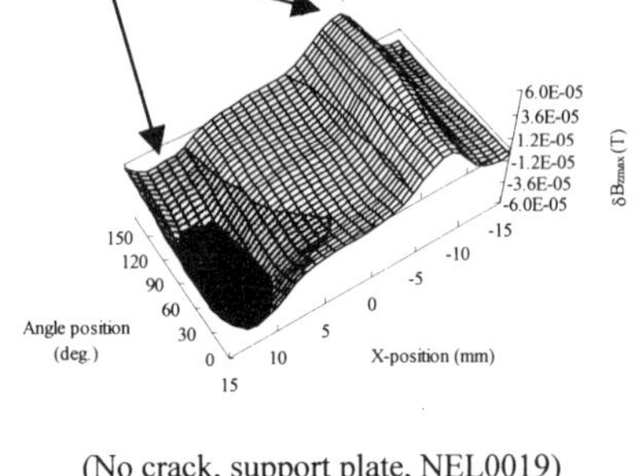

Fig. 12   Assembly drawing of a pipe holder with a round hole.

(No crack, support plate, NEL0019)
Fig. 13 B_{zmax} vs. probe position.

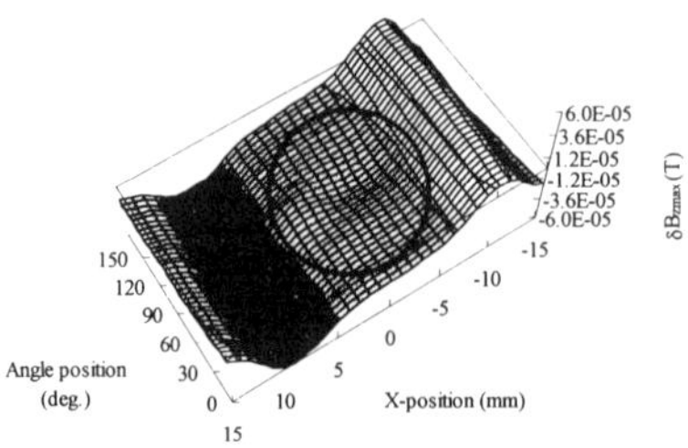

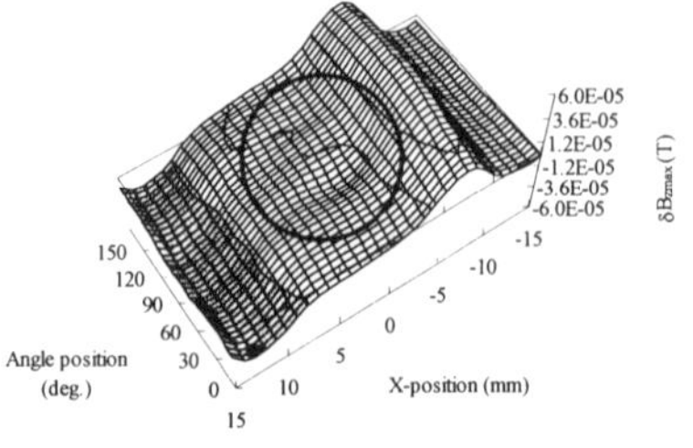

(Axial crack, support plate, NEL0017)
Fig. 14 B_{zmax} vs. probe position.

(Circumferential crack, support plate, NEL0018)
Fig. 15 B_{zmax} vs. probe position.

The influence caused by the tube plate.

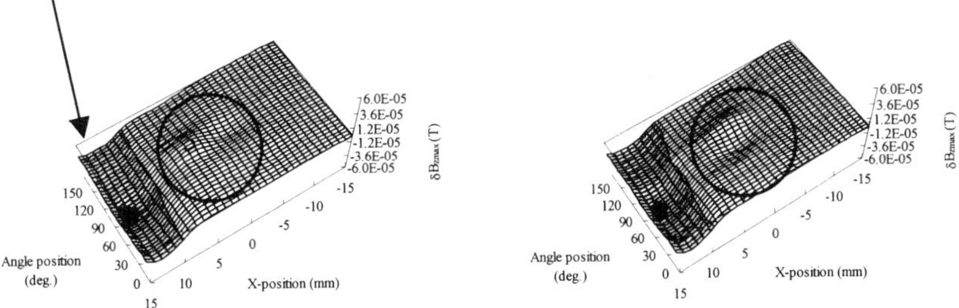

(Axial crack, tube plate, NEL0017)
Fig. 16 B_{zmax} vs. probe position.

(Circumferential crack, tube plate, NEL0018)
Fig. 17 B_{zmax} vs. probe position.

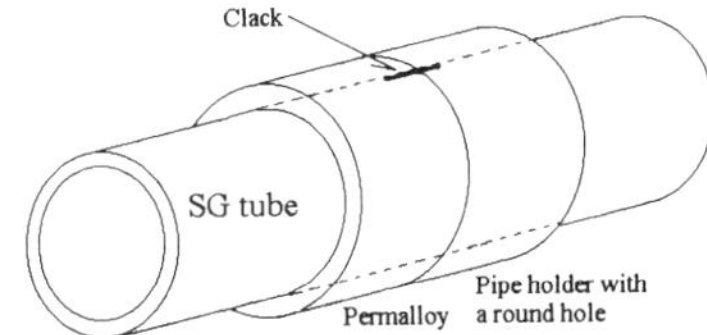

Fig. 18   Assembly drawing of a pipe holder
with a round hole and permalloy.

The influence caused by the pipe holder
with a round hole and permalloy.

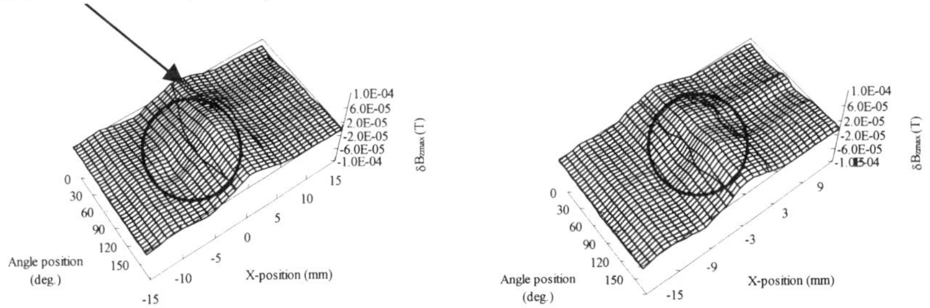

(Axial crack, round hole, permalloy, NEL0017)
Fig. 19 B_{zmax} vs. probe position.

(Circumferential crack, round hole, permalloy, NEL0018)
Fig. 20 δB_{zmax} vs. probe position.

5. Conclusions

In this paper, we present the rotational magnetic flux type ECT probe for detection of small outer cracks on SG tubes under the noisy environment. From results of experiments, the following findings are obtained.

(1) We proposed the RMF- ECT probe to detect effectively small outer cracks on SG tubes with the source of noise such as a pipe holder and sludge.

(2) If the source of noise is only a pipe holder with a round hole or a tube plate, this RMF-ECT probe can detect a small outer crack.

(3) But, when the source of noise is a pipe holder with a 1 mm BEC, this RMF-ECT probe cannot detect a small outer crack.

(3) The signal from this probe is not influenced from sludge (45 parmalloy) so much.

(4) It is thought that the influence of the source of noise becomes small due to the skin effect if the higher excitation frequency is used.

References

[1] M. Kurokawa, R. Miyauchi and K. Enami, "New Eddy Current Probe for NDE of Steam Generator", Electromagnetic Nondestructive Evaluation (III), IOS Press, Vol. 15, pp. 57-64, 1999
[2] P. Y. Joubert, D. Miller, D. Plack and E. Savin, "Multi-Detector Eddy Current Probe for the Non-Destructive Evaluation of Steam Generator Tubes, Designed for an Imaging Approach", Electromagnetic Nondestructive Evaluation (III), IOS Press, Vol. 15, pp. 34-44, 1999
[3] M. Oka and M. Enokizono, "Rotational Magnetic Flux Sensor with Three Axis Search Coil for Non-Destructive Testing", Nonlinear Electromagnetic Systems, IOS Press, pp. 357-360, 1998
[4] M. Oka and M. Enokizono, "Eddy Current Inspection by Gradio-Magneto Sensor Using MI Device", Journal of the Magnetics Society of Japan, Vol. 24, No. 4-2, pp. 859-862, 2000
[5] M. Oka and M. Enokizono, "Eddy Current Testing Using Rotational Magnetic Flux Type Probe for SG Tube", Studies in Applied Electromagnetics and Mechanics 21, ELECTROMAGNETIC NONDESTRUCTIVE EVALUATION (V), *IOS Press*, Vol. 21, pp. 234-241, 2001

Electromagnetic Nondestructive Evaluation (VI)
F. Kojima et al. (Eds.)
IOS Press, 2002

Evaluation of near-surface material properties using planar mesh type coils with post-processing from neural network model

S.C. Mukhopadhyay, S.Yamada* and M.Iwahara*

IIST, Massey University, Palmerston North, New Zealand

** Faculty of Engineering, Kanazawa University, Kanazawa, Japan.*

Abstract : The possibility of employing a planar mesh type probe for the evaluation of near-surface material properties as well as to inspect the presence of defects has been investigated in this paper. The impedance of a planar type coil in proximity of any metal surface is a complex function of many parameters including near-surface material properties. A two-dimensional model of a planar mesh type probe has been developed for the analytical calculation of magnetic vector potential., flux-linkage and impedance. The impedance of the probe is used for the evaluation of the near-surface material properties. A simple neural network model has been developed for the post-processing of output parameters from the measured impedance data.

1. Introduction

This paper is concerned with the research required to assess and inspect near-surface material properties using a planar type probe having mesh configuration. The impedance of the coil can be used to detect most kind of flaws and to evaluate degradation of materials. There are many different causes of degradation including environmental stress, corrosion and thermal treatment etc. In some circumstances this leads to dangerous situation. For example, material degradation of aircraft's outer surface may result in a catastrophic accident. In industries it is a major concern to be able to assess the coating quality, strength, contamination and hidden corrosion in a simple, versatile, effective and efficient way.

It has been demonstrated that the physical properties of a material undergo deterioration due to fatigue and aging [1, 2] so it is necessary to be able to assess the actual state of the material. Planar type meander coils have been used for the evaluation of near-surface properties and are reported in [3, 4]. The use of meander type probe for the inspection of printed circuit boards has been described in [5]. The aim is to extend the modeling technique to planar mesh type probe and to investigate the feasibility of applying it to estimate the near-surface material properties. A simple neural network model has been developed for the post-processing of output parameters from the measured impedance data.

2. *Modeling of planar mesh type probe*

The probe consists of two coils: exciting coil carries a high frequency current and generates the electromagnetic field. The induced magnetic field in the surrounding material modifies the resultant magnetic field distribution which is picked up by the sensing coil. Both coils employ mesh type configuration as shown in Fig. 1. The impedance of a coil in proximity with a metal surface varies in a complex way with different parameters such as permeability and conductivity of near-surface of the material, lift-off and pitch of the coil, operating frequency etc. and is usually used for the inspection of near-surface material properties. It is very difficult to get a simple expression to determine the near surface material properties from the measured impedance of the probe. For the derivation of the impedance characteristics of the coil an analytical model has been developed. In the analytical calculation only one pitch both along the X and Y-axis are considered. The 2-dimensional representation along the Y-axis is shown in Fig. 2 and this corresponds to the situation that the probe is used for the inspection of electroplated material. The conductivities and permeabilities of different sub-regions are shown in Table 1. Another model is used to model the system along the X-axis in which the parameters along the Y-axis are replaced by the parameters along the X-axis. The pitch along the X-axis may or may not be similar in size depending on the configuration of the coil.

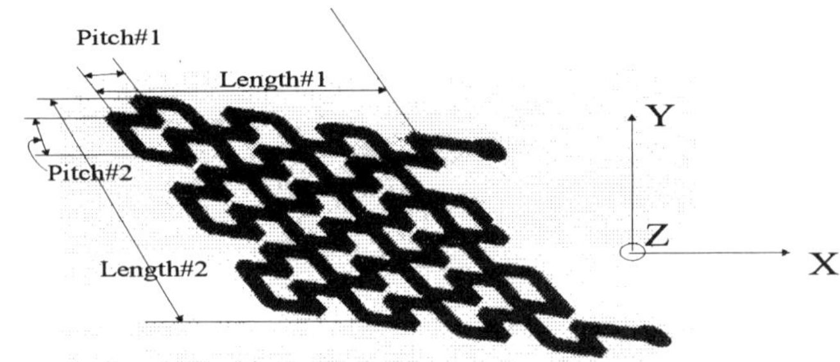

Fig. 1 Configuration of planar type mesh coil

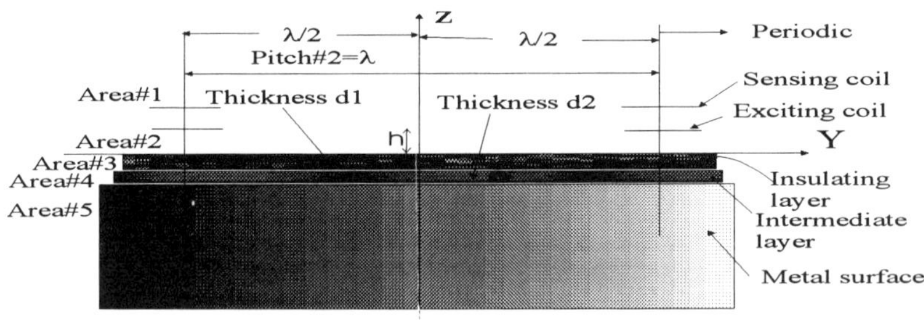

Fig. 2 2-D modeling along the Y-axis

The lift-off is designated by h. Details of the analysis for meander type probes has been described in [6]. Using the same procedure the modeling for the mesh type probe is done. The governing equation for area#1 and area#2 is given by,

$$\frac{\partial^2 A(y,z)}{\partial y^2} + \frac{\partial^2 A(y,z)}{\partial z^2} = 0 \tag{1}$$

The governing equation for area#3, #4 and #5 is given by,

$$\frac{\partial^2 A(y,z)}{\partial y^2} + \frac{\partial^2 A(y,z)}{\partial z^2} = jk^2 A(y,z) \tag{2}$$

Table 1 Conductivities and permeabilities of sub-regions

Sub-region	Conductivity	Permeability
Area#1 z >=h	0	μ_0
Area#2 0<z<h	0	μ_0
Area#3 -d1<z<0	$\sigma 1$	μ_1
Area#4 -(d1+d2)< z<-d1	$\sigma 2$	μ_2
Area#5 z<-(d1+d2)	$\sigma 3$	μ_3

The vector potential $A(x,y,z,t)$ is assumed to be independent along the x-coordinate and is function of y and z variables only. Assuming the span of the meander coil is periodic and m being integer, we can write,

$$\alpha\lambda = m\pi \tag{3}$$

The method of separation of variables is used for the solution and each part is solved separately.

The equations of the magnetic vector potential for each and every region are given below.

The equation for magnetic vector potential for area#1 is given by,

$$A_1(y,z) = \sum_{m=1}^{\infty}(C_1 e^{-m\pi\frac{z}{\lambda}})Sin(m\pi\frac{y}{\lambda}) \tag{4}$$

For area#2 we have
$$A_2(y,z) = \sum_{m=1}^{\infty}(C_2 e^{-m\pi\frac{z}{\lambda}} + C_3 e^{+m\pi\frac{z}{\lambda}})Sin(m\pi\frac{y}{\lambda}) \tag{5}$$

For area#3 we can write
$$A_3(y,z) = \sum_{m=1}^{\infty}(C_4 e^{-(r_1+j\tau_1)} + C_5 e^{+(r_1+j\tau_1)})Sin(m\pi\frac{y}{\lambda}) \tag{6}$$

The magnetic vector potential for area#4 is given by

$$A_4(y,z) = \sum_{m=1}^{\infty}(C_6 e^{-(r_2+j\tau_2)} + C_7 e^{+(r_2+j\tau_2)})Sin(m\pi\frac{y}{\lambda}) \tag{7}$$

and that of area#5 is given by
$$A_5(y,z) = \sum_{m=1}^{\infty}(C_8 e^{-(r_3+j\tau_3)})Sin(m\pi\frac{y}{\lambda}) \tag{8}$$

The exciting current is shown in Fig. 3.

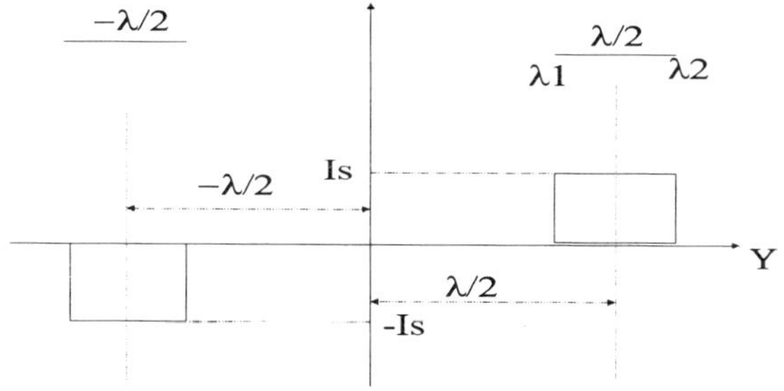

Fig. 3 Exciting current

The exciting current is expressed as
$$I(y,t) = \sum_{m=1}^{\infty}I_s Sin(m\pi.\frac{y}{\lambda}).e^{j\omega t} \tag{9}$$

The sheet current density is given by, $\qquad i(y) = I_s /(\lambda_2 - \lambda_1)$ $\qquad\qquad$ (10)

and the harmonic component current is given by,

$$I_m = \frac{2}{\lambda} \int_{-\lambda/2}^{\lambda/2} I_s(y) Sin(m\pi.\frac{y}{\lambda}) dy = \frac{4}{\lambda} \int_{\lambda_1}^{\lambda/2} I_s(y).Sin(m\pi.\frac{y}{\lambda}) dy \qquad\qquad (11)$$

All the harmonic components are separately calculated and used in the solution.

Two boundary conditions $\nabla X H = 0$ and $\nabla.B = 0$ are applied to each intersection of the sub-regions. The following conditions are obtained.

$$\begin{aligned} A_1(0, h) &= A_2 (0, h) \\ A_2 (0, 0) &= A_3 (0, 0) \\ A_3 (0, -d1) &= A_4 (0, -d1) \\ A_4 (0, -d2) &= A_5 (0, -d2) \end{aligned} \qquad\qquad (12)$$

and

$$\left.\frac{1}{\mu_0}\frac{\partial A_1}{\partial z}\right|_{z=h} - \left.\frac{1}{\mu_0}\frac{\partial A_2}{\partial z}\right|_{z=h} = i(y)$$

$$\left.\frac{1}{\mu_0}\frac{\partial A_2}{\partial z}\right|_{z=0} = \left.\frac{1}{\mu_1}\frac{\partial A_3}{\partial z}\right|_{z=0}$$

$$\left.\frac{1}{\mu_1}\frac{\partial A_3}{\partial z}\right|_{z=-d1} = \left.\frac{1}{\mu_2}\frac{\partial A_4}{\partial z}\right|_{z=-d1} \qquad\qquad (13)$$

$$\left.\frac{1}{\mu_2}\frac{\partial A_4}{\partial z}\right|_{z=-(d1+d2)} = \left.\frac{1}{\mu_3}\frac{\partial A_5}{\partial z}\right|_{z=-(d1+d2)}$$

Using the above boundary conditions all the equations are solved for the determination of the unknowns, C_1 to C_8.

Assuming the depth of the coil to be unit meter, the expression of flux is given by

$$\phi = \oint B.ds = \oint \nabla X A.ds = \int A.dl \qquad\qquad (14)$$

So the flux ϕ is given by,

$$\phi = 2A(\lambda/2, h) = 2\sum_{m=1}^{\infty}(C_1 e^{-m\pi\frac{h}{\lambda}}) Sin(m\pi/2) \qquad\qquad (15)$$

We know $\qquad Z_1 I_s = -\frac{d\phi}{dt} = 2j\omega\sum_{m=1}^{\infty}(C_1 e^{-m\pi\frac{h}{\lambda}}) Sin(m\pi/2) \qquad\qquad (16)$

So the transfer impedance per unit meter of coil depth along the Y-axis is thus given by,

$$Z_1 = 2j\omega\sum_{m=1}^{\infty}(C_1 e^{-m\pi\frac{h}{\lambda}}) Sin(m\pi/2)/I_s \qquad\qquad (17)$$

Similarly taking one coil pitch along the X-axis the expression of transfer impedance per unit meter of coil depth Z_2 can be obtained. So the total transfer impedance of the coil per unit meter is given by the vector sum of Z_1 and Z_2.

3. Results from analytical model

The transfer impedance of the probe has been calculated from the model described in the last section for different varying parameters. Fig. 4 shows the variation of the resistive part of the impedance as a function of conductivity of the intermediate layer for different coil pitch.

an operating frequency of 500 kHz. It is assumed here that total surface area covered by the probe with different pitches are kept same. It is seen from Fig. 4 that the resistive part decreases with the increase in conductivity and the value is more for larger coil pitch. Fig. 5 shows the variation of the resistive part of the impedance with the thickness of the intermediate layer at an operating frequency of 500 kHz. The thickness of top layer is maintained constant at 0.25 mm. Fig. 6 shows the variation of the resistive part of the impedance with the thickness of the top layer at an operating frequency of 500 kHz. The thickness of intermediate layer is maintained constant at 50 μm. If the coil pitch is increased beyond 2.1 mm the resistive part instead of increasing starts decreasing. So there is an optimum value of the coil pitch which is to be selected for the best results.

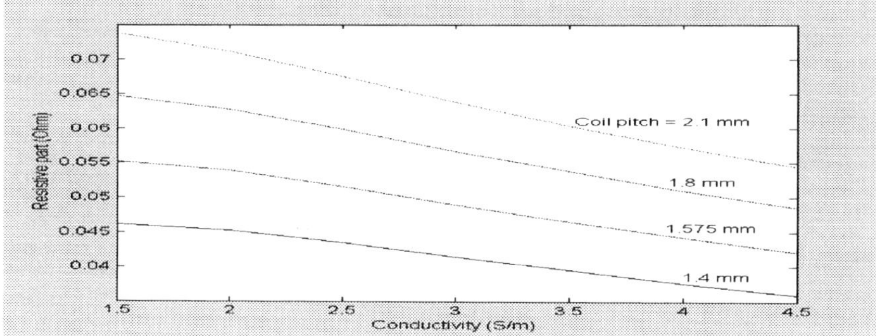

Fig. 4   Variation of resistive part of impedance with conductivity

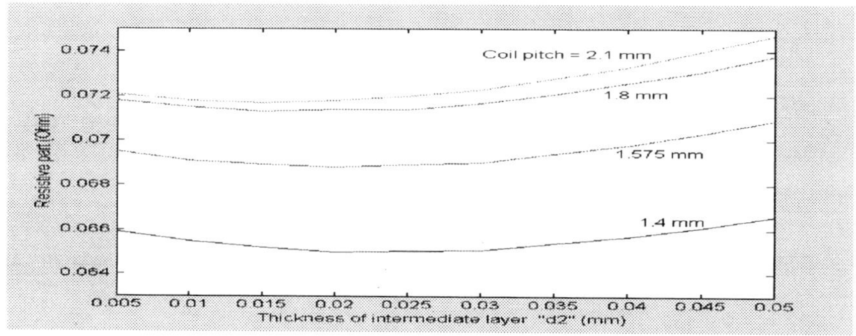

Fig. 5 Variation of resistive part of impedance with the thickness of intermediate layer

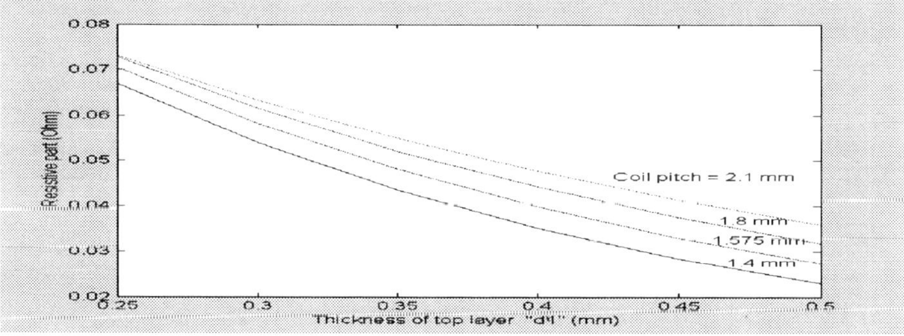

Fig. 6 Variation of resistive part of impedance with the thickness of top layer

4. Experimental set-up and results

The experimental set-up is shown in Fig. 7. The probe is placed on the electroplated material and the transfer impedance (i.e., the ratio of the voltage of the sensing coil to the current of the exciting coil) is measured with the help of Hewlet Packard make impedance analyzer HP 4194A.

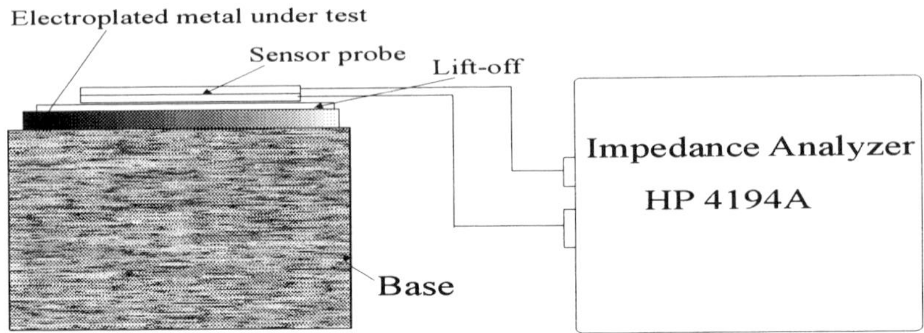

Fig. 7 Experimental set-up

In order to verify the performance of the probe the electroplated material is simulated in the laboratory and by placing the probe on that the impedance has been measured. Table 2 shows the impedance data of the meander probe at 250 kHz and 500 kHz for a base of 5 mm copper plate and the top layer of 0.25 mm insulating layer. The intermediate layer is 0.05 mm thick tungsten and molybdenum material.

Table 2 Results for different material

Freq. (kHz)	0.05 mm Tungsten		0.05 mm Molybdenum	
	Real (Z)	Imag (Z)	Real (Z)	Imag (Z)
250 kHz	0.019272	0.42921	0.020935	0.42335
500 kHz	0.052162	0.84694	0.054715	0.83488

Table 3 shows the impedance data of meander probe at 250 kHz and 500 kHz for a base of mm copper plate and the top layer of 0.25 mm insulating layer with variable thickness of tungsten material.

Table 3 Results for variable thickness

Freq. (kHz)	0.02 mm Tungsten		0.04 mm Molybdenum		0.05 mm Tungsten	
	Real (Z)	Imag (Z)	Real (Z)	Imag (Z)	Real (Z)	Imag (Z)
250 kHz	0.02082	0.40293	0.018949	0.4129	0.018928	0.41833
500 kHz	0.0513	0.80926	0.05029	0.81756	0.05101	0.83028

5. *Estimation of surface properties*

The transfer impedance of the probe depends on many parameters including the surface properties of the material in a complex way and there is no simple method to evaluate them from the impedance data. In order to make the process fast and simple, a grid system has been developed by plotting imaginary part of the impedance versus its real part for the parameters of interest at one particular frequency. A conductivity-thickness measurement grid is shown in Fig. 8 corresponds to an operating frequency of 1 MHz. The grid can be made for any two variable parameters for a 2-dimensional grid or any 3-variable parameters for a 3-dimensional grid. The data used for the generation of the grid system are obtained from the model. Each node of the grid system in Fig. 8 has fixed conductivity and thickness. The measured impedance data is plotted on the grid as shown by **O** in Fig. 8 and the surface properties are estimated by interpolation technique. The measured impedance data doesn't match very well with the results from analytical model so a correction factor is introduced while the measured data is plotted on the grid system.

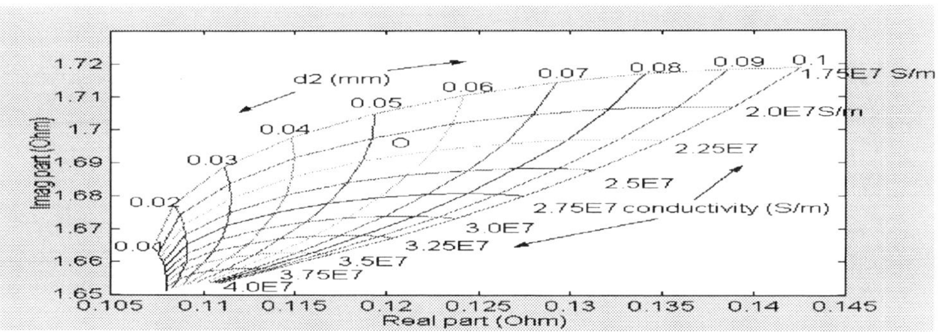

Fig. 8   Grid system for the estimation of surface properties

6. *Application of neural network model*

Instead of using the grid system the surface properties can be estimated with the help of neural network model. A schematic representation of a simple neural network model is shown in Fig. 9 in which the real and imaginary part of the impedance are used as the inputs. The conductivity and thickness (or lift-off) are considered as outputs. Matlab's neural network toolbox has been used for the solution. The data used for the generation of grid system are used to train the neural network. Depending on the situation the model can have two, three or even more number of outputs. A simple model has been developed and the results obtained from it are compared with that obtained utilizing grid system which is reported in [6] is shown in Table 4. It is seen that the level of error is consistent. The level of the error can be reduced using more data for training the network.

7. *Conclusions*

In this paper the possibility of using planar type mesh coil for the estimation of near surface material properties has been investigated. An analytical modeling has been developed from which the impedance of the coil is calculated. The impedance of the coil is used for the estimation of near-surface material properties. The approach of estimation by using a grid system has been described and an alternative to this approach is to use neural network has been proposed.

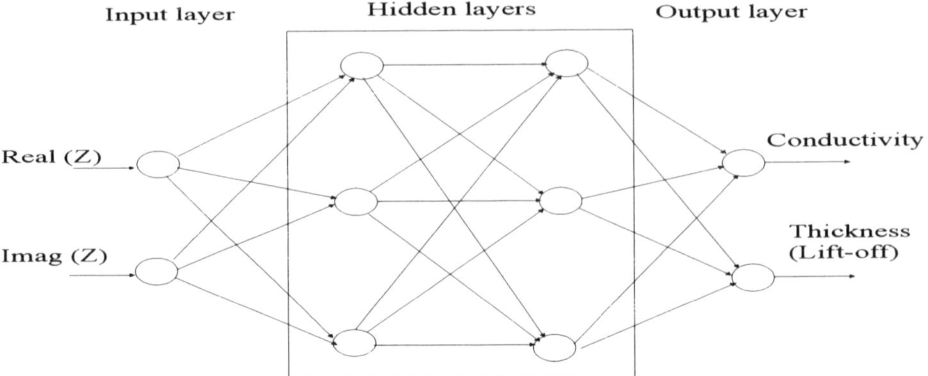

Fig. 9 Schematic representation of neural network model

Table 4 Comparison of results obtained using grid system and neural network

Conductivity of Aluminum	From Grid System (S/m)	Neural Network Model (S/m)
With No lift-off	3.6 E+7	3.53 E+7
With 0.1 mm lift-off	3.7 E+7	3.54 E+7
With 0.2 mm lift-off	3.8 E+7	3.53 E+7
With 0.3 mm lift-off	3.8 E+7	3.51 E+7
With 0.4 mm lift-off	3.4 E+7	3.49 E+7

References:

[1] Y.Shi and D.C.Jiles, "Finite element analysis of the influence of a fatigue crack on magnetic properties of steel", Journal of Applied Physics, vol. 83, no. 11, 1 June, 1998, pp. 6353-6355.

[2] N.J.Goldfine and D.Clark, "Near-surface material property profiling for determination of SCC susceptibility", 4th EPRI balance-of-plant heat exchanger NDE symposium, WY, June 1-2, 1996.

[3] N.J.Goldfine, "Magnetometers for improved material characterization in aerospace application", material evaluation, March 1993, pp. 396-405.

[4] N.J.Goldfine, D.Clark and T.Lovett, "Material characterization using model based meandering winding eddy current testing (MW-ET)", EPRI Topical workshop: Electromagnetic NDE applications in the Electric Power Inductry, Charlotte, NC, Aug 21-23, 1995.

[5] S.Yamada, H.Fujiki, M.Iwahara, S.C.Mukhopadhyay and F.P.Dawson, "Investigation of Printed Wiring Board Testing by using Planar Coil Type ECT Probe", IEEE transaction on magnetics, vol. 33, no. 5, Sep. 1997, pp. 3376-3378.

[6] S.C. Mukhopadhyay, S.Yamada and M.Iwahara, "Investigation of Near-Surface Material Properties Using Planar Type Meander Coil", presented in JAS, March 16-17, 2000, Australia, will be published in JSAEM Studies in Applied Electromagnetics and Mechanics.

Electromagnetic Nondestructive Evaluation (VI)
F. Kojima et al. (Eds.)
IOS Press, 2002

Nondestructive Detection of Inside Cracks and Delamination in Carbon-Fiber-Reinforced Plastics Using Superconducting Quantum Interference Device

Yoshimi HATSUKADE[1], Naoko KASAI[2], Hiroshi TAKASHIMA[2], Masayuki KUROSAWA[1], and Atsushi ISHIYAMA[1]
Waseda University[1], 3-4-1 Ohkubo, Shinjukuk-ku, Tokyo 169-8555, Japan
Nanoelectronics Research Inst., AIST[2], 1-1-1 Umezono, Tsukuba, Ibaraki 305-8568, Japan

Abstract. Carbon-fiber-reinforced plastic (CFRP) is a complex material, which is used in aircraft and space structures. The establishment of nondestructive evaluation (NDE) method for such structures is the urgent necessity. The potential of the NDE method using superconducting quantum interference device (SQUID) to detect inside defects in CFRP such as deep-lying cracks and inside shallow delamination was experimentally investigated using the current injection method. The results clarified that the method could be applied to such inside defects. The detection ability for the deep-lying delamination is discussed.

1. Introduction

Carbon-fiber-reinforced plastic (CFRP) is a complex material, which is recently used in many fields. Especially, the thick CFRP materials are used in aircraft and space structures. Due to increase in use of CFRP, the nondestructive evaluation (NDE) of mechanical reliability of CFRP becomes important. Moreover, the establishment of NDE method for such structures is the urgent necessity. Eddy current testing (ECT) using induction coil is a conventional NDE method for CFRP [1][2]. The ECT usually can be applied only to surface or subsurface defects due to the skin effect. However, cracks or micro cracks occur inside CFRP due to the pressure by external force. Other than cracks, delamination also often occurs inside CFRP by external impacts or during the process, since CFRP is a laminated composite. So new NDE method to detect such inside, particularly deep-lying, defects is required.

For such detection, low frequency current must flow in CFRP. Superconducting quantum interference device (SQUID) is a very high sensitive magnetic sensor even in the low frequency range [3]. So NDE method using SQUID (SQUID-NDE) has potential to detect deep-lying defects inside CFRP. Since the electric conductivity of CFRP is relatively lower than metals, a direct current injection method is more suitable to generate enough current density in CFRP than the eddy current method. So we adopted the current injection method in this study.

We constructed a SQUID-NDE system in a magnetically shielded room [4]. The SQUID-NDE system was applied to detect artificial cracks in thin CFRP plates using the current injection method [5]. For practical use of SQUID-NDE, a SQUID-NDE system is required to work in the normal environment. So we have constructed a new unshielded NDE system.

In this study, we have investigated the potential of the SQUID-NDE method to detect inside defects using the new system. Artificial deep-lying slots in CFRP planks and artificial shallow delamination in CFRP plate were used in the experiments. We discussed potential of this method for the detection of deep-lying delamination in CFRP plank.

2. CFRP Samples

Three 20-mm-thick CFRP planks with hidden slots were prepared as the crack sample. The width and length of each sample were 250 mm and 200 mm, respectively. Each sample has two slots. The slots were artificially made on the back surface of the samples. Each slot has a different depth; the depth means the distance between the sample surface and the slot. The depths of the slots are 5.0, 7.5, 10, 12.5, 15 and17.5 mm. The width and length of all the slots are 40 mm and 1 mm, respectively. The schematic figure of the crack sample is shown in Fig. 1(a).

A 2-mm-thick CFRP plate that included an insulator film inside the plate was prepared as delamination sample. The width and length of the sample were 100 mm and 200 mm, respectively. The insulator film made of polyethylene terephthalate (PET) was inserted between the central layers during the process. The film is in the shape of an ellipse. The long diameter, short diameter and the thickness of the film is 20 mm, 10 mm and 12 μm, respectively. The schematic figure of the delamination sample is shown in Fig. 1(b).

The carbon-fiber is of the cross texture type, and the packing factor of the carbon-fiber is 0.6 in this study.

3. SQUID-NDE System

The new SQUID-NDE system was constructed in a normal laboratory in National Institute of Advanced Industrial Science and Technology (AIST). The system is composed of a low temperature superconductor (LTS-) SQUID, SQUID electronics, a dewar with a 3-mm-thick bottom, a scanning stage with two stepper-motors, current supplier, lock-in amplifier and PC. The schematic diagram of the system is shown in Fig. 2. The sample is set on a table of the scanning stage, and moved under the SQUID. AC current is injected into the sample, and SQUID measures the magnetic field induced by the current flowing in the sample. The PC controls the stepper-motors through RS232C interface. The PC also controls the lock-in amplifier through GPIB interface. The SQUID output was gathered through the lock-in amplifier.

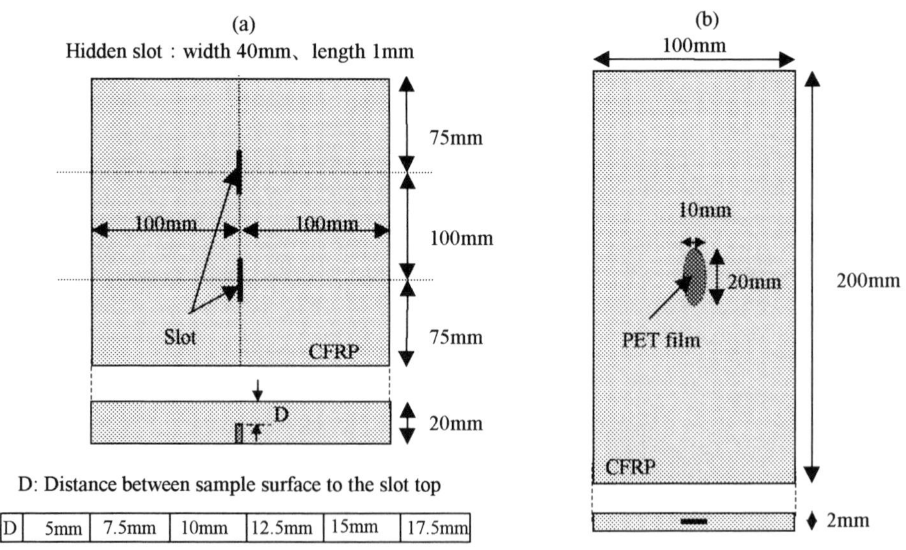

D	5mm	7.5mm	10mm	12.5mm	15mm	17.5mm

Fig. 1 Schematic figures of the samples. (a) Crack samples, (b) Delamination sample.

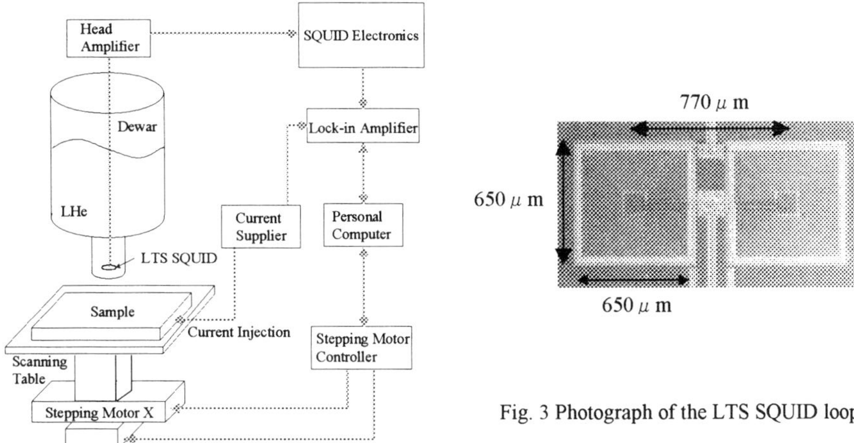

Fig. 3 Photograph of the LTS SQUID loop.

Fig.2 Diagram of the new unshielded system.

The amplitude and phase difference of measured magnetic field are stored in the PC. A data acquisition and stage-control program was developed using LabVIEW. The minimum distance between measuring plane and sample surface (standoff distance) is 3.5mm.

In this work, we used a small low-T_c SQUID loop that formed a first order gradiometeric loop in order to reduce the ambient magnetic noise [6]. Two washer coils of the SQUID loop are connected in series and reverse. The microscopic photograph of the SQUID is shown in Fig. 3. Two washer coils have the same dimension. The outer and the inner dimension of one single washer coil are 650 x 650 μm^2 and 50 x 50 μm^2, respectively. The distance between the centers of each washer coil (base line) is 770 μm. The magnetic flux noise of the SQUID measured by a flux locked loop (FLL) operation in the normal laboratory was about 5 x 10^{-6} $\Phi_0/\sqrt{}$ Hz (0.2 pT/m/$\sqrt{}$ Hz) in the white noise region (5Hz-1200Hz).

4. Method

The direct current injection method was adopted to induce current flow in the samples. For the crack samples, two electric terminals made of copper were prepared to inject current into the samples. The electric terminals were stuck on both ends of the sample surface by electrically conductive adhesive tape. AC current was injected into the sample from the one terminal and flowed out from the other terminal. The schematic view of the method is shown in Fig. 4(a).

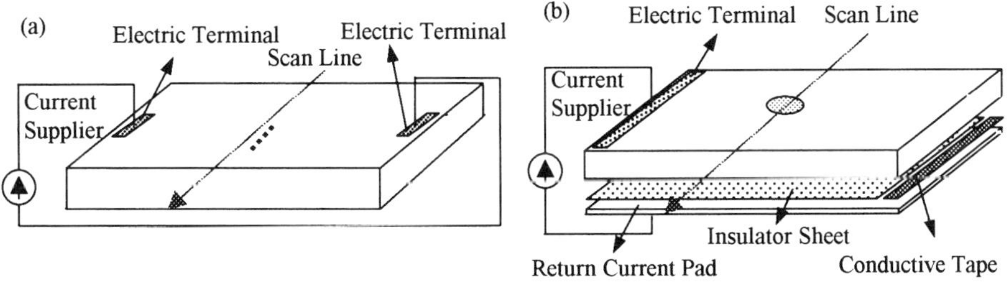

Fig.4 Injected current method; (a) without and (b) with a return current pad.

For the delamination sample, we used a thin copper pad for return-current to extract the anomalous magnetic signal due to the delamination in the sample [5]. The pad was stuck by the adhesive tape under the sample inserting a thin insulator sheet between the sample and the pad at one end of the sample. An electric terminal was also stuck on the sample surface at the opposite end. The schematic view of the method is shown in Fig. 4(b).

5. Measurements

In the detection of the slots, the crack sample was set on the scanning table so as to make the slot be parallel to the x-direction. Currents of 300 mA and 300 Hz were injected into the samples to induce current flow toward the y-direction. The distance between the sensor and the surface of the sample was set about 5.5 mm in the experiments. The SQUID was set to measure the first order gradient component in the x-direction of the z-component of the magnetic flux density, dB_z/dx. We measured the dB_z/dx along line parallel to the x-direction above the slots. The sampling space was 2 mm in the x-direction.

In the detection of the delamination, current of 60 mA and 700 Hz was injected into the sample to induce current flow toward the y-direction. The standoff distance was about 5.5 mm. First, the sample was scanned along line above the center of the delamination in the x-direction. Next, the area around the delamination was scanned. The sampling space was 1.5 mm in the x-direction and 6 mm in the y-direction.

6. Results

All results on the measurements of the crack samples are shown in Fig. 5 together. The section of the sample is also shown together. Anomalous downward signal peaks due to the slots appears above the slots. The signal amplitude due to the slot decreases as the depth of the slot increases. The slots below up to 10mm from the sample surface were successfully detected. We could not detect signal peaks due to the slot below 12.5mm.

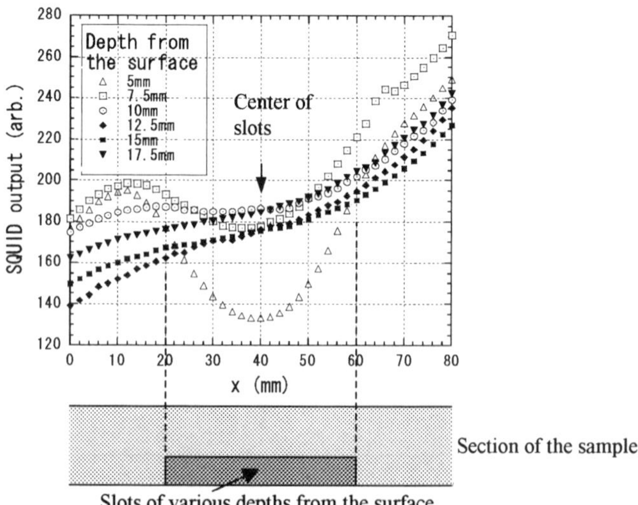

Fig. 5 All results on the hidden slots. The signal peaks due to slots below up to 10mm were successfully detected. The schematic shape of the slot is represented by dark square.

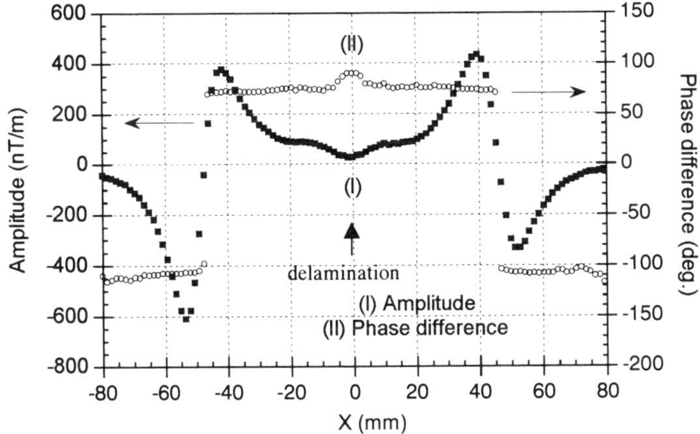

Fig.6 Result on the delamination. Signal peaks due to the delamination appear above the delamination in both curve (I) and (II). The upward and downward signals on both sides in the curve (I) and jumps in (II) are due to the edge effect.

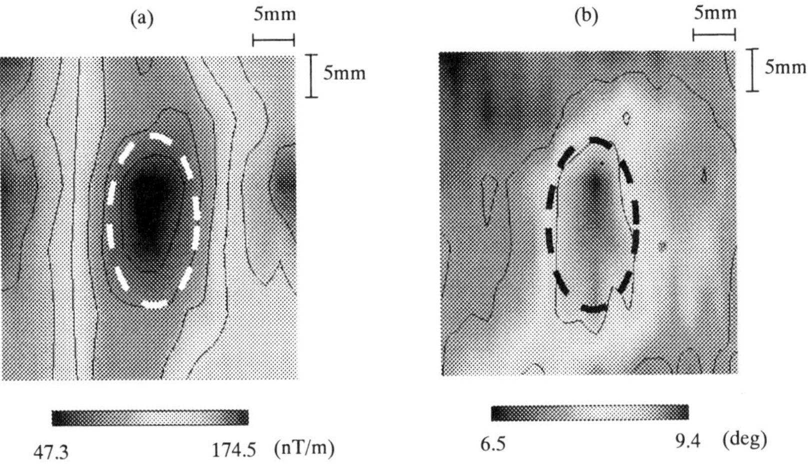

Fig. 7 Contour maps of magnetic signals dB_z/dx due to the delamination. (a) Amplitude, (b) Phase difference. The position and size of the PET film in the sample is shown together by perforated line in both figures.

The amplitude and the phase difference of the dB_z/dx measured along the line above the delamination are shown in Fig. 6 as a curve (I) and a curve (II), respectively. Signal peaks due to the delamination appeared above the delamination in both curves. The pair of large upward and downward peaks at the sample edges in (I) and the jumps on both sides in (II) are due to the edge effects. Figures 7 (a) and 7(b) show the contour maps of the amplitude and the phase difference measured around the delamination, respectively. The area where the edge effects appeared is excluded. The center peaks shown in Fig. 6 appear to form a shape of an ellipse in both figures. The position and size of these ellipses approximately agree with those of the PET film in the CFRP sample plate. These results show that the SQUID-NDE method can be applied to the detection of the delamination in CFRP.

7. Discussion

The potential of SQUID-NDE method to detect the inside shallow delamination in CFRP plate is experimentally clarified. We discuss the detection ability for deep-lying delamination.

The penetration depth and the sensitivity of a SQUID determine the detection ability for deep-lying defects. We roughly estimated the required frequency and amplitude of injection current and SQUID parameters to detect a delamination, which located 10mm below from the surface in a thick CFRP plank. In this case, injection current of lower frequency than 1200 Hz is necessary because of the skin effect. We experimentally investigated the frequency dependence of the magnetic field signal due to the delamination by measuring the delamination sample while changing the frequency of the injection current from 20 Hz to 1000 Hz. The amplitude of the signal exhibited little dependence on the frequency. However, the phase difference exhibited dependence as shown in Fig. 8. The phase displacement due to the delamination increases as the frequency increases. It shows that measuring the phase difference is effective in the frequency range above about 300 Hz. Then, current of near 1000 Hz is appropriate to detect the delamination below 10 mm using the phase difference.

The current density at 10 mm from sample surface reduces by a factor of e compared to that on the surface due to the skin effect while injecting the current of 1200 Hz into CFRP. If it is assumed that the disturbed current due to the delamination resemble to a current dipole, the signal peak amplitude of the dB_z/dx is in proportion to $1/h^3$, where h is the distance between the measuring plane and the delamination. Using the minimum standoff distance 3.5mm, the signal peak amplitude is reduced by a factor of about 10 compared to that in this study. So, the injection current of 8 A and 1200 Hz should be sufficient to detect the same size delamination located below 10 mm with same signal-to-noise ratio as in this study, using the amplitude of the magnetic field signal. In this estimation, we assumed a CFRP plank of 10 mm thickness and the same width as used in this study. The amplitude of the injection current is rather large and not so practical.

There is other way to reduce the amplitude of the current by enlarging the area of the pick-up coil. In this study, we used a small first order gradient SQUID loop because we only had that. However, it is better to use larger pick-up coil, for example 10mm x 10mm, to detect deep-lying delamination. The sensitivity of a SQUID with the large pick-up coil will increase about 40 times on assumption that the pick-up coil inductance is 5 nH and the input coil inductance is the Ketchen type and 5 nH [3]. Then the injection current of 200 mA and 1200 Hz is enough to detect the delamination below 10mm using both the amplitude and phase difference of the magnetic signal. The condition of the injection current is reasonable in practice. It suggests that the SQUID-NDE method has a possibility to detect the deep-lying delamination in thick CFRP materials.

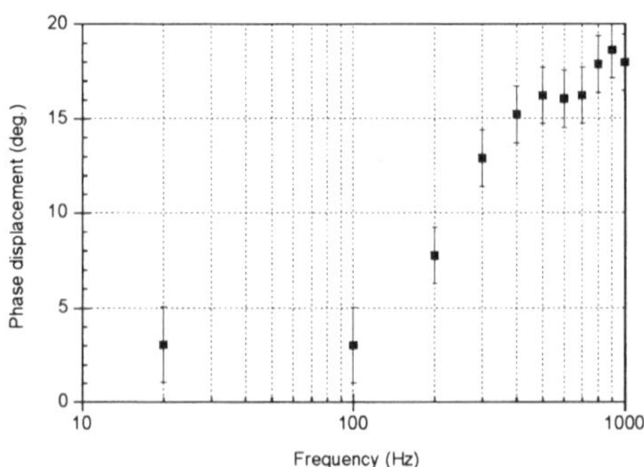

Fig. 8 Frequency dependence of the phase displacement due to the delamination.

The electric conductivity σ of CFRP is about 100 - 1000 times lower than metals such as copper or aluminium. The current injection method is easier method to induce much current flow in CFRP than a method with an induction coil. However in the practical NDE of aircraft and space structures, non-contact method is desired in many cases. We will develop the non-contact SQUID-NDE method for such cases. We have to resolve some problems with the standoff distance and imbalance of pick-up coils for the development.

8. Summary

We investigated the potential of non-destructively detecting inside defects in CFRP using an unshielded SQUID-NDE system with a low-T_c first order gradient SQUID loop. Artificial deep-lying hidden slots in 20-mm-thick CFRP planks and artificial 12-μm-thick delamination in 2-mm-thick CFRP plate were investigated using the current injection method. The slots that locate below 10 mm from sample surface were clearly detected by measuring the anomalous magnetic signals due to the slots. The position and size of the delamination was also successfully detected by mapping the amplitude and phase difference of magnetic field due to the delamination. We clarified that the SQUID-NDE method could be applied for inside and deep-lying defects in CFRP.

Acknowledgement

The authors would like to thank Dr. K. Chinone of Seiko Instruments Inc. for the low T_c SQUID.

References

[1] J. C. Treece, H. A. Sabbagh, Computed electromagnetic interactions with multi-layer advanced compositeds, IEEE Trans. Magnetics, Vol. 26, No.5, pp.1762-1764, 1990.

[2] H. A. S abbagh, L.D. S abbagh, T. M . Roberts, A n Eddy C urrent Model a nd Algorithm f or T hree-Dimensional N ondestructive E valuation o f A dvanced C omposites, I EEE T rans. M agnetics, Vol.24, No.6, pp.3201-3212, 1988.

[3] A. Barone *et al.*, Principles and Applications of Superconducting Quantum Interference Devices, World Scientific, Singapore, 1992.

[4] N. Kasai, D. Suzuki, H. Takashima, M. Koyanagi, and Y. Hatsukade, HTS-dcSQUID Gradiometer for Nondestructive Evaluation, IEEE Trans. Applied Supercond., Vol.9, No.2, 4393-4396, 1999.

[5] N. Kasai, D. Suzuki, Y. Hatsukade, and M. Koyanagi, Nondestructive Detection of Flaw in Carbon-Fiber-Reinforced Plastics Using High-T_c Superconducting Quantum Interference Device, Jpn. J. Appl. Phys. Vol.39, pp.1399-1404, Part 1, No. 3A, 2000.

[6] Y. Hatsukade, N. Kasai, H. Takashima, R. Kawai, F. Kojima, and A. Ishiyama, Development of an NDE Method Using SQUIDs for the Reconstruction of Defect Shape, to be published in IEEE Trans. Applied Supercond., 2001.

Electromagnetic Nondestructive Evaluation (VI)
F. Kojima et al. (Eds.)
IOS Press, 2002

Change of magnetic properties due to plastic deformation in $Ni_{50}Cu_{50}$ alloy

H. Satoh and S. Takahashi
Faculty of Engineering, Iwate University, 020-8551 Morioka, Japan

Abstract. The influence of plastic deformation on the magnetic properties has been studied in $Ni_{50}Cu_{50}$ alloy. The magnetism of this alloy before and after plastic deformation is analyzed by the self-consistent renormalization (SCR) theory of spin fluctuation. The spontaneous magnetization at 0K decreases considerably as a result of plastic deformation. At the same time the Curie temperature also becomes lowes. This phenomenon shows the existence of atomically ordered state in this alloy. The ferromagnetic state changes to paramagnetic along the antiphase boundary between superpartial dislocations even at 4.2K. The paramagnetic state increases in proportion to the dislocation density. The paramagnetic state extent in the long distance from the APB. The paramagnetic APB ribbon exerts an influence to the Curie temperature. The magnetism in the $Ni_{50}Cu_{50}$ alloy, including the plastically deformed ones, can be explained by the SCR theory of spin fluctuation. The dislocations are observed by the electron microscopy.

1 Introduction

Several remarkable magnetic transitions due to plastic deformation were discovered in intermetallic compounds and atomically ordered alloys. In the Ni_3Al intermetallic compound, for a example, the introduction of dislocations by plastic deformation causes the susceptibility as well as the Saturation magnetization to reduce considerably [1]. In Pt_3Fe ordered alloy, which is antiferromagnetic before plastic deformation, the application of plastic deformation causes it to become quite strongly ferromagnetic even at room temperature [2]. These magnetic transitions due to plastic deformation were explained from the viewpoint of the atomic rearrangement in the antiphase boundary (APB) between superpartial dislocations [3].

The magnetic transition due to plastic deformation includes several interests, which connect the fundamental problem of the ferromagnetic origin. The atomic arrangement in the APB is an important factor of the magnetic transition. We can get the valuable information on the relation between the magnetic ordered state and the atomic arrangement. The magnetic state of the APB ribbon enhances the neighbouring state and changes to the same state as the APB ribbon in the long range distance. The atomic arrangement of the APB ribbon is paramagnetic in Ni_3Al intermetallic compound and the paramagnetic region spreads around the APB ribbon at least as far as 100 atomic distances. An other example is Fe-Al intermetallic compounds, which are paramagnetic above 35at % Al concentration. The Fe-Al compounds change to ferromagnetic by plastic deformation [4]. The ferromagnetic APB ribbon changes neighbouring magnetism to ferromagnetic in the long distance and the extent depends on the Al concentration [5]. The magnetic extent suggests that two magnetic states exist near the APB ribbon for the same atomic structure in Ni-Cu alloy. The magnetic ordered state is explained by the two different models in general, the localized electron model and the itinerant electron model. It has been metallurgically believed that the Ni-Cu alloys form a substitutional solid solution with a face-centered cubic structure over the entire range of concentration, even in the stoichiometric composition. The atomically ordered state has not been ascertained experimentally in $Ni_{50}Cu_{50}$ alloy. It is difficult to indicate the atomic order-disorder change by X-ray and the electron diffraction, since the difference between atomic scattering factors of Ni and Cu atoms is too small. The existence of the atomically ordered state is controversial. The decrease of $M_S(0)$ and Curie temperature by plastic deformation in $Ni_{50}Cu_{50}$ alloy should be explained by the same model as the other ordered alloys and intermetallic compounds.

Our group has discovered a decrease in the values of $M_S(T)$ and T_C before and after plastic deformation in Ni-Cu alloy [6]. This peculiar phenomenon can be explained by the introduction of superpatial dislocations in Ni-Cu alloy. We confirmed it from the viewpoint of the existence of APB in order Ni-Cu alloy. In the present study, firstly, our interest is to examine the validity of the self-consistent renormalization (SCR) theory of spin fluctuation [7][8] and the change of $M_S(T)$, T_C due to plastic deformation. Secondly we aimed to examine the magnetic transition due to plastic deformation in view of the existence of APB before and after plastic deformation by magnetic influence.

2   Experimental Procedure

$Ni_{50}Cu_{50}$ alloy was prepared by cold crucible levitation melting at first and the alloy is melted by arc melting again to obtain chemical homogeneity on a water cooled copper hearth in an argon gas atmosphere at a pressure of approximately 93 kPa.

The buttons were homogenized at 1073K for 72 h and annealed at 653K for 120 h to get the atomically ordered state. We could not obtain higher uniformity by melting the samples many times in the arc furnace: the different values of magnetic properties were measured in the different parts of a button. We could prepare rather uniform composition by melting it in two processes.

The samples dimensions of approximately $2.5 \times 2.5 \times 5.0mm^3$ were cut from the button by Servomet spark cutting machine. The samples were tested in compression at room temperature by Instron-type machine. Samples for the magnetic measurement dimensions of $2.5 \times 2.5 \times 2.5mm^3$ were cut from the spark cutting machine and were chemically polished to remove surface damage. The steady-field magnetization was measured with a SQUID magnetic fluxmeter(Quantum Design) in the temperature range of 4.5K to 100K.

3   Experimental results

The magnetization curves of $Ni_{50}Cu_{50}$ alloy with different plastic strains were measured in the temperature range of 4.5 to 100K. Figure1 shows the isothermal magnetization curves for the samples with plastic strain $\varepsilon =0$ %. The $Ni_{50}Cu_{50}$ alloy is ferromagnetic at low temperatures and the spontaneous magnetization at 4.5K decreases slightly with plastic deformation. The spontaneous magnetization, $M_S(0)$, at 0K and the Curie temperature T_C have been obtained from the magnetization curves. The analysis of $M_S(0)$ and T_C is given as follows: Figure2 shows that the data fit nicely on straight lines in $M_S(T)^2$ versus H/M in $Ni_{50}Cu_{50}$

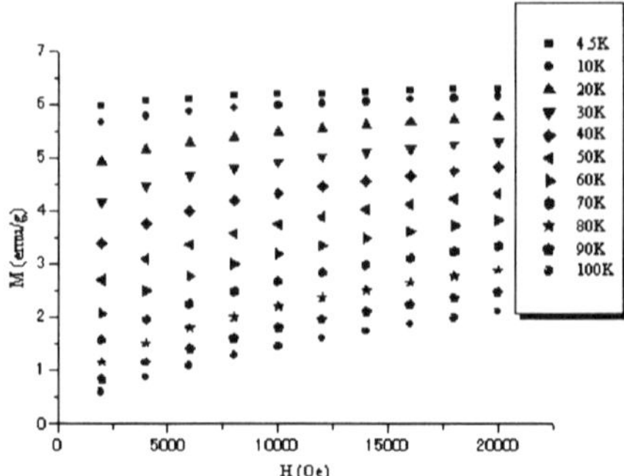

Figure 1: Magnetic curve

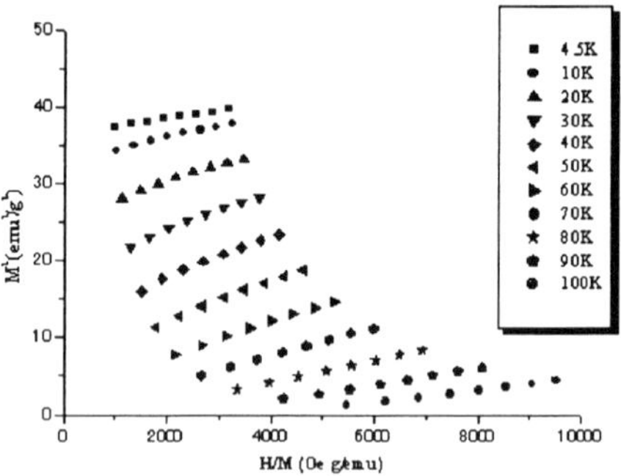

Figure 2: Arrotplot

alloy, which are commonly used in T_C (Arrot plots). The data for all the plastically deformed samples were represented well in the form of $M(H,T)^2 = M_S(T) + bH/M(H,T)$ at all temperature, except in the region of weak magnetic field. The spontaneous magnetization at TK, $M_S(T)$, can be obtained by the extrapolation of linear relation to H=0. The Curie temperature T_C shows a slight decrease with plastic deformation. The spontaneous magnetization, $M_S(T)$, in the plastically deformed $Ni_{50}Cu_{50}$ alloy is analyzed was the SCR theory of spin fluctuation. In the SCR theory the temperature dependence of $M_S(T)$ has the form of $M_S(T)^2 = M(0)^2 - \eta T^2$ at low temperatures and $M_S(T)^2 = \zeta(T_C^{4/3} - T^{4/3})$ in a fairly wide temperature range below T_C. Figure3 shows the temperature dependence of the square of the spontaneous magnetization $M_S(T)^2$ against T^2 in the $Ni_{50}Cu_{50}$ samples with $\varepsilon = 0 \sim$ 30 % strain. All the samples have the simple relationship that $M_S(0)^2$ is proportional to T^2 at low temperature. The spontaneous magnetization at T=0K, $M_S(0)$, is determined by the extrapolation of this linear relation of $M_S(T)^2$ against T^2 to T=0. The values of $M_S(0)$ for $Ni_{50}Cu_{50}$ alloy are given in table1. Figure4 shows the temperature dependence of $M_S(0)^2$

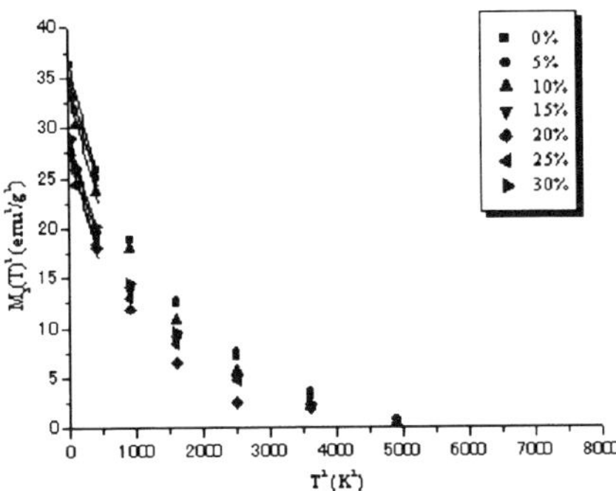

Figure 3: The temperature dependence of the spontaneous magnetization in $M_S(0)^2$ against T^2

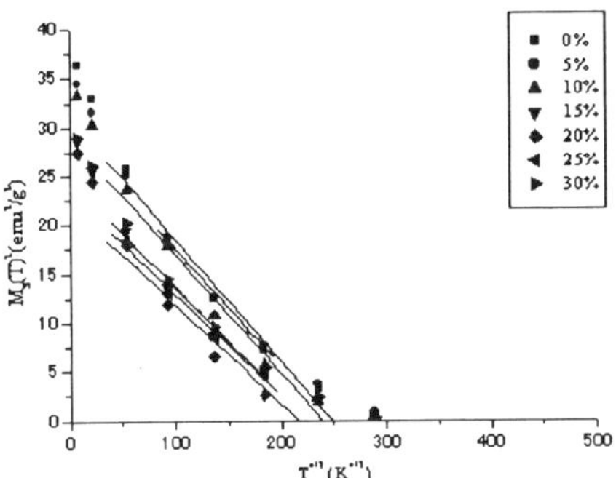

Figure 4: The temperature dependence of the spontaneous magnetization in $M_S(0)^2$ against $T^{4/3}$

on $T^{4/3}$ in $Ni_{50}Cu_{50}$ alloy with ε =0~30 %. $M_S(T)^2$ has a $T^{4/3}$ dependence in the wide temperature range below T_C for each sample. The value of the Curie temperature is obtained by the extrapolation of the linear relation to $M_S(T) = 0$. The value of T_C is nearly the same as that obtained through the Arrot plots. The valued of T_C are shown in table1. The temperature dependence of $M_S(T)$ of $Ni_{50}Cu_{50}$ alloy including the plastically deformed ones, is expressed well in the form of

$$M_S(T)^2 = M(0)^2 - \eta T^2 \tag{1}$$

at low temperature and

$$M_S(T)^2 = \zeta(T_C^{4/3} - T^{4/3}) \tag{2}$$

in a wide temperature range below T_C.

The values of η and ζ are given in table1. The strain dependence of $M_S(0)$ and T_C is shown in figure5 and 6, respectively. $M_S(0)$ and T_C decrease as the strain increases, though

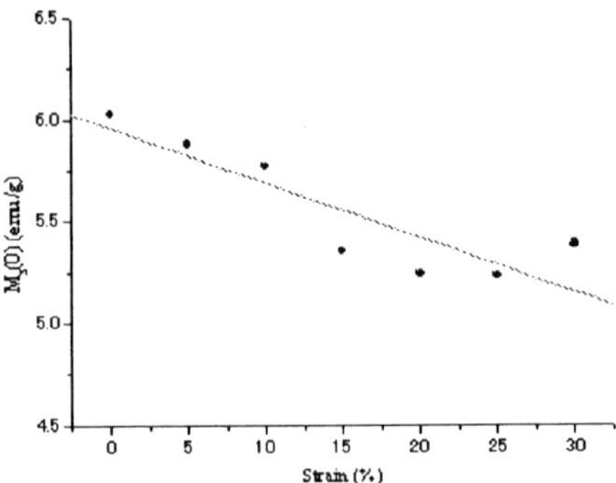

Figure 5: The temperature dependence of the spontaneous magnetization in $M_S(0)^2$ against T^2

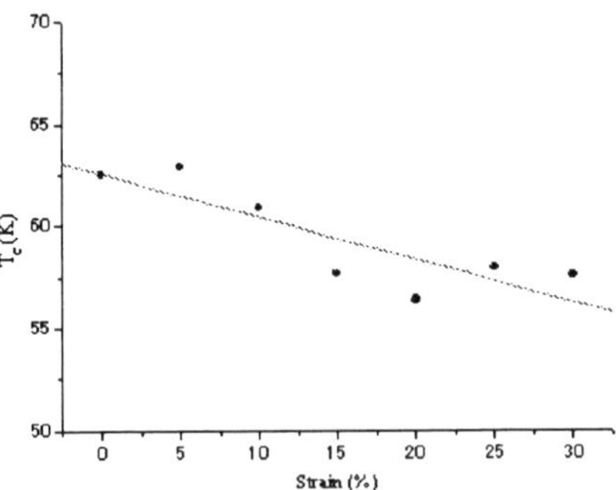

Figure 6: The temperature dependence of the spontaneous magnetization in $M_S(0)^2$ against $T^{4/3}$

some scattering exists. The scattering would be caused by the lack of uniformity of the samples.

4 Discussion

The experimental results indicated that the magnetism changed remarkably, In particuler the spontaneous magnetization and Curie temperature decrease as the strain increases. The ordered structure is $L1_0$-type in Ni-Cu alloy. The change of magnetism was explained according to the formation of APB in order-state ($L1_0$-type) by plastic deformation. It changes from ferromagnetic to paramagnetic to destroy the cycle of atomic arrangement. The experimental results of temperature dependence of $M_S(T)$ are consistent with the results of the SCR theory of spin fluctuations in $Ni_{50}Cu_{50}$ alloy with plastic deformation. According to the SCR theory

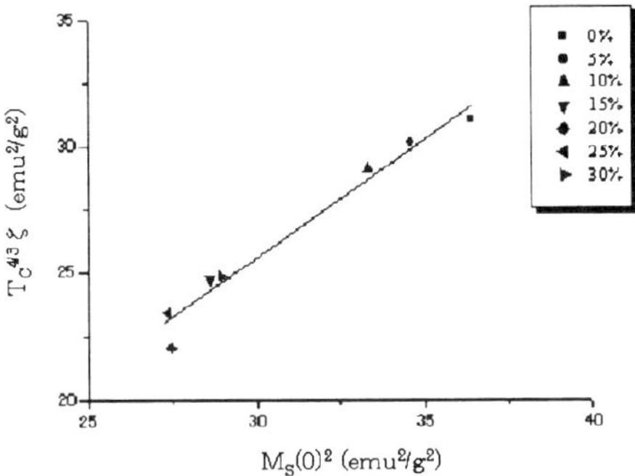

Figure 7: ζ is the coefficient of the $T_C^{4/3}$ term in the $M_S(0)^2$ versus $T^{4/3}\zeta$ plot

of spin fluctuation, the temperature dependence of $M_S(T)$ is given by

$$M_S(T)^2 \propto \Gamma_1/R(T_C^{4/3} - T^{4/3}) \tag{3}$$

where Γ_1 and R are parameters depending on the band structure, and the Curie temperature is represented as

$$T_C^{4/3} \propto (\alpha - 1)/\Gamma_1 \tag{4}$$

where $\alpha = IN(EF)$, i.e. the product of the intra-atomic exchange constant, I, and the density of states, $N(EF)$ at the Fermi level, EF. As both $M_S(0)$ and T_C decreases with plastic deformation, it seems a plausible conclusion that the factor $(\alpha - 1)$ decreases and the parameter R increases with plastic deformation, keeping Γ_1 nearly constant.ζ in equation (3) corresponds to Γ_1/R. ζ decreases with increasing plastic strain in table1. The increase of R is consistent with the decrease of ζ. The increase of R due to plastic deformation and the constant Γ_1 are obtained in $Ni_{75+X}Al_{25-X}$ intermetallic compounds. As the T_C decreases with plastic deformation, $\alpha = IN(EF)$ decreases as ϵ increases. R can be represented by $N(EF)$ and the first and second derivative of $N(EF)$ by.

$$R = \dot{N}(E_F)^2/N(E_F)^4 - \ddot{N}(E_F)/3N(E_F)^3 \tag{5}$$

Plastic deformation would exert some influence on the derivative of $N(EF)$ as well as $N(EF)$. The decrease of α is consistent with the increase of R.

It gave experimentally a simple relationship between T_C and $M_S(0)$ in $Ni_{75+X}Al_{25-X}$ compounds, i.e.

$$T_C^{4/3}\zeta = KM_S(0)^2 \tag{6}$$

where K is a constant. Equation (6) can be obtained by supposing that $(\alpha - 1)/R = (\alpha_0 - 1)/R_0$, where α_0 and R_0 are the values at $T = 0K$. The relationship of equation (6) has been explained in the present study. Figure7 shows the relationship between $T_C^{4/3}\zeta$ and $M_S(0)^2$. The relationship indicates that the magnetism in the $Ni_{50}Cu_{50}$ alloy, including the plastically deformed ones, can be explained by the SCR theory of spin fluctuation. The values of $M_S(0)$ are a function of the number of atomic configurations in the vicinity of APB, which increases proportionally to the dislocation density. The change of the spontaneous magnetization due to plastic deformation can be written as

$$M_S(0) = (N_0 - N_1)\mu_{Ni} \tag{7}$$

where N_0 and μNi are the total number of Ni atoms and their magnetic moment before plastic deformation, respectively. N_1 is the number of Ni atom near the APB. N_1 can be written as

$$N_1 = CnS^2/a^2\bar{r}\rho \tag{8}$$

where $r\rho$ is the size of APB ribbon between superpartill dislocations. The number of these atomic moment increases as the strain increases. This shows the relationship of size of APB ribbon. The relationship between spontaneous magnetization and dislocation density is shown theoretically in comparison with the experimental results. The value of dislocation density has not been obtained by direct method such as the electron microscope. We can estimate the dislocation density of the present samples theoretically to be $\rho = 10^{12}cm^{-2}$. The number of Ni atoms, which carry no moment near the APB, is ten times of that constituting the APB. The different atomic configuration from the ordered state is restricted in the APB ribbon. There exist paramagnetic and ferromagnetic states for the same atomic configuration near the APB ribbon. The two different states for the same atomic structure near the APB were confirmed in the other alloys and intermetallic compounds.

Thus, there exist paramagnetic state near the APB ribbons in $Ni_{50}Cu_{50}$ alloy. The relationships between the spontaneous magnetization and dislocation are depending on localized electron model. The exchange interaction in long distance cannot be explained in Ni-Cu alloy. Further, the itinerent electron model of Ni-Cu alloy cannot be adapted directly. The results are consistent with SCR theory.

References

[1] De Boer FR,Shinkel CJ,Biesterbos J and Proost S 1969 J.Appl.Phys.40 1049

[2] Crangle J 1959 J.Phys.Radium 20 435

[3] Takahashi S and Ikeda K 1983 Phys.Rev.B 28 5225

[4] Taylor A and Jones RM 1958 J.Phys.Chem.Solids 6 16

[5] Takahashi S,Umakoshi Y 1991 J Phys Condens Matter 3 5805

[6] Takahashi S,Li XG,Chiba A 1996 J Phys Condens Matter 8 5101

[7] Moriya T,Kawabata A 1973 J.Phys.Soc.Japan 34 639;35 669

[8] Sasakura H,Suzuki K,Masuda Y 1984 J Phys Soc Japan 53 754

Electromagnetic Nondestructive Evaluation (VI)
F. Kojima et al. (Eds.)
IOS Press, 2002

Optimization of Pipeline Inspection Tool using Taguchi Method

Zhiwei Zeng, Pradeep Ramuhalli, Lalita Udpa and Satish Udpa
Material Assessment Research Group, Dept. of Electrical and Computer Engineering
Iowa State University, Ames, IA 50010, USA

Abstract: Taguchi analysis is a well-established technique for optimizing the design parameters in experiments. In pipeline inspection this technique can be used to optimize the inspection tool. This paper first introduces the basic idea underlying the Taguchi method. The factors and levels being considered in the Taguchi study are also described. Procedures for designing and simulating the experiments are given along with simulation results, which are analyzed using the main effects and ANOVA. Conclusive remarks of the analysis are finally summarized.

1. Introduction

Natural gas transmission pipeline inspection employs the magnetic flux leakage (MFL) method, where the pipe wall is magnetized axially by permanent magnets and leakage fields due to defects are detected by an array of Hall sensors. Two major parameters in this experiment that affect the detection capability are magnetization level and sensor spacing in the array. A systematic way of quantifying the influence of these factors and optimizing the tool design is provided by the Taguchi method.

In pipeline inspection, the design quality characteristic of the inspection tool is defined by the probability of detection (POD) of a critical defect. Thus for the tool to be optimum the POD should be as large as possible. The optimum condition can be found by conducting experiments (simulations) and determining the contribution of each experimental factor to the POD of critical defect. The study presented in this paper follows the standard approaches described in [1] and [2]. A model based POD evaluation method [3] is employed in calculating PODs.

2. Taguchi Method

The Taguchi method is a statistical analysis technique that is used in quality improvement and design of experiments. In this technique, Taguchi simplifies the conventional statistical tools by identifying a set of stringent guidelines for experiment layout and analysis of results. The approach ensures quality by optimizing the design of product/process and making the design insensitive to influence of uncontrollable factors (robustness).

The Taguchi method for experiment design minimizes the number of experiments that need to be performed to obtain the most information by employing specially constructed tables known as orthogonal arrays (OA) [1], which makes the design of experiments easy and consistent.

Table 1 shows a standard OA $L_4(2^3)$. This OA is designed for experiments with three factors and two levels of each factor. The number of trials using this OA is four. In Table 1, each column contains two level 1 and two level 2 conditions for the factor assigned to the

Table 1 Orthogonal Array $L_4(2^3)$

Trial \ Column	$L_4(2^3)$		
	1	2	3
1	1	1	1
2	1	2	2
3	2	1	2
4	2	2	1

column. Two 2-level factors can be combined in four possible ways, (1,1), (1,2), (2,1), and (2,2). When any two columns of an array contain these combinations the same number of times, the columns are said to be orthogonal or balanced. For example, column 1 and column 2 in Table 1 have all the above combinations once. This also holds for the other pairs of columns.

Uncontrollable factors (noise factors) are included in a second OA (outer array), which is used in conjunction with the array of controllable factors (inner array).

The Taguchi approach involves listing the factors and the levels of each factor that need to be considered, designing the experiment, conducting the experiment, analyzing the results, and determining the optimum condition and effects of each factor. A $0.5''\times0.5''$ rectangular defect, of 15% depth and 45° surface angle, is used as the critical defect for POD calculations, in this Taguchi study.

3. Factors and Levels

Four factors with two levels each were considered in this study. Based on prior experience, magnetization level, sensor spacing, radial liftoff, and circumferential shift between sensor and defect have obvious influences on the measured signal. The factors can be classified into controllable or uncontrollable (noise) factors, depending on whether or not its level can be controlled.

Magnetization level is a controllable factor. In pipeline inspection, permanent magnets are used to establish magnetic fields in the pipe wall. Magnetization level, denoted by B_r, is a design parameter characterizing the strength of the permanent magnets. Intuitively, the MFL signal at a high magnetization level is stronger than that obtained at low magnetization level. The two magnetization levels used in this Taguchi analysis are 0.8T and 1.2T.

Radial Liftoff is defined as the distance from a sensor to the inner surface of the pipe wall. It is an uncontrollable factor. When the liftoff distance increases, the MFL signal decreases monotonically. Radial liftoff is a random variable whose level represents its mean value. The two levels selected are $0.05''$ and $0.1''$.

Sensor spacing in the circumferential sensor array is a controllable factor. Ideally one would wish to reduce the number of sensors used in the tool without compromising the POD of a critical flaw. Circumferential shift, identified as the lateral or circumferential distance between sensor and defect, is an uncontrollable factor. A combination of sensor spacing and circumferential shift is used to determine the positions of sensors in the circumferential direction, as shown in Figure 1. In the FEM prdeiction, signal is discrete corresponding to a sensor spacing S_0, specified by the element size in the circumferential direction, and the maximum circumferential shift d equals $S_0/2$. Signal magnitudes for different circumferential shifts can be obtained by interpolating the FEM signal. A Gaussian distribution of circumferential shifts with an assumed mean (level) chosen as a fraction of the sensor spacing is used to calculate the POD.

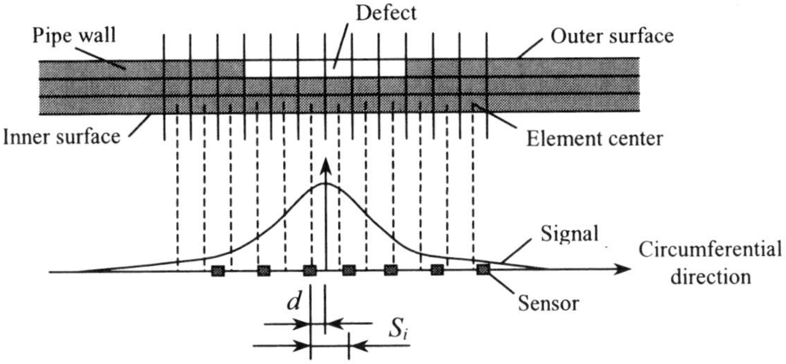

Fig 1 Illustration of Sensor Spacing (S_i) and circumferential shift (d)

Table 2 Factors and levels

Factors		Levels	
		Level 1	Level 2
Controllable	Magnetization level (A)	0.8T	1.2T
	Sensor spacing (B)	1″	2″
Uncontrollable	Radial liftoff	0.05″	0.1″
	Circumferential shift	0	25%

Table 3 Experiment design and results

Outer Array

Noise Factors	Col	1 (1 1 0)	2 (1 2 0)	3 (2 1 0)	4 (2 2 0)	
Inner Array						
Controllable Factors		Results (PODs)				
Trial — Column	1 2 3	1	2	3	4	S/N
1	1 1 0	0.90141	0.88436	0.77364	0.74519	-1.7471
2	1 2 0	0.90143	0.80468	0.76795	0.66378	-2.2662
3	2 1 0	0.99867	0.99743	0.98491	0.97839	-0.0896
4	2 2 0	0.99857	0.98534	0.98492	0.93857	-0.2107

Two levels of sensor spacing, 1″ and 2″, and two levels of circumferential shifts, 0 and $S_0/4$, are used in the Taguchi study. The factors and levels are summarized in Table 2.

4. Experiment Design and Results

Table 3 shows the inner and outer orthogonal arrays used in the Taguchi study. Note that the last columns of both the inner array and the outer arrays in Table 3 are not used, because we have only two controllable and two uncontrollable factors. With the factors and

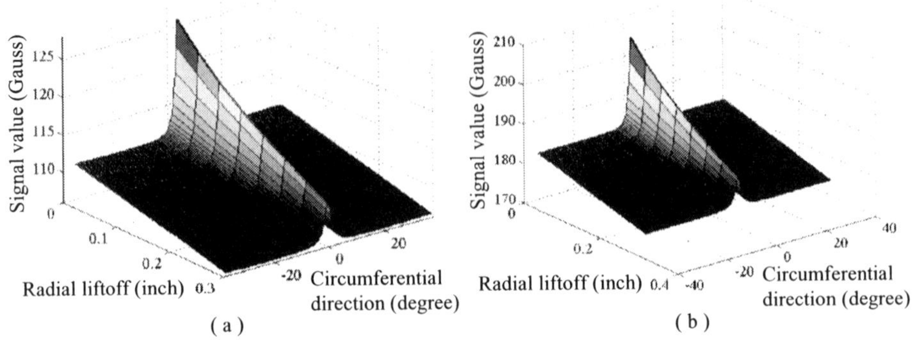

Fig 2 2-D signals in the surface perpendicular to the axial direction at the center of the flaw
(a) B_r=0.8T (b) B_r=1.2T

levels chosen as shown in Table 2, the number of experiments is the same as that of full factorial design. However, the method will be more efficient if more factors and/or more levels are considered in future research.

To obtain complete data, each trial run of the inner array must be repeated for each of the four noise combinations (in outer array). That is, the experiments consist of the computation of POD for each combination of magnetization level, sensor spacing, and distributions of radial liftoffs and circumferential shifts.

FEM simulations are performed to predict signals. With each simulation of the FEM at a particular magnetization level, we obtain a signal in the 3-D space (radial, circumferential, and axial). Since we are interested in the peak value of the signal, we can consider only the signal at the center (in the axial direction) of the flaw. The 2-D signals perpendicular to the axial direction with axial coordinate at the center of the flaw obtained with B_r=0.8T and 1.2T are shown in Figures 2 (a) and (b).

For each run of the inspection tool, we have a random radial liftoff and a random circumferential shift. In order to account for the effects of the uncontrollable factors, multiple simulations are performed assuming Gaussian distributions for radial liftoffs and circumferential shifts. Using 2-D interpolation, we can find signal magnitudes at the corresponding radial liftoffs and circumferential shifts, and thus find the probability density function of the signal magnitudes.

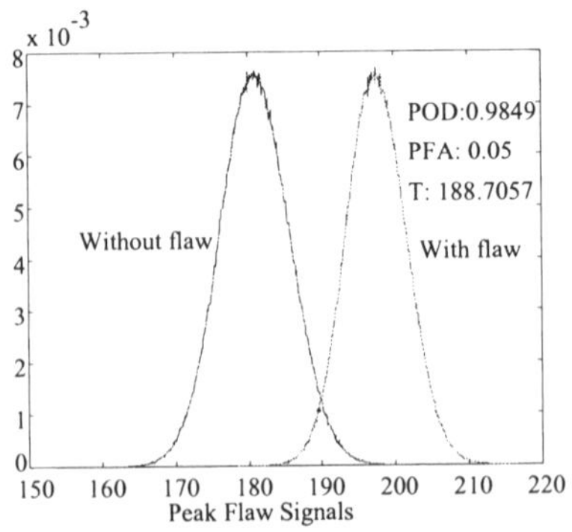

Fig 3 pdf of signal magnitudes with and without flaw
when B_r=1.2T, sensor spacing is 1″, mean radial liftoff equals 0.1″,
and zero mean circumferential shift

Besides random variations in radial liftoff and circumferential shift, there is also measurement noise present in the signal, which is independent of any of the four factors being considered. This common measurement noise should be taken into account when calculating the 16 PODs. Typically, we assume the noise to be Gaussian with zero mean and variance 25 Gauss.

Conducting each of the 16 simulations in Table 3, pdf's of measurement outputs with and without flaw were found. Figure 3 presents the pdf's of measurement outputs when B_r = 1.2T, sensor spacing = 1″, mean radial liftoff = 0.1″, and mean circumferential shift = 0. The POD and PFA are calculated as

$$POD = \int_T^\infty p(y \mid x_1)dy \qquad (1)$$

and

$$PFA = \int_T^\infty p(y \mid x_0)dy \qquad (2)$$

where T is the threshold, $p(y \mid x_1)$ and $p(y \mid x_0)$ are the pdf's of signal magnitudes with and without a flaw respectively. The threshold was selected so that the PFA is 0.05. All 16 PODs are shown in Table 3.

5. Analyses

5.1 Objectives and Procedures

The results of the Taguchi experiments were analyzed using the standard steps. First, the factorial effects (main effects) were evaluated and the influences of the factors were determined in qualitative terms. The optimum design parameters and the performance under this condition were also determined from the factorial effects. In the next phase, analysis of variance (ANOVA) was performed to identify the relative influence of the factors.

5.2 S/N Analysis

The signal to noise ratio measures the sensitivity of the quality characteristic investigated in a controlled manner, with respect to uncontrolled (noise factors). The S/N ratio is defined as

$$S/N = -10\log_{10}(MSD) \qquad (3)$$

where MSD is mean squared deviation from the target value of the quality characteristic. Consistent with its application in engineering and science, the value of S/N is desired to be large; hence the value of MSD should be small. In our case, we want the quality characteristic to be as large as possible. In this case, the MSD is defined as

$$MSD = \left(1/y_1^2 + 1/y_2^2 + 1/y_3^2 + \cdots\right)/n \qquad (4)$$

In (4), y_i is the result of the ith repetition, and n is the number of repetitions. The S/N ratio calculation is based on data from all observations of a trial condition. The set of S/N ratios can then be considered as trial results without repetitions. The four S/N ratios under each trial condition (controllable) are shown in Table 3.

Table 4. Main Effects for S/N analysis

Factors	Level 1 (d_1)	Level 2 (d_2)	d_2-d_1
Factor A	-2.007	-0.150	1.857
Factor B	-0.918	-1.238	-0.320

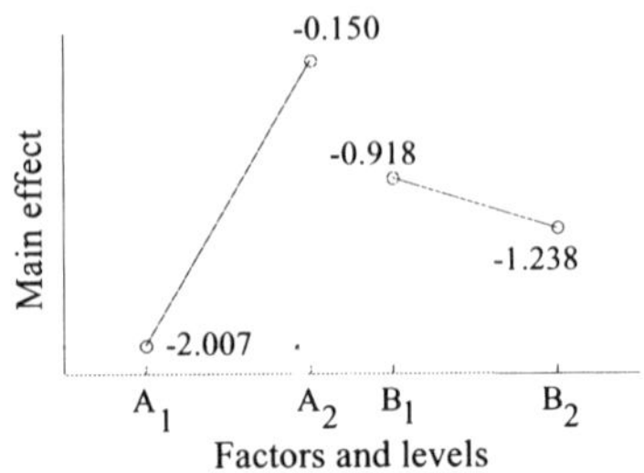

Fig 4 Main effects for S/N analysis

5.3 Main Effects

In Taguchi analysis, the optimum condition is identified by studying the main effects of each of the factors. The main effects indicate the general trend of the influence of the factors.

To compute the main effect of each factor at each level, we add results (S/N ratios) for trials including the factor at this level, and then divide by the number of such trials. For example, to compute the main effect A_1 of Factor A (magnetization level) at level 1 (0.8T), we look in the column for A in the inner array in Table 3 and find that level 1 occurs in experiment numbers 1 and 2. Thus the main effect of A_1 is

$$\overline{A}_1 = \frac{1}{2}\left((S/N)_1 + (S/N)_2\right) = \frac{1}{2}\left((-1.7471) + (-2.2662)\right) = -2.007 \qquad (5)$$

The average effects of other factors are computed in a similar manner and listed in Table 4 and plotted in Figure 4.

From these results, we can conclude that in the four trials studied the combination A_2B_1 leads to maximum POD, and Factor A (magnetization level) has more influence than Factor B (sensor spacing) on the POD.

5.4 ANOVA Study

The analysis of variance (ANOVA) is a statistical technique most commonly applied to the results of experiments for determining the percent contribution of each factor. Study of the ANOVA table for a given analysis helps determine which of the factors need control and which do not. The following analyses follow the standard steps described in [1].

The first important concept in ANOVA is sum of squares. The total sum of squares is expressed as

$$S_T = \sum_{i=1}^{N}(S/N)_i^2 - \frac{T^2}{N} = 3.589 \qquad (6)$$

where $(S/N)_i$ is the S/N ratio of the ith trial; N is the number of trials; $T = \sum_{i=1}^{N}(S/N)_i$ is the summation of all results. The term T^2/N in (6) is called the correction factor, C.F. The sum of squares for a factor (e.g. Factor A) can be calculated from:

$$S_A = \sum_{k=1}^{L} \frac{1}{n_k}\left(\overline{A_k} \cdot n_k\right)^2 - \frac{T^2}{N} \tag{7}$$

where L is the number of levels; n_k is the number of trials for Factor A at level k; $\overline{A_k}$ is the main effect of Factor A at level k and hence $\left(\overline{A_k} \cdot n_k\right)$ is the summation of S/N ratios when the level of Factor A is k. Subtracting all the factor sums of squares from the total sum of squares, the residual is referred to as error sum of squares, expressed as

$$S_e = S_T - S_A - S_B \tag{8}$$

Degree of freedom (DOF) is an important and useful concept that is defined as a measure of the amount of information that can be uniquely determined from a given set of data. DOF of a factor equals one less than the number of levels. For a factor A with two levels, A_1 can be compared with only A_2, not with itself. Thus a two level factor has DOF=1, i.e., $f_A = 1$. DOF of an experiment (total DOF or f_T) is one less than the number of trials. DOF of the error term is given by

$$f_e = f_T - f_A - f_B \tag{9}$$

Variance measures the distribution of the data about the target value. Since the data is representative of only a part of all possible data, DOF rather than the number of observations is used in the calculation.

$$\text{Variance} = \frac{\text{Sum of Squares}}{\text{Degrees of Freedom}} \quad \text{or } V = \frac{S}{f} \tag{10}$$

The variance ratio, commonly called the F statistic, is the ratio of variance due to the effect of a factor to the variance due to the error term. This ratio is used to measure the significance of the factor under investigation with respect to the variance of all factors included in the error term. The F values obtained in this study are F_A=87.055, F_B=2.589, F_e=1. Assuming a confidence level of 90%, the F values determined from the F-table (available in most statistical handbooks) for both Factors A and B are $F_{0.1}(1,1)$=39.864, where the two parameters in the parentheses are DOF of Factor A and DOF of error term, respectively. Since $F_A > F_{0.1}(1,1) > F_B$, we say that Factor A (magnetization level) does contribute to the sum of squares at the confidence level of 90%, while Factor B (sensor spacing) does not contribute to the sum of squares at this confidence level.

The pure sum of squares is defined as the sum minus the degrees of freedom times the error variance:

$$S_A' = S_A - f_A \cdot V_e \tag{11}$$

$$S_e' = S_e + \left(f_A + f_B\right) \cdot V_e \tag{12}$$

The percent contribution for any factor is obtained by dividing the pure sum of squares for that factor by S_T and multiplying the result by 100.

Results of the above analyses are listed in Table 5 under the part "before pooling", where f, S, V, F, and P stand for DOF, sum of squares, variance, variance ratio, and percent contribution, respectively.

When the contribution of a factor is small, the sum of squares for that factor is combined with the error S_e. This process of disregarding the contribution of a selected factor and subsequently adjusting the contributions of other factors is known as pooling. In Table 5, we find that the percent contribution of Factor B (sensor spacing) is only 1.753%. Thus Factor B should be pooled. The new error variance is computed as:

Table 5. ANOVA Table

Factor	Before pooling					After pooling				
	f	S	V	F	P (%)	f	S	V	F	P (%)
A	1	3.45	3.45	87.06	94.94	1	3.45	3.45	48.52	94.06
B	1	0.10	0.10	2.59	1.75	(1)	(0.10)	Pooled		
error	1	0.04	0.04		3.31	2	0.14	0.07		5.94
Total	3	3.59			100.00	3	3.59			100.00

$$V_e = \frac{S_B + S_e}{f_B + f_e} \tag{13}$$

With a pooled V_e, all S' values must be modified to reflect pooling. The new analysis results with Factor B being pooled are also listed in Table 5. The percent contribution of Factor A is now 94.061%. The contribution of errors is 5.94%.

6. Conclusions

The Taguchi method is an effective way of optimizing an experiment setup. This study considered two factors each with two levels and, the optimum combination A_2B_1, that is, 1.2T magnetization level and $1''$ sensor spacing was determined to be optimal with respective to the probability of detecting a critical flaw.

Analyzing the variance ratios, we find that the magnetization level contributes to the sum of squares within 90% confidence level, while sensor spacing does not.

In the ANOVA table without pooling, the percent contribution of sensor spacing is considered to be too small and is therefore pooled. The percent contribution of magnetization level is 94.061% after pooling.

These results were obtained by considering a $.5''\times.5''\times15\%$ deep critical defect. This defect was chosen on the basis of the smallest element in the FE mesh. If the critical defect is smaller, the sensor spacing will have a greater effect on the POD. Also considering more factors and levels can increase the effectiveness of the study.

References

[1] R. K. Roy, "A primer on the Taguchi method", Van Nostrand Reinhold, 1990.
[2] K. Dehnad, "Quality control, robust design, and the Taguchi method", Pacific Grove, Calif., 1989.
[3] Z. Zhang, Y. Zhang, L. Udpa, S. S. Udpa, "Probability of detection model for Pipeline inspection", QNDE, Vol 16, 1997, pp. 1307-1314.

Electromagnetic Nondestructive Evaluation (VI)
F. Kojima et al. (Eds.)
IOS Press, 2002

Profile reconstruction of conductive cracks from eddy current signals by means of a neuro-fuzzy system

Noritaka YUSA and Kenzo MIYA

International Institute of Universality
1-4-6-SB801, Nezu, Bunkyo, Tokyo, 113-0031 Japan

Abstract: The present paper proposes a numerical scheme that reconstructs profile of conductive cracks from eddy current signals. What is considered here is not electro-discharge machined notches that have zero conductivity but conductive cracks that allow eddy current to flow across their surfaces. Only boundary shape of the crack is explicitly reconstructed, which enables to deal with complicated cracks. The scheme is based upon an artificial neural network and a fuzzy system. The neural network is trained to simulate mapping between crack profile and eddy current signals using simulated signals. In order to enhance credibility of result, a parameter is proposed and the fuzzy system gives degree of accuracy of reconstruction by calculating the proposed parameter. The scheme is validated by reconstructing crack profile from simulated signals.

1   Introduction

Eddy current testing has been applied to non-destructive inspections of steam generator tubes that are one of the most important parts of pressurized water nuclear power plants[1]. Although severe demands on safety and reliability of nuclear power plants have rejected presence of cracks so far, it is necessary to accept harmless cracks in order to optimize maintenance[2]. From this aspect, profile of the crack is indispensable knowledge, and, therefore, importance of study on eddy current inversion scheme is increasing. While most of eddy current studies have focused on artificial electro-discharge machined notches, it has been widely known that there are significant discrepancies between artificial electro-discharge machined notches and natural cracks that occur actual steam generator tubes[3][4]. One of the most important characteristics of natural cracks is considered as their non-zero conductivity. Namely, natural cracks are not electrical insulator with zero thickness and eddy current flows across their surfaces.

The authors proposed an innovative numerical scheme to reconstruct profile of conductive cracks from eddy current signals in the 6th international workshop on electromagnetic nondestructive evaluation, E'NDE2000, held in Budapest[5]. The scheme was based upon an artificial neural network that was trained to simulate mapping between crack profile and eddy current signals. Figure 1 illustrates the crack model and geometrical configuration adopted in the previous study. The biggest advantage of the scheme is that it can deal with cracks that have distributed conductivity as shown in left-hand side of Fig. 1. Although an artificial neural network has ability to simulate any kinds of non-linear function, it is very difficult to simulate exact mapping between two spaces, especially in case of inverse mapping. Key idea of the scheme is that it is not necessary to take all crack parameters into consideration for the reason that only size of cracks is important in view of practical inspections. Thus, mapping that the scheme dealt with reduced into much more simple one, and accurate reconstruction became possible. It was revealed that the scheme is robust against noises, and profiles of natural cracks occurred in a steam generator tobe was successfully reconstructed by the scheme[7]. On the other hand, the scheme has significant drawback; no criterion that indicates accuracy of the reconstruction is available unlike reconstruction schemes using model-based approaches[8], because it is impossible to calculate eddy current signals from a reconstructed crack.

In the present paper, the authors enhance the scheme by means of a neuro-fuzzy system in order to overcome the drawback. An overview of the scheme is illustrated in Fig. 2. The difference between the scheme proposed here and the one in the previous study is a fuzzy system that processes output of the neural network. The fuzzy system computes a parameter that will be proposed in the present paper and gives degree of credibility on the output.

2 Simulation of mapping between eddy current signals and crack profile by means of a neural network

The scheme proposed in the present paper is mainly based upon an artificial neural network that simulates mapping between eddy current signals and crack profile. The neural network is trained by simulated signals and a database for training contains no measured signals or artificial noise. Crack model and geometrical configuration adopted here is the same ones as the previous study[5], which are shown in Fig. 1. A crack is modeled as an assembly of cells that have uniform conductivity in order to model natural cracks. A Crack is present on a wall of a steam generator tube and eddy current signals are measured by a pancake probe, whose parameters are

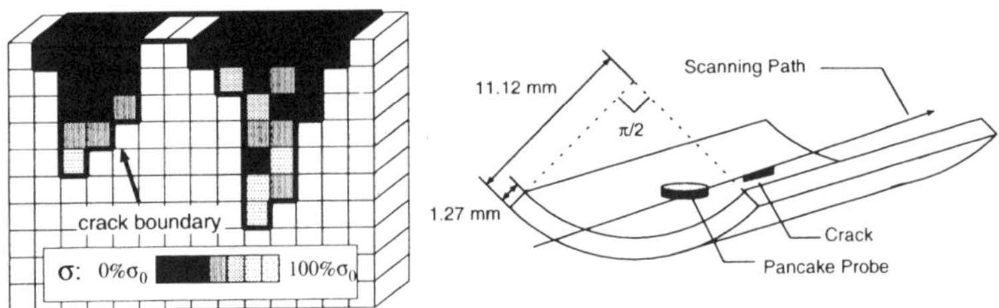

Figure 1: The adopted crack model (left) and geometrical configuration (right); crack is modeled as an assembly of cells that have uniform conductivity, and eddy current signals are obtained by using a pancake probe that scans directly above an axial crack.

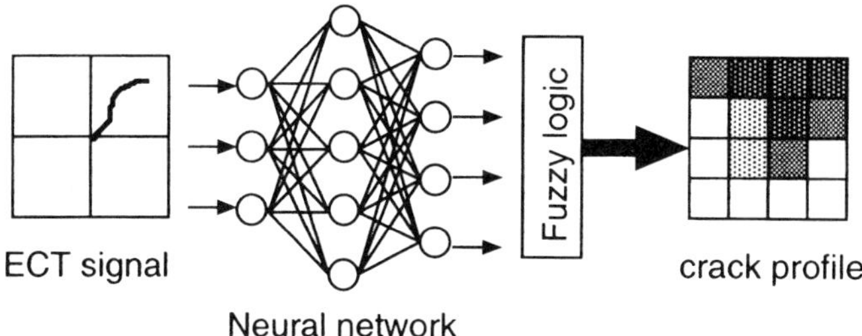

Figure 2: An overview of the scheme; a neural network trained to simulate mapping between crack profile and eddy current signals evaluates crack profile from unknown signals, and a fuzzy system gives credibility on the output of the neural network system.

listed in Table 1. Signals are obtained by means of simulations using a FEM-BEM coupling code[6] accelerated by a fast forward solver[8]. Details of forward analyses are not discussed here, since they are carried out only to establish the database and independent from the proposed scheme.

A crack is parameterized as a set of integers that indicates depth of boundary profile with 10% pitch. and conductivity of crack is not explicitly parameterized. For instance, a parameter vector that represents the crack shown in Fig. 1 is {0 1 5 4 3 0 0 1 2 7 6 3 1 0}. The neural network is a feed-forward type neural network whose transfer function is hyperbolic tangent. Structure of the neural network is modified during training procedure so that the neuron that corresponds to crack parameter excites most in the output layer. Since it is assumed in the present study that depth of crack equals to or less than 80% thickness of tube wall. nine neurons are necessary in the output layer to deal with unit element of the crack parameter vector. A statistical signal processing and a shifting aperture method are applied to enhance accuracy of the scheme. For further details, one can refer to the reference [5].

Table 1: Parameters in the simulation.

	material	INCONEL 600		shape	pancake
	conductivity	1.0×10^6 S/m		inner radius	0.8 mm
	permeability	$1.0\ \mu_0$		outer radius	1.8 mm
specimen	shape	1/4 tube	probe	height	0.8 mm
	thickness	1.27 mm		frequency	300kHz
	length	40 mm		lift-off	0.5 mm
	outer radius	11.12 mm		current	1 A $\times$ 140 turn

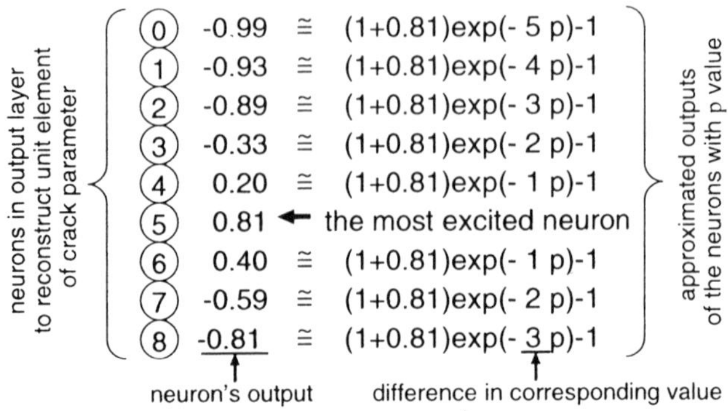

Figure 3: A typical response of a trained neural network to unknown signals. A circle with a value, n, illustrates a neuron in output layer of a neural network whose excitation indicates crack parameter is n. Transfer functions of the neurons are hyperbolic tangent functions whose output is in the range of -1 to 1. While the sixth neuron is excited most, other neurons are also excited to some extent.

3 Processing of output of the neural network system by means of a fuzzy system

The role of the fuzzy system is to process output of the neural network and give credibility on the output without computing eddy current signals from reconstructed cracks. Ideally, only the neuron that corresponds to the value of the crack parameter is excited and other neurons keep calm when a trained network receives unknown signals. However, the mapping of eddy current inversion is so complicated that it is impossible for neural networks to exactly simulate it. It sometimes happens that a neuron different from the desired one is excited most, or all neurons keep calm. Increasing size of database for training would solve the problem to some extent, which will make the training much more difficult and cause other problems.

Figure 3 illustrates a typical response of a trained neural network to unknown signals. Since the sixth neuron that indicates the value of the crack parameter is 5 is excited most, a value of 5 is taken as the reconstructed crack parameter according to the strategy explained in the previous section and the reference [5]. What should be noted is that other neurons are also excited to some extent. If the neural network was trained well and the unknown signals are very similar to one of the signals with which the neural network was trained. all other neurons would not be excited at all. Therefore, it is reasonable to suppose that the difference of the most excited neuron from the other ones in the degree of excitement can be a criterion of credibility of the output.

Here, the authors propose a parameter p that is computed as follows: (1) find the most excited neuron and let y equal its value, (2) suppose that output values of other neurons are given by $(y + 1) \times \exp\{-pD\} - 1$, where D is absolute value of difference from the most excited neuron in their corresponding values. (3) obtain p by means of the least square method. The parameter p represents uniqueness of the most excited neuron's value. Small value of p indicates that all neurons are almost equally excited, while large vlaue of p indicates that only one neuron is excited and other neurons keep calm. Figure 3 also illustrates a scheme to compute the p-value. Since the sixth neuron is excited most and its output value is 0.81, output value of the third neuron is, for instance, supposed to be $(1 + 0.81) \exp\{-p|5 - 2|\} - 1$. Six examples of a neural network's responses to unknown signals and calculated p-values are listed in Table 2. A glance at Table 2 reveals that the p parameter well represents degree of "confidence" that the trained neural network had in its response.

4 Reconstruction of crack profile from simulated signals

The scheme is validated by means of reconstruction using simulated eddy current signals. Whether a crack is present on inner or outer tube wall is assumed to be known in advance. Number of training, validation and test data are respectively 200. Size of cell is 0.5 mm in length, 0.127 mm (10% of specimen's thickness) in depth and 0.2 mm in width. In order to avoid too difficult problem, it is supposed here that depth of a crack is equal to or less than 80% specimen's thickness, and conductivity is less than 50% of base material's one. After several preliminary calculations, number of layers of the neural network was set to 3 and number of neurons in the hidden layer was 30.

Four reconstruction results of outer defects are shown in Fig. 4 with corresponding p-value. In the each figure, hatched region shows true profile of the crack and black solid line indicates reconstructed boundary. Digits in the cells inside true crack region represent percentage of their conductivity to the base material's one, which are not reconstructed here. Calculated p-value is shown together in the figure. Figure 4 reveals that reconstructed boundary profile with small p tends to be different from the true one, and one with large p is correct. It should be noted that Table 2 shows output of a neural network is highly reliable if calculated p-value is larger than 2.

Table 2: Examples of neural network's output and its p-value

	Case 1	Case 2	Case 3	Case 4	Case 5	Case 6
output of neuron 1	0.9100	0.7323	-0.0231	-0.4882	0.0424	0.5809
output of neuron 2	-0.7828	-0.7662	-0.7221	-0.6373	-0.6246	-0.6582
output of neuron 3	-0.9035	-0.8276	-0.7023	-0.6448	-0.7521	-0.7980
output of neuron 4	-0.9799	-0.9635	-0.9217	-0.9030	-0.9290	-0.9228
output of neuron 5	-0.9826	-0.9731	-0.9466	-0.9096	-0.9167	-0.9347
output of neuron 6	-0.9741	-0.9640	-0.9516	-0.9514	-0.9387	-0.8938
output of neuron 7	-0.9860	-0.9841	-0.9765	-0.9699	-0.9745	-0.9791
output of neuron 8	-0.9943	-0.9939	-0.9916	-0.9905	-0.9934	-0.9946
output of neuron 9	-0.9917	-0.9920	-0.9938	-0.9938	-0.9953	-0.9970
p-value	2.0965	1.8310	0.8820	0.3881	0.8524	1.3541

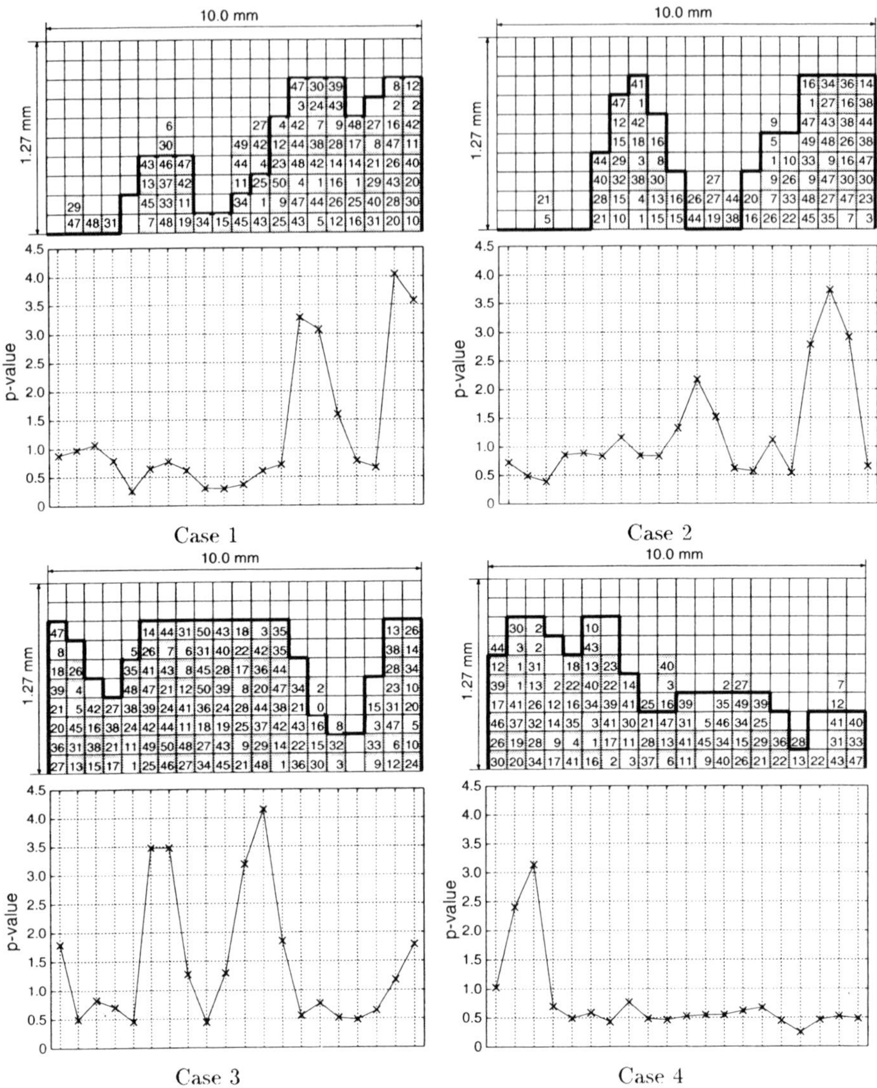

Figure 4: True and reconstructed profiles of cracks with computed p-value.

5   Conclusion

The present study have developed an inversion scheme that reconstructs crack profile from eddy current signals by means of a neuro-fuzzy system. A neural network simulated inverse mapping and a fuzzy system gave a criterion that indicated to what extent the neural network had confidence in its outputs. The p parameter was proposed in order to overcome the drawback of the original scheme. Cracks with distributed conductivity were reconstructed from simulated eddy current signals by means of the inversion scheme. The paper revealed that the parameter gives criterion of credibility of reconstructed profile. It is concluded that the scheme can deal with modeled natural cracks and its reliability is greatly enhanced by the fuzzy system that calculates the parameter without obtaining eddy current signals from the reconstructed crack.

References

[1] T. Takagi and K. Miya, ECT round-robin test for steam generator tubes, Journal of the Japan Society of Applied Electromagnetics and Mechanics, Vol. 8, No. 1, pp. 121-129, 2000.

[2] Y. Tsujikura, T. Nakamura, K. Matsueda and S. Masamori, Current maintenance activities at nuclear power plants and in other industries, Journal of the Japan Society of Applied Electromagnetics and Mechanics, Vol. 8, No. 2, pp. 159-172, 2000. (in Japanese)

[3] W. R. Randle, B. D. Woody, Caution about Simulated Cracks in Steel for Eddy Current Testing, Material Evaluation, January 1989, pp. 44-48.

[4] Z. Badics, Y. Matsumoto, K. Aoki, F. Nakayasu, A. Kurokawa, Finite element models of stress corrosion cracks (SCC) in 3-D eddy current NDE problems, Nondestructive testing of materials, IOS Press, Amsterdam, pp. 21-29, 1995.

[5] N. Yusa and K. Miya, Numerical analysis of JSAEM round robin test, Electromagnetic Nondestructive Evaluation(V), IOS Press, Amsterdam, pp. 325-332, 2001.

[6] F. Matsuoka and A. Kameari, Calculation of three dimensional eddy current by FEM-BEM coupling method, IEEE trans. on Mag. Vol. 24, No. 1, pp. 182-185, 1988.

[7] N. Yusa, Z. Chen and K. Miya, Quantitative profile evaluation of natural cracks in a steam generator tube from eddy current signals, International Journal of Applied Electromagnetics and Mechanics, (to appear).

[8] Z. Chen, K. Miya and M. Kurokawa, Rapid prediction of eddy current testing signals using A-ϕ method and database, NDT&E International, Vol. 32, pp. 29-36, 1999.

Electromagnetic Nondestructive Evaluation (VI)
F. Kojima et al. (Eds.)
IOS Press, 2002

Inverse Analysis of Steam Generator Tube ECT Signals with Noise Sources Outside

Haoyu Huang and Toshiyuki Takagi
Institute of Fluid Science, Tohoku University, Katahira 2-1-1, Aoba-ku, Sendai 980-8577,
JAPAN

Abstract. This paper describes an algorithm of the crack reconstruction from noisy signals. In eddy current testing of steam generator tubes, signals are sometimes rendered noisy by the presence of other structures, such as support plates and tube sheets outside of test materials. Moreover, copper and magnetic deposits will also adhere outside the tube near the support plates and tube sheets. A round-robin test of SG tubes with noise source outsides is introduced by JSAEM. Experimental results of a transmit-receive type ECT probe are shown. Noisy signals are processed by a multi-frequency method but residual noise remains. Some other signal processing methods are discussed. Several artificial cracks are reconstructed and inverse results are shown in this paper.

1. Introduction

In eddy current testing (ECT) of steam generator (SG) tubes, signals are sometimes rendered noisy by the presence of other structures, such as support plates and tube sheets outside of test materials. Moreover, copper and magnetic deposits will also adhere to the outer surface of the tube near the support plates and tube sheets. The support plates and tube sheets here are sometimes made of ferromagnetic materials. To solve ECT inverse problems, a fast forward solver is required because the ECT signals must be accurately computed many times. Some fast pre-computed database approaches [1,2] have been proposed, but a problem still exists because it is impossible to treat the ferromagnetic materials. By the extension of these database approaches, a new method [3] is proposed to solve the ECT problems including ferromagnetic materials as noise sources.

A round-robin test of SG tubes with noise source outsides is introduced by the Japan Society of Applied Electromagnetics and Mechanics (JSAEM). The noise sources here include support plates, tube sheets, copper and magnetic deposits, as well as their combination. Crack shape reconstruction problems with support plate noise have been solved [4] by a multi-frequency method and inverse analysis. However, the multiple noise source problems have never been discussed as far as authors' knowledge. Experimental results of the round-robin test using a transmit-receive type ECT probe are shown. Noisy signals are processed by a multi-frequency method [5] and other methods as well. Several artificial cracks are reconstructed and inverse results are shown in this paper.

2. Experimental results of round-robin test

The round-robin test includes eight kinds of test samples, as shown in Table 1. The arrangement of tubes and noise sources is shown in Fig. 1. An ECT probe used here is

Table 1　Test Samples

Sample Name	Tube Type	Crack Type	Depth	Length
NEL-0017	Straight	EDM Axial	OD20%	5mm
NEL-0018	Straight	EDM Circ.	OD20%	5mm
NEL-0019	Straight	Crack-free	0	0
NEL-0021	Expansion	EDM Axial	ID20%	5mm
NEL-0022	Expansion	EDM Circ.	OD20%	5mm
NEL-0023	Expansion	Crack-free	0	0
NEL-93114	Straight	SCC Circ.	N/A	N/A
NEL-95205	Expansion	SCC Circ.	N/A	N/A

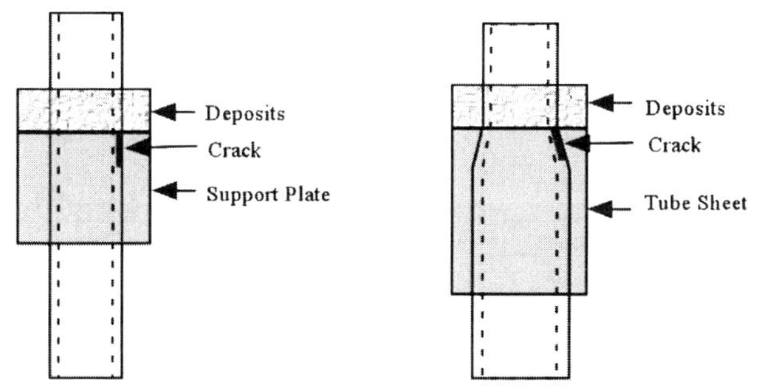

(a) Straight tube and a support plate　　　(b) Expansion zone and a tube sheet

Fig.1　Arrangement of tube and noise sources

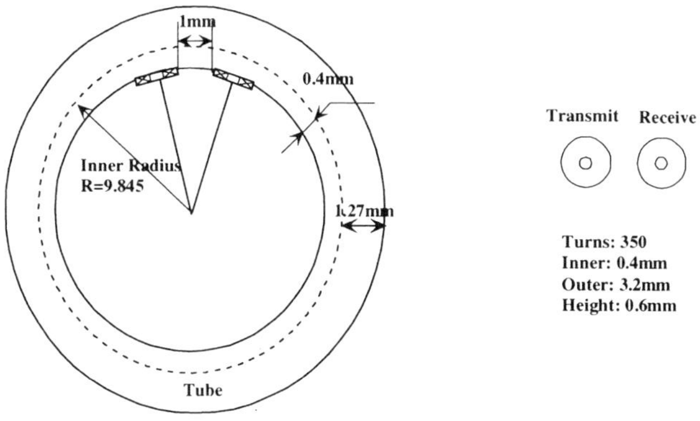

(a) Cross-section of the probe and the tube　　　(b) Exciting and Pickup Coils

Fig.2　Transmit-receive type ECT probe

developed by Nuclear Engineering Ltd, which is a transmit-receive type, with same size round shape exciting and pickup coils. Shape and sizes of the probe are shown in Fig. 2.

Axial and circumferential cracks are detected separately by axial (coils arrange in the axial direction) and circumferential channels (coils arrange in the circumferential direction). Both 400kHz and 200kHz are used in order to apply the multi-frequency method. As an example, ECT signal of sample NEL-0017 with a support plate and magnetic deposits is shown in Fig. 3. A peak can be found in the middle that shows the position of the crack, left side of it is the magnetic deposits noises, and the right side is the support plate noises. Signal shown here is the Y component of the complex measured signal. As a processing method of ECT, the support plate noises are rotated to X direction. Much larger noises are due to the magnetic deposits in Y component graph.

3. Discussion of signal processing

The multi-frequency method is well known as an effective signal processing to decrease the noises by the presence of structures outside tube such as support plates. It is based on the phenomenon that phase of the noises is almost the same, and on the other hand, phases of crack signals are quite different. However, when two kinds of noise sources are combined together, it is impossible to decrease both noises at the same time by a multi-frequency method. Signal of a support plate only and signals from both support plate and magnetic deposits are shown in Fig.4. In order to see the phase of the signals more clearly, B-scan (along the axial direction) graphs are used. The ECT signals in the complex plane start from the start point, get the peak value and then go back to start point.

Other signal processing methods can be considered, because of the properties of revolving symmetry of the noise sources. One is the method to subtract a base line where cracks does not exist, such the line where x=1 in Fig.3. It is called here a base-line subtraction method. Another method is called a differential filtering method. Equations used in these two methods are shown as follows:

$$S'(x=i) = S(x=i) - S(x=1) \tag{1}$$
$$S'(x=i) = S(x=i) - 0.5*(S(x=i-1) + S(x=i+1)) \tag{2}$$

where S is the raw signals and S' is the processed signals.

Signal processing results from raw signals shown in Fig.3 are compared using above three methods, as shown in Fig. 5. It is clear that the multi-frequency method mainly decreases the magnetic deposits noises in Fig.3, but at the same time, support plate noises become larger. The other two results are nearly the same, where the differential filtering method reduces noises more effectively, and crack signals as well. Considering the reason that the fast signal prediction method using pre-computed database [3] is able to compute the signals due to cracks directly, it will be suitable to use the base-line subtraction method as the signal processing method. Anyway, it can only be applied to the case that noises in the circumferential direction are uniform. If noise sources are not uniform, residual noises exist where noise sources change. In this case, residual noise must be separate from crack signals by phase or other messages somehow. Otherwise cracks can't be recognized.

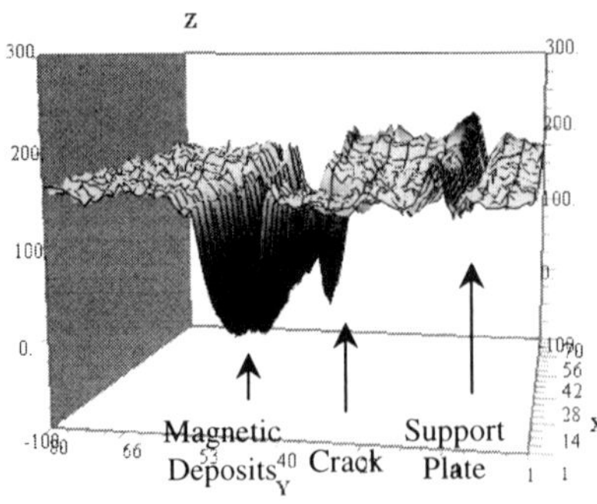

Fig.3   Signal of NEL-0017 with a support plate
and magnetic deposits (400kHz)

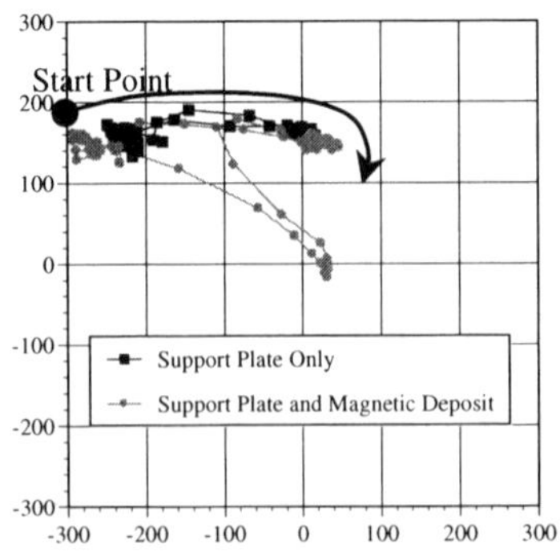

Fig.4   Support plate and magnetic deposits
noises

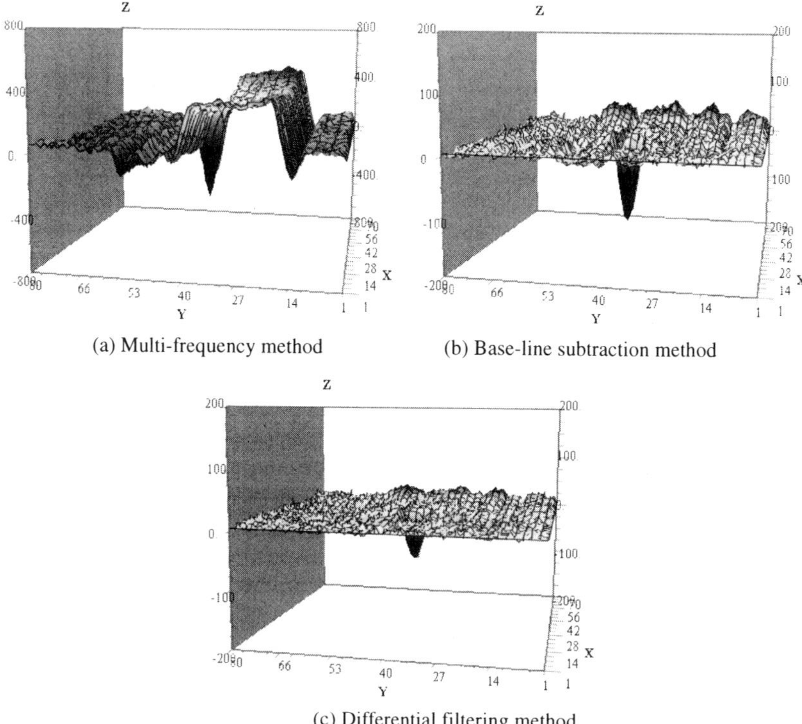

(a) Multi-frequency method
(b) Base-line subtraction method

(c) Differential filtering method

Fig.5   Comparison of three signal processing results

4.   Inverse analysis results

Inverse analysis is performed from the signals processed by the base-line subtraction method. Experimental results of noise-free cases of this probe are also available and the transform coefficients from measurement data to numerical data are obtained using OD60% as a standard. B-scan data are used, 21 scan points along the crack direction. Numerical data and the transformed measurement data are compared in Fig. 6. When signals of OD60% are used as standards, signals of OD20% also fit each other very well except some residual noises.

Using the processed signals for the inverse scheme needs more time to compute the forward problem, but there are no difficulties when using our fast signal estimator. Combining the signal processing method with normal inverse approaches, the actually depth of a crack can be predicted. It is a normal approach [4] to solve the inverse problem by minimizing the least square error between the estimated signals and observation signals. The following evaluation function is used in our studies:

$$J(x) = \sum_{i=1}^{N_{pos}} \left| S_{comp}^i(x) - S_{obs}^i \right|^2 \left/ \sum_{i=1}^{N_{pos}} \left| S_{obs}^i \right|^2 \right. , \tag{3}$$

where J is an evaluation function, x is the vector characterizing the shape of the cracks, $S_{comp}^i(x)$ are the predicted signals related to the vector x, and S_{obs}^i are the observation signals. $N_{pos} = 21$ is the total observation points and i is the point number. The steepest descent method is applied to minimize the least square error.

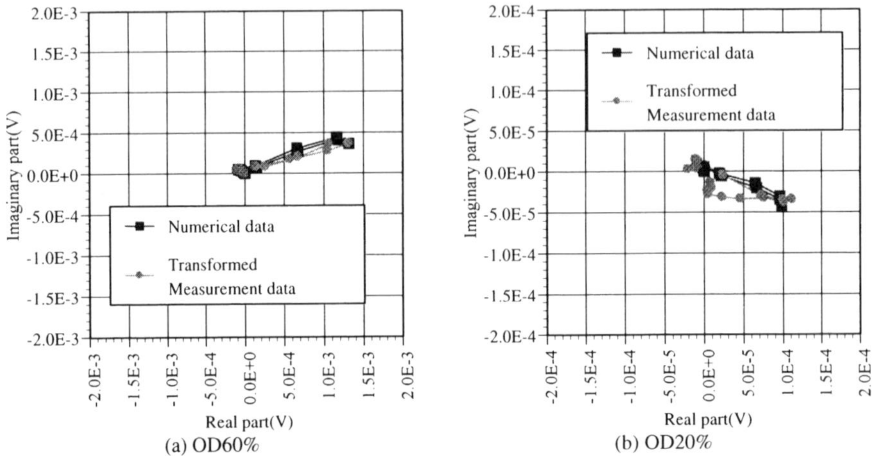

(a) OD60%   (b) OD20%

Fig.6   Comparison of numerical signals and transformed measurement signals (NEL-0017)

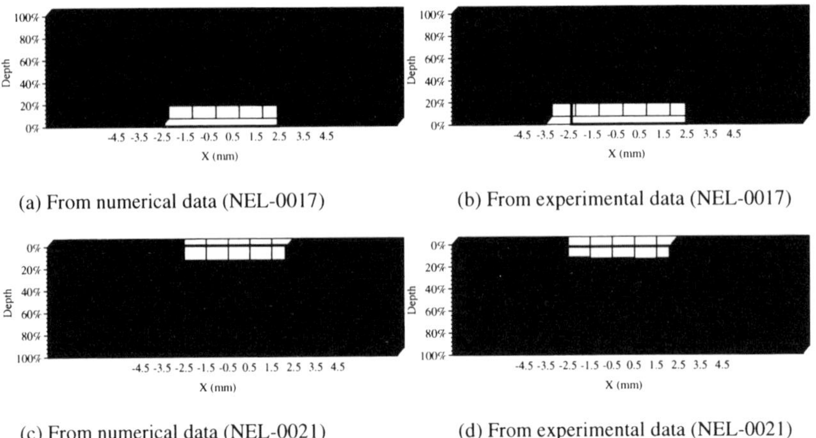

(a) From numerical data (NEL-0017)   (b) From experimental data (NEL-0017)

(c) From numerical data (NEL-0021)   (d) From experimental data (NEL-0021)

Fig.7   Reconstructed results

To apply the fast forward ECT signal simulator [3], a "suspect region" which includes the possible crack region should be defined. For reconstructions of axial cracks, a suspect region used here is a $9*1.27*0.2mm^3$ block, which is divided into $9*5*1$ elements. Parameters are the depths of each 1mm column, along the crack direction. Shape reconstructions of the sample NEL-0017 and NEL-0021 are shown in Fig. 7. Comparison of reconstruct results and true results is shown in Table 2. Crack depth reconstructed from experimental data agrees with the true depth well, and the predicted length of OD20% (NEL-0017) is 1mm longer than the real value because of the residual noises. Database used here is for axial cracks only. To reconstruct the circumferential crack shapes of sample NEL-0018, NEL-0022, NEL-93114, NEL-95205, databases of circumferential cracks are needed. After preparing a database for circumferential cracks, inverse problems of other tube samples are also possible to be solved.

Table 2   Comparison of reconstructed results and true results

Sample Name		Reconstructed Results		True Results	
		Depth (%)	Length (mm)	Depth (%)	Length (mm)
NEL-0017	Numerical Data	OD19.7	5	OD20	5
	Experimental Data	OD20.5	6	OD20	5
NEL-0018	Numerical Data	ID20.2	5	ID20	5
	Experimental Data	ID20.5	5	ID20	5

5. Summary

In eddy current testing of steam generator tubes, signals are sometimes rendered noisy by the presence of other structures, such as deposits, support plates and tube sheets outside of test materials. Crack shape reconstruction methods from ECT signals with combined noise sources are discussed in this paper.

1. Round-robin test of SG tubes with noise sources outside is introduced, arrangement of straight tube, expansion zone of tube, deposits, support plates and plate sheets are shown.
2. Signal processing methods of the noised experimental signals are discussed. Multi-frequency method is not able to reduced two or more noises at the same time. Therefore, base-line subtraction method is used in this study.
3. Inverse analysis method of noised ECT signals is given. The reconstructed results agreed with real shapes, especially the depths of cracks. Length of OD20% is reconstructed 1mm longer than true one because of the residual noises.

Acknowledgment

This study was supported in part by the Research Committee on Nondestructive Evaluation Technology by Eddy Current Testing of the Japan Society of Applied Electromagnetics and Mechanics through a grant from 5 PWR utilities and Nuclear Engineering Ltd. This study was also supported in part by the Grant-in-Aid for COE Research (11CE2003) by the Ministry of Education, Culture, Sports, Science and technology.

References

[1]   Z. Chen and K. Miya, "ECT inversion using a knowledge-based forward solver," J. Nondestructive Evaluation, Vol.17, No.3, pp.167-175, 1998.
[2]   T. Takagi, H. Huang, H. Fukutomi and J. Tani, "Numerical evaluation of correlation between crack size and eddy current testing signal by a very fast simulator," IEEE Trans. Magnetics, Vol.34, No.5, pp.2581-2584, 1998.
[3]   H. Huang, T. Takagi and H. Fukutomi, "Fast signal prediction of noised signals in eddy current testing," IEEE Trans. Magnetics, Vol. 36, No. 4, pp. 1719-1723, 2000.
[4]   H. Huang, T. Takagi and H. Fukutomi, "Crack reconstruction from noisy signals using a novel ECT probe," Electromagnetic Nondestructive Evaluation (IV), IOS press, 2000, pp. 143-150.
[5]   D. E. Bray and R. K. Stanley, "Nondestructive evaluation: A tool in design, manufacturing and service," CRC Press, 1997, pp.418-419.

Electromagnetic Nondestructive Evaluation (VI)
F. Kojima et al. (Eds.)
IOS Press, 2002

Crack Shape Identification of Steam Generator Tubes by Approximate Output Least Square Problems

Nobuyuki OKAJIMA, Fumio KOJIMA, Futoshi KOBAYASHI
Graduate School of Science and Technology, Kobe University
1-1, Rokkodai, Nada-ku, Kobe 657-8501, JAPAN

Abstract.   A computational method is considered for the fast recovery of a crack shape of steam generator tubes of nuclear power plants. A numerical model of the inspection process is given with the hybrid use of finite element and boundary element methods. The impedance change due to a defect is calculated by building the difference of the impedance values with the flaw present and absent. In order to achieve the high reduction of the computational costs, the solution space of the model output is reconstructed using multivariate B-spline series. A very fast recovering scheme is performed by solving the output least square problems for the approximated observation space.

1   Introduction

The detection of a crack on steam generator tubes is a critical issue for the structural integrity of the plants. Eddy current testing (ECT) is used for in-service inspection of tubes in steam generators because of high detectability and rapid scanning process. A transmitter-receiver pancake type probe coil is scanned along the inner surface of steam generator tubes. Defects can be detected as a probe impedance or voltage trajectory from the receiver by exciting the transmitter. The inversion techniques by parameter estimation problems have been studied computationally and experimentally by many authors (e.g. [1]). However, those are computationally intensive iterative procedures in which the accurate forward problems must be solved numerous times. Recently, most efforts in computational savings are directed to the use of database for an ensemble of the forward solutions. Huang et al [2] and Chen et al [3] have been developed the current dipole method based on the reciprocity theorem. In this method, once the forward model can be simulated for the material without flaws, the ECT signals can be obtained by recalculating the corresponding model only defined on the suspect region. Those method are very effective if the size of the suspect region become too small. Banks et al [4] has been proposed the Proper Orthogonal Decomposition methodology which makes it possible to produce the reduced order forward model. In this paper, a new approach for parameter estimation is proposed for the sensitivity evaluation of the output least square method. In order to cut the computational costs drastically, the model output is directly reconstructed by data sets of the forward solutions. Thus the sensitivities of the cost can be directly evaluated from the approximated observation space.

2 Mathematical Modeling of Inspection

The eddy current problems considered here involve a transmitter-receiver pancake type probe coil as shown in Fig. 1. The force current density vector is applied to the transmitter, and then the impedance of the receiver is affected due to the existence of the eddy current excited by the transmitter coil. Let (t, x) be the time $t \in [0, \infty)$ and the location $\mathbf{x} = (x_1, x_2, x_3)$ in R^3. Let V_0 be the inspection area of the steam generator tube defined by

$$V_0 = \left\{ \mathbf{x} = (x_1, x_2, x_3) \, \middle| \, \left(\frac{d_i}{2}\right)^2 < x_1^2 + x_2^2 < \left(\frac{d_o}{2}\right)^2, 0 < x_3 < d_l \right\} \tag{1}$$

where d_i, d_0 and d_l denote the inner diameter, the outer diameter and the length of the tube, respectively. The mathematical model of eddy current testing can be described by the magnetic vector potential $\mathbf{A} = (A_1, A_2, A_3)$ and the electric scalar potential ϕ. The mathematical model can be expressed as

$$\frac{1}{\mu_0} \nabla^2 \mathbf{A} = j\omega\sigma(\mathbf{A} + \nabla\Phi) \quad \text{in } V_0 \tag{2}$$

$$\nabla \cdot j\omega\sigma(\mathbf{A} + \nabla\Phi) = 0 \quad \text{in } V_0 \tag{3}$$

$$-\frac{1}{\mu_0} \nabla^2 \mathbf{A} = \chi_T \mathbf{J}_T \quad \text{in } R^3 - V_0 \tag{4}$$

where Φ denotes the time integration of the scalar potential ϕ, $\mathbf{J}_T$ denotes the force current density in transmitter coil, χ_T denotes the characteristic function on the transmitter coil, μ_0 denotes a space permeability, and σ denotes an electric conductivity. Eqs.(2-4) is numerically solved by the hybrid use of finite element and boundary element methods[1],

$$[P + j\omega Q + j\omega R + K] \left\{ \begin{array}{c} A \\ \Phi \end{array} \right\} = \left\{ \begin{array}{c} [M][G]^{-1}\{F\} \\ 0 \end{array} \right\} . \tag{5}$$

The corresponding ECT signal can be evaluated as

$$\Delta Z = \frac{1}{I^2} \iiint_{\Omega_c} \mathbf{E}^u \cdot (\sigma_0 - \sigma_c)(\mathbf{E}^u + \mathbf{E}^f)^* dV \tag{6}$$

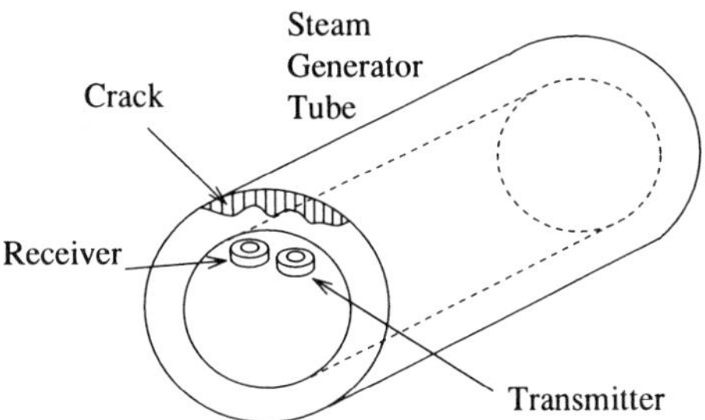

Fig. 1. Tube with stress corrosion crack

where I is an electric current of the coils, E^u and E^f are the electric field with flaw absent and the difference between the field with the flaw present and with the flaw absent, expressed by

$$\mathbf{E^u} = -j\omega\mathbf{A^u} - \nabla\Phi^u \tag{7}$$

$$\mathbf{E^f} = -j\omega\mathbf{A}^\delta - \nabla\Phi^\delta. \tag{8}$$

(See [3] for more details.)

3   Approximated Output Least Square Problems

The defect considered here is assumed to be circumferential cracks outside the SG tube and the profile is represented as shown in Fig. 2 where N_c denotes the number of division of crack length and where $\mathbf{q} = \{q_i\}_{i=1}^{N_c}$ denote the crack profile parameters (the ratio of crack depth to the nominal thickness of the SG tube). The admissible class of parameter $\mathbf{q}$ is assumed to be given by the set

$$Q_{ad} = \{\mathbf{q} \in R^{N_c} \mid 0 \leq q_i < \bar{q}_U < t_c < \infty \quad i = 1, 2, \cdots, N_c\} \tag{9}$$

where t_c and $\bar{q}_U$ denote the thickness of the SG tube and the maximum depth of the crack, respectively. Our aim is to recover defect geometries using probe impedance trajectories. Parameter estimation problems are applied to our inversion technique. The usual procedure can be formulated as a minimization of the fit-to-data functional:

$$E_r(\mathbf{q}) = \frac{1}{2}|\Delta Z(\mathbf{q}) - \Delta Z_d|^2 \tag{10}$$

with respect to the geometrical parameters $\mathbf{q}$. Here ΔZ_d denotes the measurement impedance data, while $\Delta Z(\mathbf{q})$ implies the model output corresponding to the defect parameter $\mathbf{q}$. The model output can be computed by applying the hybrid use of FEM-BEM model (5) combined with the current dipole method. As is well-known, the minimizer of the fit-to-data functional $E_r(\mathbf{q})$ is to implement

$$\mathbf{q}^{(l+1)} = \mathbf{q}^{(l)} - \lambda^{(l)}\mathbf{d}^{(l)} \quad \text{for } l = 0, 1, 2, \cdots \tag{11}$$

where $\mathbf{d}^{(l)}$ denotes the feasible direction vector and where

$$E_r(\mathbf{q}^{(l)} + \lambda^{(l)}\mathbf{d}^{(l)}) = \min_{\eta\in[0,\infty)} E_r(\mathbf{q}^{(l)} + \eta\mathbf{d}^{(l)}). \tag{12}$$

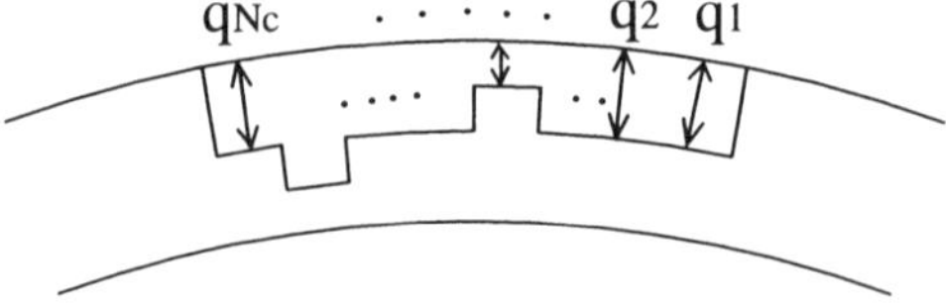

Fig. 2. Parametrization of crack profile

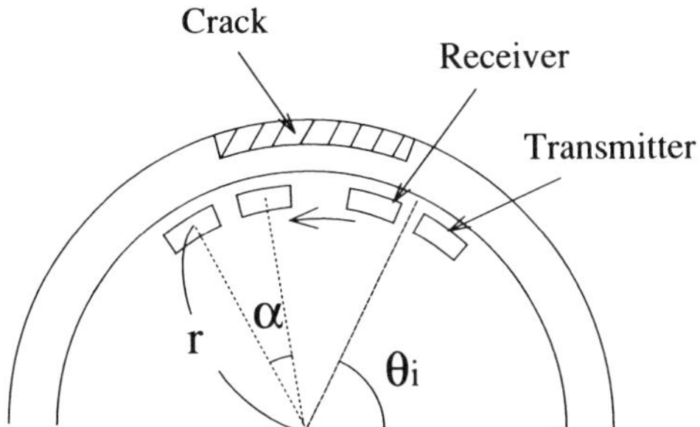

Fig. 3. Scanning position of transmitter-receiver probe coil

The practical implementations of the algorithm (11) involve the crucial part of computational volumes. Using the multivariate bilinear spline functions[6], the model output can be approximated by

$$\Delta Z(\mathbf{q}) \approx \sum_{i=1}^{N_d+1} \Delta Z_i B_i^{\Delta}(\mathbf{q}). \tag{13}$$

Thus, instead of solving the output least square problem for Eq.(10), we solve the following approximate parameter estimation problem.

Approximate Identification Problem:
Find the optimal $\mathbf{q} = \hat{\mathbf{q}}$ which minimizes the approximate fit-to-data functional

$$E_r^{\text{approx}}(\mathbf{q}) = \frac{1}{2} \left| \sum_{i=1}^{N_d+1} \Delta Z_i B_i^{\Delta}(\mathbf{q}) - \Delta Z_d \right|^2 \tag{14}$$

with respect to $\mathbf{q} \in Q_{ad}$.

4   Computational Experiments

In our numerical experiments, the defect was assumed to be the circumferential crack (width: 0.2 [*mm*], length: 4.77 [*mm*]) outside SG tube (inner diameter: $d_i = 19.69$ [*mm*] , thickness: $t_c = 1.27$ [*mm*]) made from INCONEL600 (electric conductivity: $\sigma_0 = 1.0 \times 10^6$ [*S/m*], magnetic permeability: $\mu_0 = 1.0 \times 10^{-7}$ [*H/m*]). Two crack patterns (a stair-step crack and multiple cracks) were tested for computationbal experiments. The number of finite elements and boundary elements were set as 1040 and 706, respectively. The frequency of transmitter coil was $400[kHz]$. The coils were moved along the crack as shown in Fig. 3. The scanning positions of coils were set as

$$x^P = (r\cos\theta_i, r\sin\theta_i, x_3^P) \quad (i = 1, 2, \cdots, m_p)$$

where $r - 9.145[mm]$, $m_p - 20$, and θ_i were set by a unit of $1.75[deg]$ from $73.41[deg]$ to $106.59[deg]$. The difference of angle between transmitter and receiver is $\alpha = 25[deg]$. Then, the solutions were added random noises, and used as 'data' for our inverse algorithm. The

Table 1. True and estimated values of **q** at iteration 500 (Stair-step Crack)

	$\mathbf{q}^{\text{true}}$	Noise Free	1% Noise	2% Noise	5% Noise	10% Noise
q_1	80	80.05	79.68	80.06	79.60	78.25
q_2	60	60.08	60.18	60.18	61.40	63.04
q_3	60	60.10	58.94	58.94	54.10	49.86
q_4	40	40.00	40.55	40.55	43.27	47.39
q_5	40	40.01	39.99	39.99	40.00	39.86
q_6	20	19.10	19.76	19.76	18.53	16.75
$\Delta E_r^{\text{approx}}$		1.148E-3	2.316E-3	8.816E-2	5.435E-1	2.169
Accuracy		936.3	105.2	55.97	21.33	10.06

Table 2. True and estimated values of **q** at iteration 500 (Multiple Cracks)

	$\mathbf{q}^{\text{true}}$	Noise Free	1% Noise	2% Noise	5% Noise	10% Noise
q_1	60	60.31	60.49	59.10	58.96	62.52
q_2	60	59.97	59.71	60.25	60.71	55.73
q_3	0	0.27	0.01	0.14	0.31	0.15
q_4	40	39.85	40.04	40.00	40.00	40.59
q_5	80	80.00	79.99	79.78	79.44	79.78
q_6	60	60.03	60.01	60.14	60.37	60.01
$\Delta E_r^{\text{approx}}$		7.970E-4	1.112E-2	4.487E-2	2.747E-1	84.11
Accuracy		312.1	240.1	141.7	97.04	27.42

crack shape of initial guess was chosen as no crack. For the optimization routine of Eq.(14), we used the gradient projection method. Tables 1 and 2 summarize the estimated parameter results for the data without noise and with 1, 2, 5 and 10% relative noise. The average of the elapsed time required for estimation procedures was about $84[sec]$. In Tables 1 and 2, the accuracies of the estimated values were evaluated by

$$\text{Accuracy} = \frac{|\mathbf{q}^{\text{true}}|}{|\hat{\mathbf{q}} - \mathbf{q}^{\text{true}}|}. \tag{15}$$

Figs. 4 and 5 depict the true and estimated crack shape.

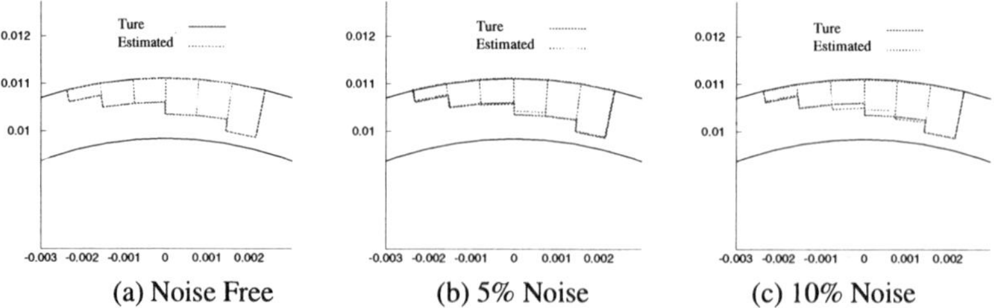

(a) Noise Free   (b) 5% Noise   (c) 10% Noise

Fig. 4. Estimated crack shape (Stair-step Crack)

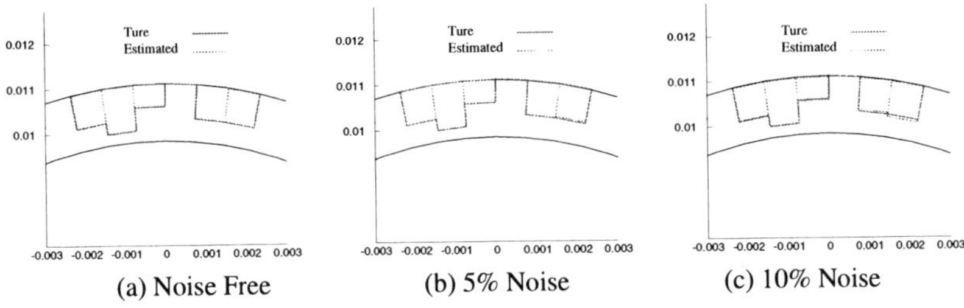

Fig. 5. Estimated crack shape (Multiple Cracks)

5   Concluding Remarks

A fast computational solver was proposed for recovering a crack shape of steam generator tubes. Computational savings have been achieved by building the approximate observation space with multivariate B-splines. A very fast recovering scheme was effectively tested with computational experiments. Current efforts are directed to show the effectiveness of the proposed method to the real experimental data.

6   Acknowledgements

This study was supported in part by the Research Committee on the Nondestructive Evaluation Technology by ECT of the Japan Society of Applied Electromagnetics and Mechanics through a grant 5 PWR utilities and Nuclear Engineering Ltd.

References

[1] F. Kojima, Computational method for crack shape reconstruction using hybrid FEM-BEM scheme based on $A - \phi$ method, *Studies in Applied Electromagnetics and Mechanics*, **12** (1997) 279-286.

[2] H. Huang, T. Takagi, and H. Fukutomi, A novel crack reconstruction method for steam generator tube ECT with sources outside, *Review of Progress in Quantitative Nondestructive Evaluation*, **20** (2000) 513-520.

[3] Z. Chen and K. Miya, ECT inversion using knowledge based forward solver, *J.Nondestructive Evaluation*, **17** (1998) 167-175.

[4] H. T. Banks, M. L. Joyner, B. Wincheski, and W. P. Winfree, Evaluation of material integrity using reduced order computational methodology, *Technical Report CRSC*, TR99-30, North Carolina State University, 1999.

[5] Z. Badics, Y. Matsumoto, K. Aoki, F. Nakayasu, M. Uesaka, and K. Miya, Accurate Probe-Response Calculation in Eddy Current NDE by Finite Element Method, *Journal of Nondestructive Evaluation*, Vol. 14, No. 4, 1995.

[6] C. K. Chui, Multivariate Splines, Vol. CBMS54, SIAM, 1988.

Electromagnetic Nondestructive Evaluation (VI)
F. Kojima et al. (Eds.)
IOS Press, 2002

A Processing Method of Eddy-Current Testing Image Using Subband Technique without Phase Distortion

T. Taniguchi*, S. Yamada**, and M. Iwahara**
** Department of Electronic Engineering, The University of Electro-Communications
1-5-1 Chofugaoka, Chofu, Tokyo, 182-8585 Japan*
*** Laboratory of Magnetic Field Control and Applications, Faculty of Engineering
Kanazawa University, 2-40-20 Kodatsu-no, Kanazawa, 920-8667 Japan*

Abstract As an application of eddy-current testing (ECT) technique, the inspection of printed circuit board (PCB) using ECT probe has been proposed, and for the actual acquisition of the existence and position of the defect, analyzing methods of ECT images have been also investigated. In those approaches, one important theme is the removal of undesired components contained in images, which disturb correct data analysis. This paper describes an image processing method for this aim using multiresolution analysis. Since filters used in this method have a linear phase, the waveforms of signals are maintained in each subband, which results in a correct manipulation about amplitude. The ability of the method to extract defect signal in undesired signals is shown through analysis of ECT images derived from samples for benchmark test and PCBs.

1. Introduction

Eddy-current testing (ECT) is a very useful tool for the detection of defects in metallic objects. It is applied in the area of inspections of steam generator pipes of atomic power plants [1], wings of aero plane [2], and printed circuit boards [3], [4], and its usefulness has been reported. But the inspection is not completed without a data processing for the analysis of features of defects, including existence, position, shape, size, and direction. What is important here is to carry out the reduction of the undesired components such as noise, offset signal and its variation which disturb the correct estimation of defect features, simultaneously with the image analysis. The wavelet-based denoising [6]-[8] is one good choice for this aim, but generally their bases have nonlinear phase and it might lose correct information on location of the signal in the wavelet region.

In this paper, we present an image processing method based on wavelet-like subband technique without this problem. Namely, since bases with complete linear phase (LP) are adopted, the decomposed signal keeps correct wave shape and phase information in the transformed domain. The advantages and problems of the proposed method are discussed through examples.

2. ECT Measurement

The defects on metallic object are examined by using ECT technique as depicted

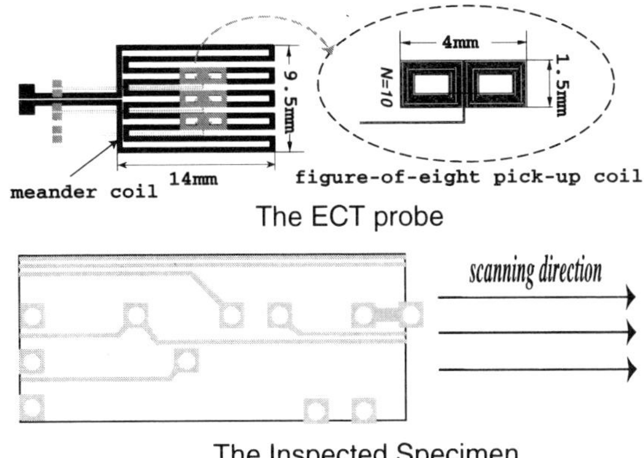

Figure 1: Inspection of metallic object by using ECT probe.

in Fig. 1. The ECT probe shown here consists of a meander type coil for the excitation of eddy-currents in the conductor and figure-of-eight (Fig. 1) or solenoid pick-up coil [4]. The ECT probe usually has a constant output, and detect disturbance of eddy-currents and generates a peak signal formed by a pair of positive and negative pulses only in the place where nonuniform distribution of conductor exists. In addition, the output signal contains measurement noise and offset component (and its variation) which have bad influences to the feature extraction of defects. Hence, we should remove those components before the analysis of defect features. As one way to cope with this problem, we present an image processing method utilizing multiresolution technique in the following section.

3. Subband Image Processing Method

For the reduction of undesired components in ECT images, many types of filter have been used. The wavelet-based method also can be interpreted as a kind of octave filter bank [9], i.e., a set of band pass filters accompanied with sampling rate conversion. But here, it is desirable that each filter has a linear phase, especially when a manipulation about amplitude is carried out in the transformed domain. In this case, if the phase distortion results in a change of waveform, we might have wrong amplitude information.

The system given in Fig.2 has a same structure as that of the discrete wavelet transform (DWT) based method [8], but using exactly LP bases for analyzer and synthesizer, it can hold correct spatial information also in the transformed domain.

The whole procedure of image processing is given as follows.

(i) The signal is decomposed into $J+1$ subbands $\{c_{-1,k}, \cdots, c_{-J,k}, d_{-J,k}\}$ through the system in Fig. 2 (a).

(ii) The subbands which contain undesired components but not defect signal are eliminated (Frequency filtering).

(iii) To reduce the undesired components occupying the same frequency band as the defect signal, the elements of the subbands not eliminated in (ii) are removed if their absolute value is smaller than threshold A (Amplitude filtering) [10].

(iv) The signal is reconstructed through the synthesizer in Fig. 2 (b).

Steps (i) to (iv) are applied to each horizontal line of ECT images considering that

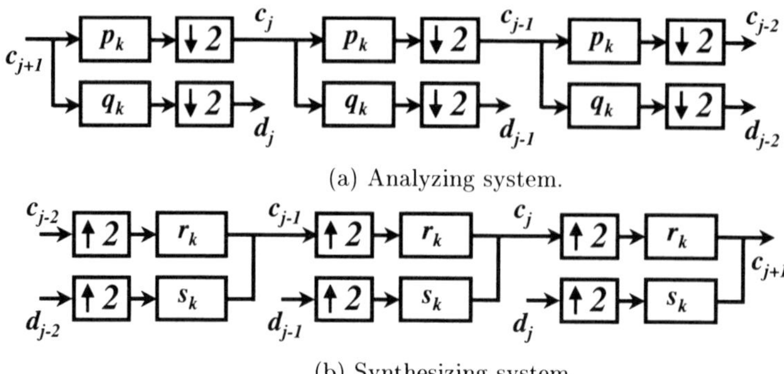

(a) Analyzing system.

(b) Synthesizing system.

Figure 2: Multiresolution analyzing system.

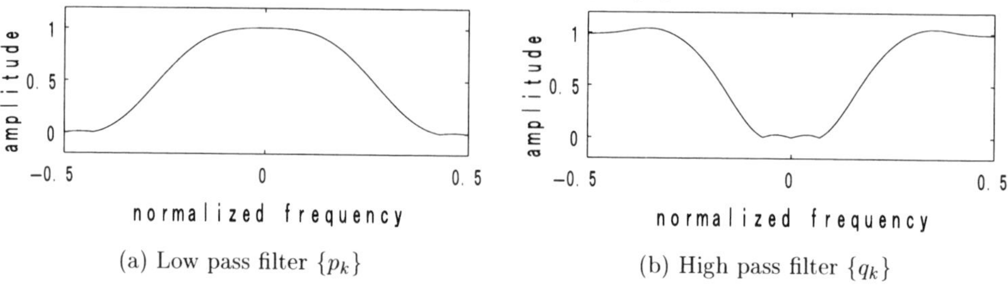

(a) Low pass filter $\{p_k\}$ (b) High pass filter $\{q_k\}$

Figure 3: Filter amplitude responses in analyzer.

the measurement is carried out along this direction and they can be dealt as independent signal in this stage. Each filter is designed through iterative min-max method [11], [12], which enables us flexible choice of frequency characteristics.

After reconstruction, the correlation between the theoretical waveform of defect (which is derived by computer simulation or actual measurement) and ECT image is calculated. This operation has an effect to emphasize defect-like signal and simplify the image, especially when the defect signal is small [5]. Then the image is compared with a reference image obtained from a sample previously known to be without defect. If a peak signal is detected only in test image, it is considered to be coming from a defect since undesired components have been already removed.

4. Examples

In this section, the effectiveness and problems of the proposed method are investigated through examples.

The benchmark data (NEL98-25) shown in Fig. 4 (a) contains defect peaks as shown inside of the broken ellipses, but due to the existence of the undesired components, the discrimination of defects is very difficult. Hence the image processing method described in the previous section is applied to this example. The filters $\{p_k\}$ and $\{q_k\}$ (amplitude responses are in Fig. 3) are designed with the order of 5 and 11, respectively. Namely, 9 multiplications are required to calculate one sample when the symmetry of them is considered. First,

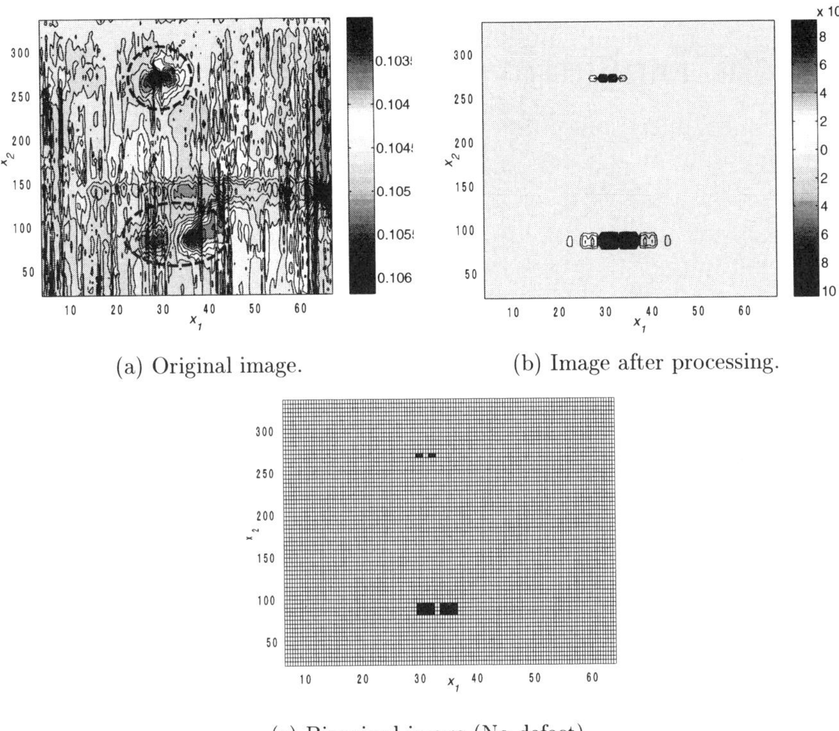

(a) Original image.　　　(b) Image after processing.

(c) Binarized image (No defect).

Figure 4: Image processing of benchmark data (NEL98-25).

the periodically extended image is decomposed into 4 levels and the signals which belongs to coefficients $\{c_{-1,k}, c_{-2,k}, d_{-4,k}\}$ or have an absolute value smaller than $A = 0.0005$ are removed. Then the image is reconstructed, and consequently, the defect peaks become clearer (Fig.4 (b)) compared to the original. The binarized image shows us the existence and position of defects.

The results of PCB data processing are also shown in Fig.5. In this case, we should carry out the comparison between reference (previously known to have no defect) and test image to discriminate the origin of the peak signal, i.e., they are coming from defect or distribution of wiring.

In spite of the effectiveness and the phase-distortion-less nature shown above, this method has a problem as shown in Fig. 6. Both wavelet processing and the proposed method clearly show the defect peak in undesired components, but residual high frequency components are still observed in Fig.6 (c). The low regularity (i.e., low smoothness) of the given bases is considered as a reason. This is the inferior point of the proposed method to wavelet processing which is designed with high regularity.

5. Conclusions

In this paper, we have presented an ECT image processing approach using subband technique for the extraction of the defect. The proposed method consists of two steps: (1) the

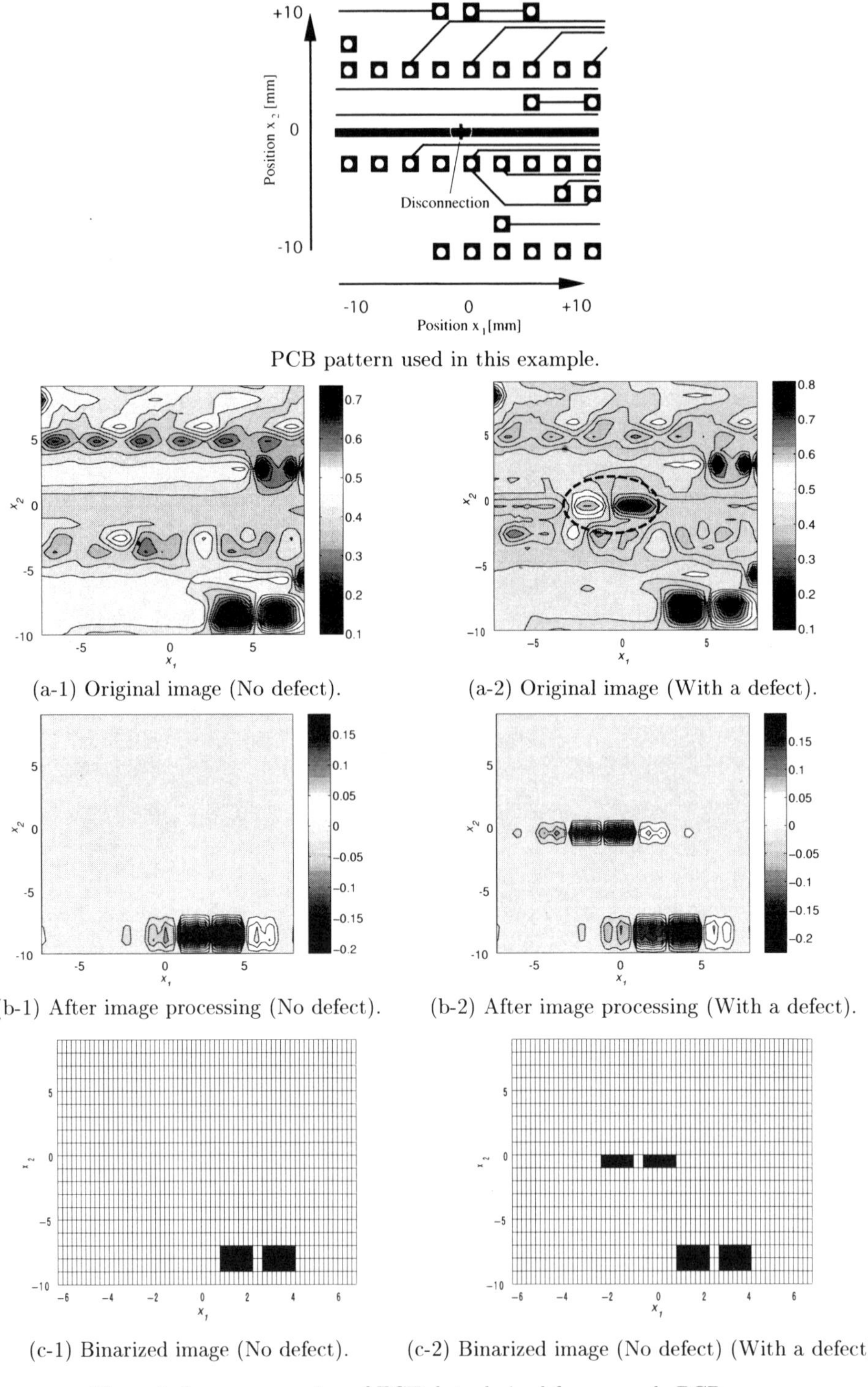

(a-1) Original image (No defect).

(a-2) Original image (With a defect).

(b-1) After image processing (No defect).

(b-2) After image processing (With a defect).

(c-1) Binarized image (No defect).

(c-2) Binarized image (No defect) (With a defect).

Figure 5: Image processing of ECT data derived from sample PCBs.

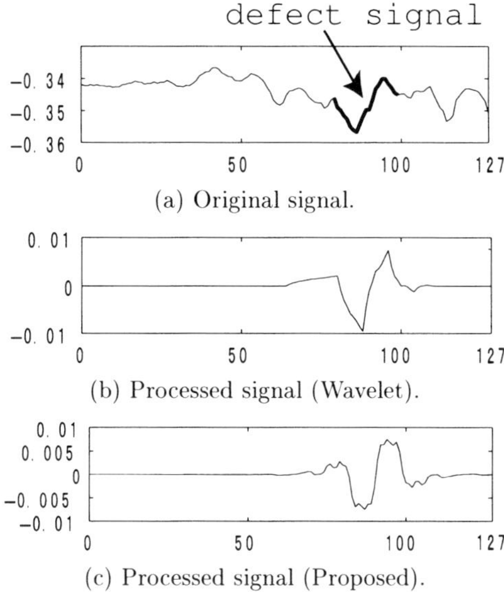

(a) Original signal.

(b) Processed signal (Wavelet).

(c) Processed signal (Proposed).

Figure 6: Results of ECT signal processing

reduction of the undesired signals using multiresolution analysis and (2) actual discrimination of defect-based signals from others, based on the comparison of two images (Step (2) is necessary for the case of PCBs). Using LP bases, the correct information on phase is maintained also in the transformed domain, and the parameters of the filters can be chosen more flexibly. Through examples, the ability of the proposed method to reduce the undesired components is demonstrated together with its problem, the poor regularity of the frequency characteristics.

As a future work, we should develop a design method of LP bases which have a higher regularity. The examples in this paper concerns only with the exinstence and position of defects. Hence another theme is the investigation of the usefulness of the proposed method for the extraction of other defect features.

References

[1] R. Mol, "Evolution of inspection tools for SG tubes," Nuclear Europe Worldscan, vol.3–4, pp.38–39, 1999.

[2] R. Leclerc, and R. Samson, "Eddy current array probe for corrosion mapping on ageing aircraft," Review of Progress in Quantitative Nondestructive Evaluation, vol.19A, pp.489–496, 1999.

[3] S. Yamada, H. Fujiki, M. Iwahara, S. C. Mukhopadhyay and F. P. Dawson, "Investigation of printed wiring board testing by using planar coil type ECT probe," IEEE Trans. on Magnetics, vol.33, no.5, pp.3376–3378, Sept. 1997.

[4] D. Kacprzak, T. Miyagoshi, S. Yamada, and M. Iwahara, "Inspection of printed circuit board by planar ECT probe," IEEE INTERMAG99, GR-04, Aug. 1998.

[5] T. Taniguchi, D. Kacprzak, S. Yamada, M. Iwahara, and T. Miyagoshi, "Defect detection of printed circuit board by using eddy-current testing technique and image processing," The 5th International Workshop on Electromagnetic Nondestructive Evaluation, 1999-30, Des Moines, Iowa, 1999.

[6] G. Chen, A. Yamaguchi, and K. Miya, "A novel signal processing technique for eddy-current testing of steam generator tubes," IEEE Trans. on Magn., vol.34, no.3, pp.642–648, March 1998.

[7] K. Kawata, S. Kumano, and M. Kurokawa "Signal processing of eddy-current signal by 2 dimensional wavelet transform," 8th MAGDA Conference, Hiroshima, 1999.

[8] T. Taniguchi, D. Kacprzak, S. Yamada, and M. Iwahara, "ECT image processing for PCB inspection by using wavelet transform," Journal of JSAEM, vol.9, no.1, March 2001 (in print).

[9] C. K. Chui, "An introduction to wavelets," San Diego, CA:Academic Press, 1992.

[10] D. L. Donoho, "De-noising by soft-thresholding," IEEE Trans. on Information Theory, vol.41, No.3, pp.613-627, 1995.

[11] M. Vetterli, "Filter banks allowing perfect reconstruction," Signal Processing, vol.10, no.3, pp.219–244, 1986.

[12] M. Ikehara, A. Yamashita, and H. Kuroda, "Design of two-channel perfect reconstruction QMF," Trans. on IEICE, vol.J75-A, no.8, pp.1333–1340, August, 1992.

Electromagnetic Nondestructive Evaluation (VI)
F. Kojima et al. (Eds.)
IOS Press, 2002

An approach for identifying current distributions by means of magnetic field sensor with high spatial resolution

Norio MASUDA
EMC Engineering Center, NEC Corporation, Kanagawa, Japan
Tatsuya DOI
Ashikaga Institute of Technology, Tochigi, Japan

Abstract. In this paper, we have carried out identification of pulse current distribution on a PCB (printed circuit board) by using the magnetic sensor with high spatial resolutions. Experimental examinations have shown that the time-varying of pulse current distributions can be identified by means of our approach, which is expected to clarify electromagnetic phenomena on a PCB.

1. Introduction

Development of high speed signal transmission techniques on electronic devices causes the electromagnetic compatibility (EMC) problems, such as leakage magnetic fields on printed circuit boards (PCB). Identifying working currents on a practical PCB is one of the important techniques for the EMC design of PCB [1,2]. Then, the working current identification problems of PCB requires any approach to identify current distribution from locally magnetic field distribution measured in both of high spatial resolution and high frequency. However, practical measured magnetic fields are under the influence of a magnetic field sensor's spatial characteristics. This means that the spatial characteristics of sensor should be taken into account in order to calculate current distribution from such measured magnetic fields.

In this paper, we have proposed an approach for identifying current distributions calculated from locally magnetic field distribution considering spatial characteristics of a magnetic field sensor. A magnetic field sensor with high spatial characteristics has been utilized for the magnetic field measurement of PCBs. At first, magnetic field distribution on simplified PCB has been measured by means of the magnetic filed sensor. Second, current distributions on the simplified PCB have been calculated by utilizing spatial characteristics of the sensor.

2. Identification of current distribution on simplified PCB

Fig. 1(a) shows the miniature magnetic field sensor having a maximum 250-μ m spatial resolution at a 6 dB degrading point, enabling the measurement of magnetic fields near LSIs and dense-PCBs. The characteristics and the general specification of the sensor are shown in table 1(a), 1(b). As shown in Fig. 1(b), at frequencies below 3 GHz, the output is approximately proportional to the frequency having the gradient of about 20dB/dec., permitting magnetic field calibration at less than 3 GHz. If the measurement target is a line on PCB, the current-calibration coefficient can be used to convert the measured magnetic field over the line into current. The design is based on a precise glass ceramic multilayer board, and the detector is 2 mm wide and 1 mm thick [3,4].

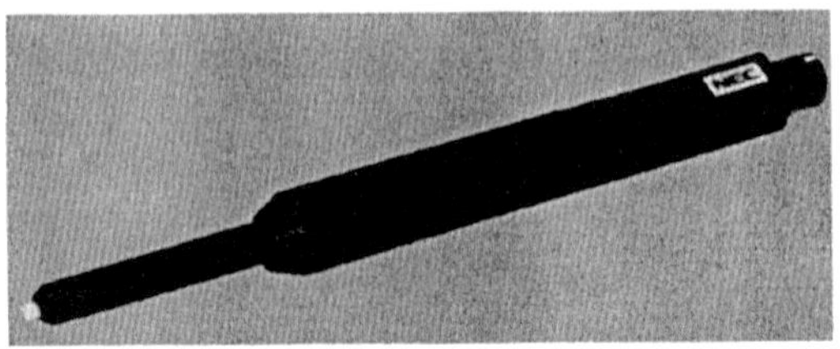

(a) External view.
(Photograph is supplied by NEC Glass Components, Ltd.)

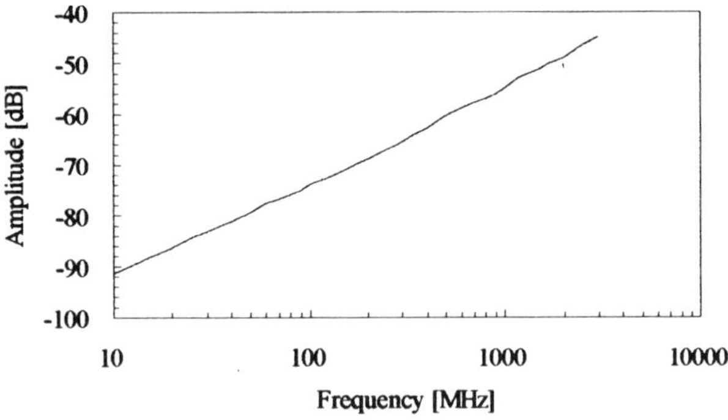

(b) The frequency characteristic measured by using a 50 ohms microstrip line as a magnetic field source. (Input power to the microstrip line is 0dBm)

Fig. 1. A miniature magnetic field sensor.

Table 1. Specifications of the magnetic field sensor.
(a) Characteristics.

Frequency range		10MHz – 3GHz
Electric field shielding		Less than -20dB
Spatial resolution	-6dB	About 0.25mm

(b) General Specifications.

Connector	SMA (receptacle type)
Impedance of the measurement receptacle	50 ohm
Size of outline (probe tip)	Length 135 mm x Dia. 12 mm (Thickness 1 mm x Width 2 mm)
Substrate material	Glass ceramics (ε_r=7.1)
Environment	Room temperature and room humidity

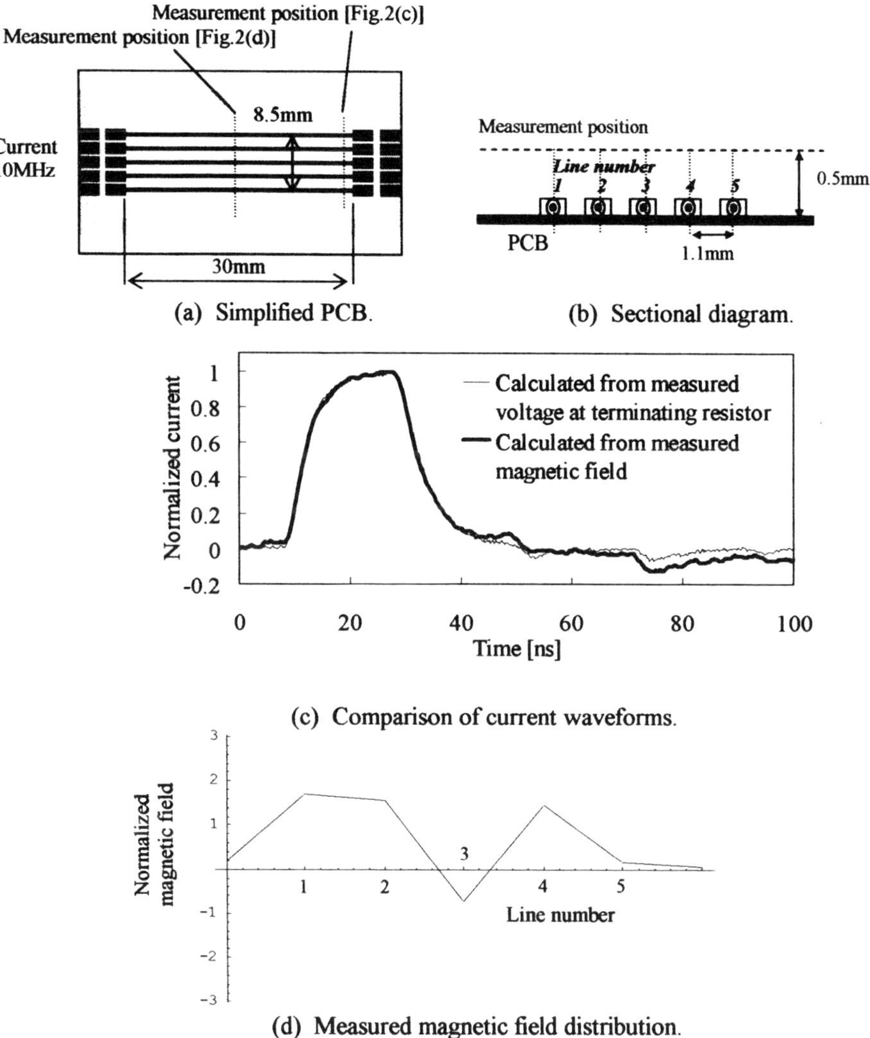

(a) Simplified PCB.

(b) Sectional diagram.

(c) Comparison of current waveforms.

(d) Measured magnetic field distribution.

Fig.2. Magnetic field measurement on PCBs with high spatial resolution.

Fig. 2(a) and 2(b) show the simplified PCB, sectional diagram, respectively. Fig. 2(c) shows the comparison of current waveform caluculated from measured voltage waveform and that of calculated from measured magnetic field by integrating output voltage waveform of magnetic field sensor. The voltage was measued at the terminating resistor (50 ohm) by a differential high impedance probe. The maginetic fields were measured at the hight of 20 μ m on the line close to the resistor. It is apparent from Fig. 2(c) that both current values are agreed well. Thus, it is shown that the current waveform is measured with high accuracy by means of the magnetic field sensor. Fig. 2(d) shows a magnetic field distribution X measured by means of the magnetic field sensor. This magnetic field distribution is obtained at the height of 5mm. Therefore the spatial resolution is degradated. Under such condition, we consider identifiation of current distribution in the lines on PCB from locally measured magnetic field.

When $\mathbf{X}'$ is magnetic field distribution to be obtained, whose elements correspond to the magnetic fields caused only by currents directly below each measured positions, system equations is

$$
\begin{pmatrix} H_1 \\ H_2 \\ \vdots \\ H_n \end{pmatrix} = \begin{pmatrix} c_{11} & c_{12} & \cdots & c_{1n} \\ c_{21} & c_{22} & & \vdots \\ \vdots & & \ddots & \vdots \\ c_{n1} & \cdots & \cdots & c_{nn} \end{pmatrix} \begin{pmatrix} H_1' \\ H_2' \\ \vdots \\ H_n' \end{pmatrix}, \qquad (1)
$$

or

$$
\mathbf{X} = C\mathbf{X}'. \qquad (2)
$$

where C is system matrix. And n is a number of measurement points.

Rows of system matrix C corresponds to the spatial characteristics of the sensor. Consequently, the system matrix C in eq.(2) becomes always a regular matrix. Then, magnetic field distribution $\mathbf{X}'$ to be obtained is obtained uniquely by

$$
\mathbf{X}' = C^{-1}\mathbf{X}. \qquad (3)
$$

Finally, currents in the lines on the PCB can be calculated only to apply the Ampere's law to the solution $\mathbf{X}'$.

Fig.3(a) is spatial characteristics of the sensor, which have practically measured under same condition in Fig.2. Each element of matrix C is obtained by substituting the value of each position in Fig. 3(a) to c_{ij} in eq. (1).

$$
c_{ij} = h_k, \quad k = i + j - 2, \quad i = 1,2,3,\ldots\ldots,n, \quad j = 1,2,3,\ldots\ldots,n. \qquad (4)
$$

Where, k indicates the position and h_k is the value in Fig.3(a).

Fig.3(b) is system matrix C calculated from eq. (4), whose diagonal elements are almost 1. Gray-code districts in the figure are correspondence with the values of c_{ij}. Inverse of the system matrix C can be obtained exactly. Fig.3(c) shows calculated current distribution by means of the solution (3).

As a result, our approach makes it possible to compute the currents on the PCB by using only the spatial characteristics of sensor. Also, the results have verified that the approach is capable of applying to identifying the current distribution on a PCB, which requires high spatial resolution.

3. Pulse waveform in wires

We have developed of a method for measuring current waveforms by integrating the ouput voltage of magnetic field sensor. In this paper, the approximate function is generated based on interpolation, analytically integrated, thus the current waveform is obtained. "Mathematica" is used for the operations of generating intepolation functions and integration.

Currents of each line is calculated from measured magnetic field distibution over PCB and spatial characteristics of sensor [Fig.3(a)]. Here, the pulse signal is impressed at 1st line in synchronism with 2nd and 4th, the signals are equal-strength. The input signal to 3rd line is dephased. The termination of 5th line is electrically opened, thus the current does not flow. Experiment conditions and input currents are shown in Table 2.

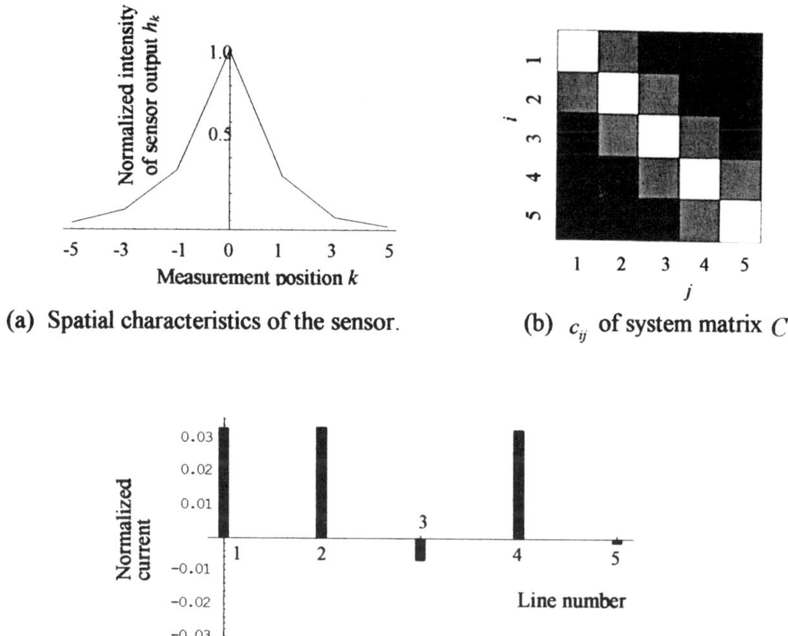

(a) Spatial characteristics of the sensor.

(b) c_{ij} of system matrix C.

(c) Calculated current distribution.

Fig.3. Identification of current distribution on the simplified PCB.

Fig. 4(a) shows an example of output voltage waveform of the magnetic field sensor. The period of pulse is 100 ns (corresponding to 10MHz) and the duration time of the pulse is 20 ns (corresponding to 20% duty). It is found out that the form of pulse is restored well which indicates the usefulness of the data-processing procedure. The magnetic fields can be converted to current by using conversion factor [3], thus waveform shown in Fig. 4(b) is regarded as current waveform. The dc component of magnetic field waveform was not close to zero. This dc component is caused by the slight voltage offset of oscilloscope. Fig. 5 shows the time varying magnetic field distributions across the PCB. The time varidation of number 3 line is different from others. Fig. 6 shows a stereoscopic 3-D view of time varying current distribution. Shades of gray in the figures are correspondence with the value of currents. There is 10.8 ns delays between the pulse of number 3 line and other lines. The value of delay time is close to that of shown in table 2.

It is clear that the working current distribution and waveform are indentified by applying proposed method in this paper. Futhermore, the current waveforms are identified as pulse shape.

Table 2. Measurement conditions and input current.

Line number	Current input
1,2,4	10MHz, 0dBm(Power), 20% Duty Pulse
3	10MHz,0dBm(Power), 20% Duty Pulse, Delay time 10.6ns (Measured)
5	Open

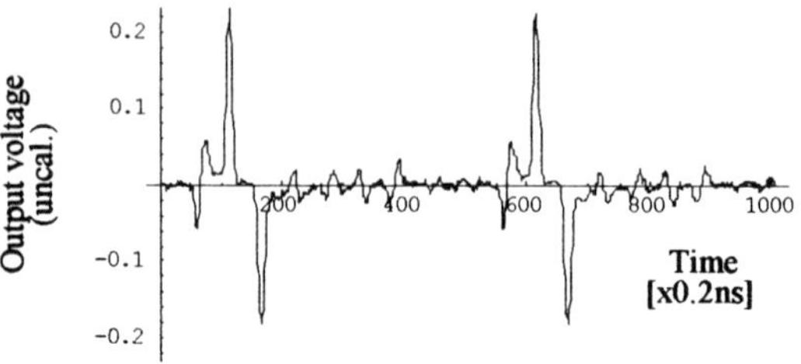

(a) Output voltage waveform of magnetic field sensor.

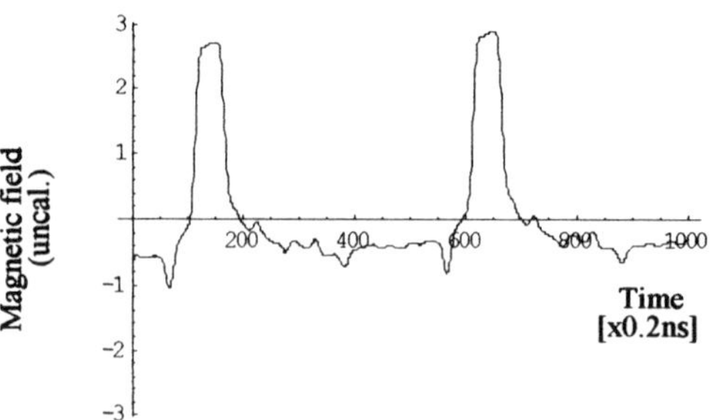

(b) Magnetic field waveform obtained by integration.

Fig.4. Waveform of magnetic field calculated from output voltage waveform of the sensor.

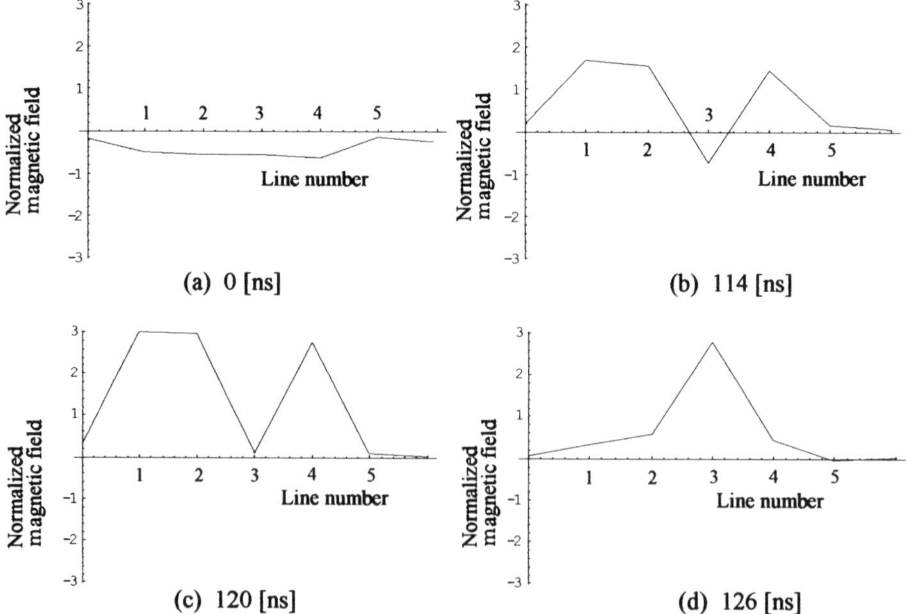

(a) 0 [ns]   (b) 114 [ns]

(c) 120 [ns]   (d) 126 [ns]

Fig.5. Magnetic field distrbutions calculated from output voltage waveform of the sensor.

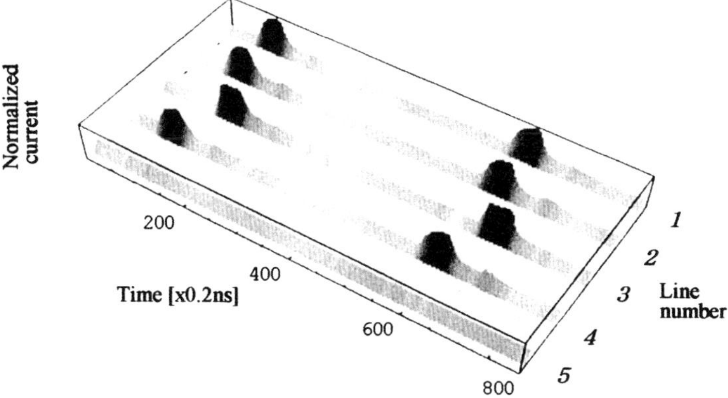

Fig.6. Stereoscopic 3-D view of time varying current distribution.

Table 3. Measurement conditions and input current.

Points, Pitch		3 points,　2.mm
Measurement gap		0.5 mm
Input signal	Number 1 and 3	50MHz, sine wave, Power : 0dBm,
Input signal	Number 2	Open

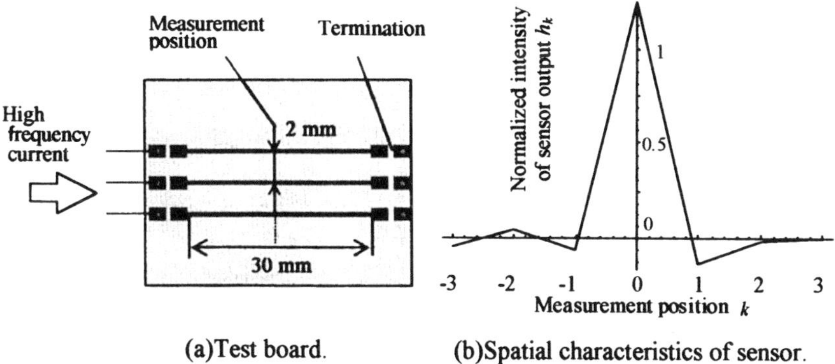

(a)Test board.　　　　(b)Spatial characteristics of sensor.

Fig.7. Microstrip lines.

4. Sinusoidal waveform in microstrip line

Mostly, PCBs of electronic equipment have a multilayer construction which inner layers is ground plane. Thus, we carried out the calculation for identifiying the current waveform in the microstrip lines constructed from a strip conductor and a ground plane. Fig. 7(a) and Fig. 7(b) show a test board, and spatial characteristics of sensor at the measurement height of 0.2 mm, respectively. The half band width of spatial characteristcs of sensor is narrower than that of Fig. 3(a), because there exists the effect of current distribution on grand plane. In other words, the magnetic fields generate from the currents distributed on ground plane are included with the measured magnetic fields over a microstrip line.

However, the currents on ground plane do not have little effect to the calculation, because the equation (3) performs operation for canceling out the effect. Fig. 8 shows the time varying current waveform distributed on the test board. It is apparent from Fig. 8 that the current distribution is identified even if the PCB has microstrip construction.

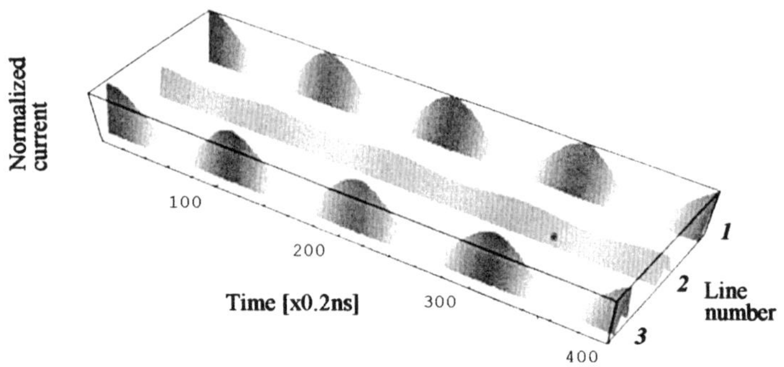

Fig.8. Stereoscopic 3-D view of time varying current distribution
in the microstrip lines.

5. Conclusions

In the present paper, we have proposed the approach for identifying high spatial resolution current distributions, which utilizes spatial characteristics of a magnetic field sensor. Identification of current distributions on a simplified printed circuit board have been carried out by means of our approach. As a result, the experimental examinations show validity of our approach to identify working current distributions on a PCB. Futhermore, we have presented that the proposed method is also useful for identification of current distribution on microstrip lines.

References

[1]T. Doi, S. Hayano and Y. Saito, "Magnetic field distribution caused by a notebook computer and its source searching", J. Appl. Phys., Vol. 79, No.8, pp.5214-5216, 1996.

[2] T. Doi and N. Masuda, "Leakage magnetic field source searching of micro processing unit on printed circuit board", Trans. IEEJ., Vol. 120-A, No.10, pp.5214-5216, 2000.(in Japanese)

[3] N. Masuda, N. Tamaki, T. Watanabe, K. Ishizaka, "A miniature high-performance magnetic-field probe for measuring high-frequency currents", NEC R&D, Vol.42, No.2, pp.246-250, 2001.

[4] T. Harada, H. Sasaki, and E. Hankui, "Time-Domain Magnetic Field Waveform Measurement Near Printed Circuit Boards ", Electrical Engineering in Japan Vol.125, No. 4, 1998, translated from Denki Gakkai Ronbunshi, Vol. 117 - A, No. 5, pp. 523-530, May 1997

Wavelet Basis Function Neural Network for Eddy Current Signal Characterization

Jaein Lim, Ping Xiang, Pradeep Ramuhalli, Satish Udpa and Lalita Udpa
Material Assessment Research Group, Department of Electrical and Computer Engineering
Iowa State University, Ames, IA 50010, USA

Abstract. A wavelet basis function (WBF) neural network is proposed to estimate the defect profile from eddy current inspection data of steam generator tubes. The calibration and phase gating schemes are first employed. Calibration is used to optimize the frequency and sensitivity settings, which are used to classifiy the actual defects relative to the reference defects. Phase gating is employed to improve the visibility of the defect after calibration. A WBF neural network is subsequently employed to map the signal appropriately to obtain the geometrical profile of the flaw. Test results showing the effectiveness of the approach are presented. A calibration curve method is also evaluated for a comparative study of results obtained with the WBF neural network.

1. Introduction

Inverse methods are used when the parameters governing the behavior of a physical process are to be estimated from a set of measured signals. However, the inversion processes are usually ill-posed. In nondestructive evaluation (NDE), the solution to the inverse problem involves prediction of defect parameters based on probe measurement signals [1]. One of the earliest and most popular approaches used in industry is the calibration curve method. An alternative approach involving the use of WBF neural networks is presented here.

WBF neural networks allow functional approximation using multiresolution wavelets. In this method, increasing or reducing the number of resolutions can produce multiple scales of approximation. Several such networks with varying architectures and employing different wavelets have been studied. Zhang and Beveniste [2] introduced wavelet networks employing the first derivative of a Gaussian function, which satisfies frames requirements, as a basis function. The locations of basis functions were determined by using a density function and then optimized by employing a stochastic gradient algorithm that is similar to the backpropagation algorithm. Zhang et al. [3] employ an orthogonal scaling function as a basis function, which provides a unique representation of the function being approximated. The locations of the basis functions are set using a dyadic scheme with the number of basis functions at resolution L. This neural network is generally not an RBF neural network since scaling functions can be radially nonsymmetrical. Lim and Bakshi et al. [4,5] presented WBF neural networks using both scaling function and wavelet as basis functions. This study employ the WBF neural network to estimate defect profiles from eddy current signals obtained in the inspection of steam generator tubes [6].

The multiresolution nature of the expansion implies that the number of resolutions in the network architecture can control the level of prediction accuracy and the confidence of the network prediction can be measured by observing the improvement in outputs.

In this paper, calibration and phase gating schemes are first employed. A calibration procedure is employed to optimize the frequency and sensitivity settings, which are used to classifiy the actual defects relative to the reference defects. A phase gating scheme is employed to improve the visibility of the defect after calibration. Data from sets of sensors are fused to estimate the defect profile as shown in Figure 1, where WBFNN represents a wavelet basis function neural network. The data was collected using a pancake coil rotating probe at four excitation frequencies; 400 kHz, 300 kHz, 200 kHz, and 100 kHz.The four signals are combined using weighted average fusion method into a signal. A WBF neural network is finally employed to map the fused signal appropriately to obtain the geometrical profile of the defect. A calibration method is also evaluated for a comparative study of results obtained with the WBF neural network.

2. Calibration

In order to provide accurate defect analysis, it is necessary to obtain a piece of tube of the same material and dimensions as the tubes to be inspected and use it for calibrating the instrument. The calibration step is essential to optimize the frequency and sensitivity settings required in order that actual defects can be classified relative to the reference defects. The calibration procedure consists of three steps; normalization, phase rotation, and voltage scaling. The normalization step suppresses the signals from structural discontinuities such as tube support plates or tube sheets. This is obtained by removing the median value along each column and then along each row. The median along each row is subtracted to remove the trend along the axial direction. In phase rotation, a reference angle is first set to the through-wall hole data (40° for MIZ30 and 26° for MIZ18). The difference between the phase angle corresponding to the through-wall hole and reference angle is subsequently calculated as shown in Figure 2. These phase differences of each channel are applied to the raw data channel subsequently. In voltage scaling, the magnitude of signal corresponding to a through-wall hole is first scaled and set at a fixed value (20 volts). The raw data is then scaled, channel by channel.

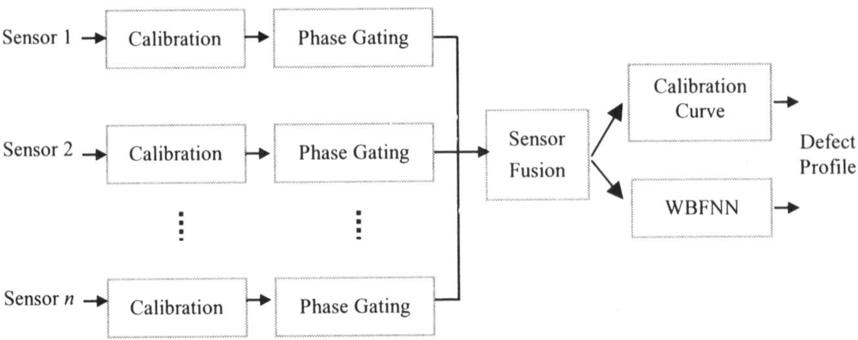

Figure 1. Defect Characterization Scheme

3. Phase Gating

Although the calibration procedure improves the visibility of the defect, additional artifacts are introduced due to phase wrapping. In order to minimize the artifacts, we employ a thresholding approach to highlight the defect indication more clearly. The threshold is computed dynamically based on the statistical properties of the signals within a swath of data.

The phase gating step includes two thresholding procesures; one for magnitude and the other for phase. In magnitude thresholding, the signals whose magnitude are less than the thresdhold level are treated as noise and replaced with zeros both in the horizontal and vertical components of eddy current signal. In phase thresholding, the signals with phase in the wrong quadrant (3 and 4) are eliminated. Signals whose phase lie between 5° and 175° are retained.

4. Weighted Average Data Fusion

The weighted average is one of the simplest and most intuitive methods of data fusion. This technique takes a weighted average of redundant information provided by a group of sensors [7]. The weighted average of n sensor measurements x_i is defined as

$$\bar{x} = \sum_{i=1}^{n} w_i x_i \quad , \text{where} \sum_{i=1}^{n} w_i = 1 \text{ and } 0 \le w_i \le 1 \tag{4-1}$$

The weights account for the differences in accuracy between sensor measurements and are estimated using the variances of the signals as follows:

$$w_i = \frac{\dfrac{1}{\sigma_i^2}}{\sum_{j=1}^{n} \dfrac{1}{\sigma_j^2}} \tag{4-2}$$

where σ_i^2 is the variance of sensor measurements x_i.

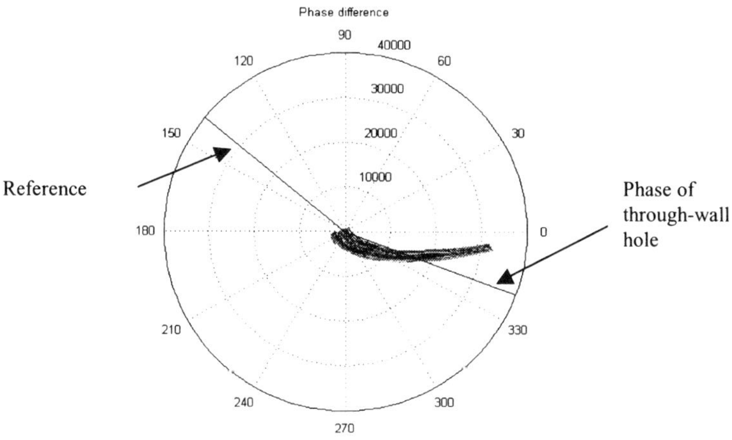

Figure 2. Phase difference

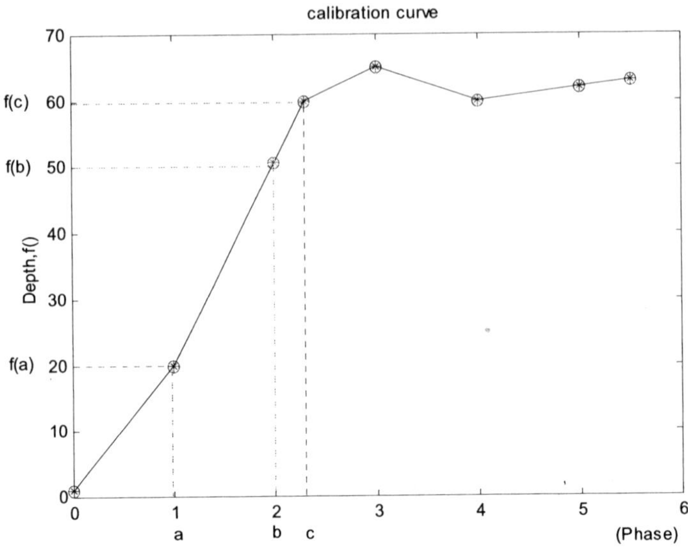

Figure 3. Calibration curve based on linear interpolation scheme.

5. Calibration Curve Method

The calibration curve relates the phase of the eddy current signal with the depth of the defect. We employ linear interpolation techniques to estimate the depth, when a signal whose phase is intermediate between sample values is to be analyzed. For example, we first assume that the calibration curve f is piece-wise linear between each of the known mapping points. The relationship used to find the depth when the signal phase is b is given by:

$$f(b) = f(a) + \frac{b-a}{c-a} \cdot [f(c) - f(a)] \qquad (5\text{-}1)$$

Figure 3 shows a typical calibration curve.

6. Wavelet Basis Function Neural Network

WBF neural networks allow functional approximation based on multiresolution wavelets using a limited number of wavelets, ψ, and scaling functions as follows [5]:

$$f(\mathbf{x}) = \sum_{k=1}^{N_L} s_k^L \phi_k^L \left(\| \mathbf{x} - c_k \| \right) + \sum_{j=1}^{L} \sum_{k=1}^{N_j} d_k^j \psi_k^j \left(\| \mathbf{x} - c_k^j \| \right) \qquad (6\text{-}1)$$

where L is the number of resolutions, N_j is the number of translated basis functions at resolution j, k is the translated point, ψ_k^j is a dilated and translated version of a mother wavelet, and c_k is the center of basis function at translation point k. The architecture of a

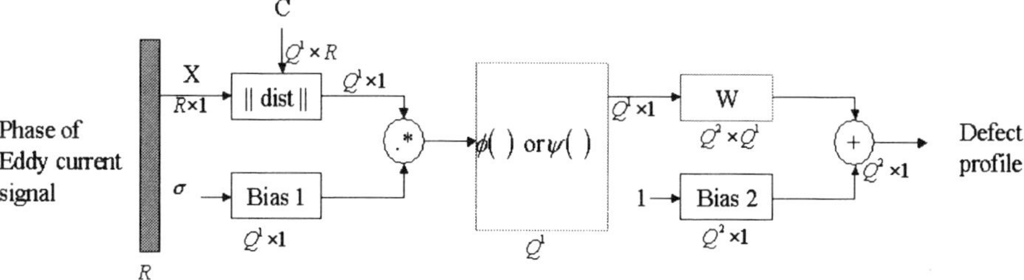

Figure 4. Wavelet Basis Function (WBF) Neural Network

typical WBF neural network implementing the above equation is similar to radial basis function (RBF) neural networks in that both neural networks have a single hidden layer. In contrast to RBF neural networks, a WBF neural network employs a family of wavelets as basis functions and has sets of wavelet function nodes depending on the number of resolutions. In this method, the number of resolutions is either increased or decreased to control the level of prediction accuracy. The number and location of the basis function centers C is determined using the well known ISODATA clustering scheme. σ is the standard deviation that is used as a bias of network, R is the length of input eddy current signal, Q^1 is the number of training data set, Q^2 is the length of defect profile, and W is the weight matrix relating the input phase and the output defect profile.

6.1. Scaling and Wavelet Basis Functions

Scaling and wavelet functions are employed as basis functions so that the norm error of the function approximation is minimized using a minimum number of functions [5]. An intuitive procedure is used to select these functions. Wavelet basis function with a high degree of smoothness are usually chosen for obtaining good generalization in the case of multidimensional problems with limited training data. In this paper, the Mexican hat function is employed as a wavelet.

$$\psi(x) = (1 - x^2) \exp\left(-\frac{x^2}{2\sigma^2}\right) \tag{6-2}$$

where σ is the standard deviation. This wavelet decays rapidly in the time and frequency domains and the frame constant is very near unity. A Gaussian radial basis function is used as a scaling function.

$$\phi(x) = \exp\left(-\frac{x^2}{2\sigma^2}\right) \tag{6-3}$$

6.2. Centers Selection

The centers of the basis functions can be estimated using the k-means clustering algorithm or one of its several variants. The number of centers at each resolution can be determined heuristically. Previous studies have shown that the performance of the network is highly dependent on the location and number of centers. Therefore, an LMS algorithm is typically employed in order to obtain better performance instead of using the k-means clustering algorithm not only for determining the optimal locations, but also for determining the widths of the centers. However, such an algorithm is unsuitable for training a multiresolution network.

In this paper, we use a dyadic center selection scheme for calculating the location and number of centers [5]. Since the task at the lowest resolution is to coarsely approximate the function, the locations of basis functions are chosen using the ISODATA clustering algorithm. This corresponds to the first term in the approximation given by Equation (6-1). Subsequently, new centers, which correspond to the second term in Equation (6-1), are calculated using the dyadic selection scheme to obtain finer approximations. The dyadic selection scheme is similar to the dyadic dilation scheme in wavelet theory. The centers at the next resolution are expanded by dividing the Euclidean distance between adjacent centers at the previous resolution. If too may basis functions are used and overfitting has occurred, a center pruning algorithm is used in order to eliminate unnecessary centers.

7. Results and Conclusions

The algorithms were evaluated with eddy current data obtained from defects machined in tube samples. The data was collected using a three coil rotating probe (two pancake coils and a plus-point coil) at several excitation frequencies. In this paper, data from pancake coil rotating probe was used. Figure 5 and Figure 6 show the outputs of the calibration and phase gating stages respectively. Figure 7 indicates the calibration curve obtained from calibration data.

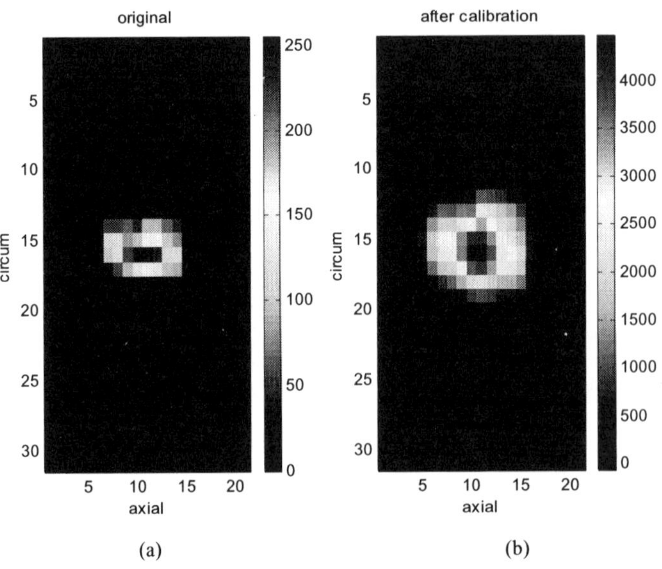

Figure 5. Calibration result of vertical component; (a) original, (b) after calibration.

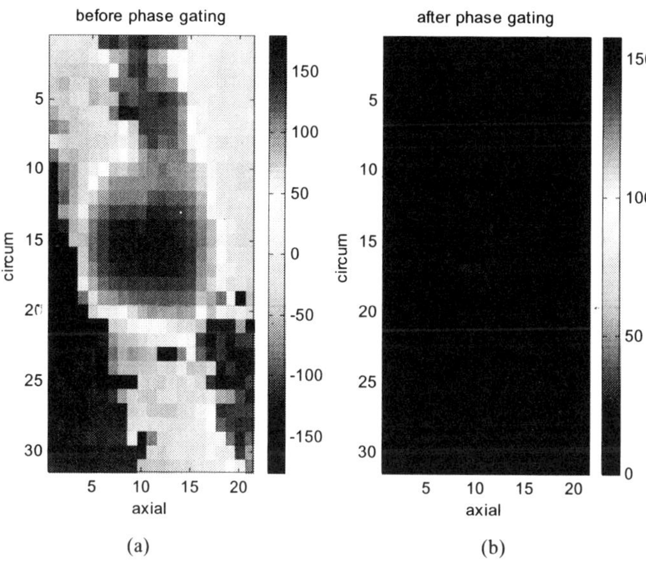

Figure 6. Phase gating result of eddy current signal; (a) before phase gating, (b) after phase gating

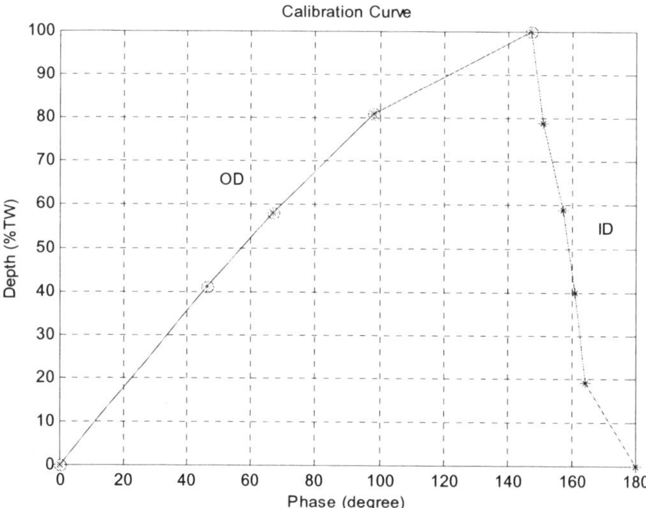

Figure 7. Calibration curve.

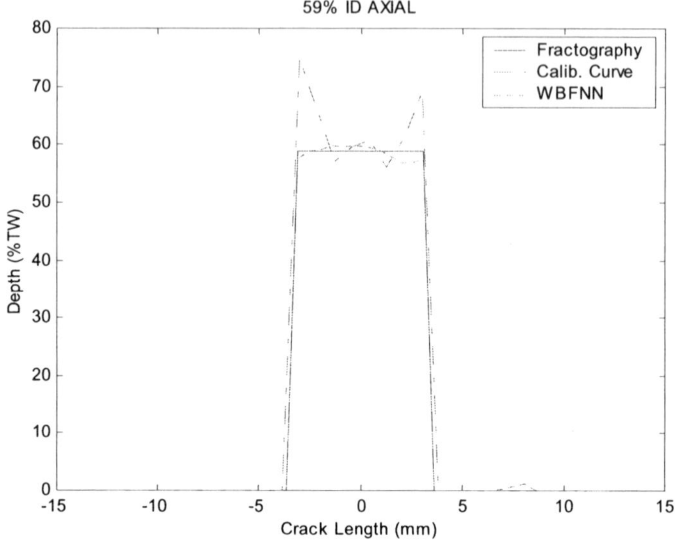

Figure 8. Defect characterization result based on calibration curve method and WBFNN.

Figure 8 describes the defect characterization result based on calibration curve and WBF neural network. The test data was a deep 59% ID axial defect. The result shows that WBF neural network shows considerable promise in predicting defect profiles. Additional improvement can be obtained using an adaptive boundary extraction algorithm to locate the defect area and using data fusion techniques to improve the characterization result.

References

[1] L. Udpa and W. Lord, A Discussion of the Inverse Problem in Electromagnetic NDT, in Review of Progress in Quantitative Nondestructive Evaluation, vol. T, New York, 1986, pp. 375-382.

[2] Q. Zhang and A. Benveniste, Wavelet networks, in IEEE Transactions on Neural Networks, **3**, no. 6, 1992, pp. 889-898.

[3] J. Zhang, G. G. Walter, Y. Miao, and W. N. W. Lee, Wavelet neural networks for function learning, in IEEE Transactions on signal processing, 43, no. 6, 1995, pp.1485-1497.

[4] B. R. Bakshi and G. Stephanopoulos, Wavelets as basis function for localized learning in a multiresolution hierarchy," *AICheE J.*, vol. 39, no. 1, 1992, pp. 57-81.

[5] J. Lim, S. S. Udpa, L. Udpa, and M. Afzal, Multisensor Fusion for 3-D Defect Characterization using Wavelet Basis Function Neural Network, To appear in Review of Progress in Quantitative Nondestructive Evaluation, vol. 20, 2001.

[6] J. Stolte, L. Udpa, and W. Lord, Multifrequency Eddy Current Testing of Steam Generator Tubes using Optimal Affine Transform, in Review of Progress in Quantitative Nondestructive Evaluation, vol. 7A, New York, 1988, pp. 821-830.

[7] M. A. Abidi and R. C. Gonzalez, Data fusion in robotics and machine intelligence, Academic presss inc., San Diego, 1992.

Electromagnetic Nondestructive Evaluation (VI)
F. Kojima et al. (Eds.)
IOS Press, 2002

Magnetic Transition due to Tensile Deformation in SUS 304 stainless steel

Y. Okino, T. Ueda, S. Takahashi, J. Echigoya, and K. Mumtaz
Faculty of Engineering, Iwate University, 020-8551 Morioka, Japan

Abstract. The martensite phase, such as ferromagnetic, can be created in SUS304 stainless steel by plastic deformation near the part with high dislocation density. The information about dislocation density can be obtained by the magnetic measurement techniques accurately. The occurence of martensite transformation however, depends on various factors such as dislocation density, grain boundaries, chemical composition, external stress and temperature. The first three factors are material related and the last two are external factor. The relationship between these factors should be made clear.

The role of grain boundaries are investigated in the present study. The grain boundaries are obstacles for the dislocation motion. At the same time the grain boundary makes the austenite state stable. The coercive force provides us the information on the shape of the martensitic phase.

1. Introduction

NDE of nuclear reactors is the most pressing problem. The generic interests are focus on the heat transfer tubes and pressure vessels. The currently employed NDE methods are eddy currents, ultrasonic testing , X-ray measurements etc., but the subject of investigation is the crack; the discovery of micro-cracks and monitoring of their development. The new NDE method is introduced recently and the subject is the degradation before the cracks appear by the use of relationship between magnetism and lattice defects in A533B steel [1]. The mechanisms of magnetic changes due to metal fatigue are different in low carbon steels such as A533B, and SUS 304 stainless steels; the former is due to the magneto-elastic coupling energy and strain field of lattice defects and the latter is the magnetic transition due to martensitic transformation.

Another interest of NDE in stainless steels is detection of stress corrosion cracking (SCC). The SCC is induced by three factors: the environment, the internal stress and the chemical in homogeneity. The internal stress is important in SCC as well as metal fatigue. NDE of the internal stress and the chemical inhomogeneity would be made possible by using the magnetic methods in connection with the martensitic transformation. The NDE of internal stress before the appearance of cracks is of our interest and the purpose of the present study is to investigate the relation between magnetic properties and lattice defects due to plastic deformation in SUS 304 stainless steel. The magnetic transition due to plastic deformation has been investigated in SUS 304 stainless steel for a long time. The data are inconsistent. The reason is that the magnetic transition depends on the experimental conditions, such as Ni content, test temperature and the method of introducing plastic

Table 1 Chemical composition , annealing temperature and grain size number of samples

Sample number	Chemical composition								Annealing temperature	Grain size number
1	C	Si	Mn	P	Ni	Cr	S	Fe	1323K	6
2	0.06	0.43	1.16	0.33	8.33	18.44	0.009	Bal.	1533K	1

deformation. For the construction of the data base under the precise experimental condition, especially, the initial condition is important. In the present study we investigated the influence of the grain size on the martensitic transformation induced by plastic deformation.

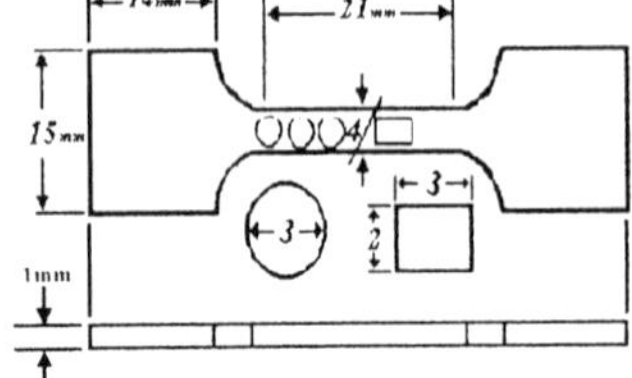

Fig.1 Shape and size of sample

2. Experimental Procedure

The chemical composition and the grain size of SUS-304 stainless steel samples are shown in Table 1. Shape and size of sample is shown in Fig.1. In order to change the grain size, various annealing temperatures (1323K and 1533K) were used after keeping the samples for 1 hour in the furnace and followed by water quenching. Grain sizes of samples were measured using a optical microscope. The grain size of sample 1 is about 40 μm and is about 100 μm for sample 2.

The samples were plastically deformed using Instron tensile machine at room temperature with a strain rate of 0.5 mm/min to examine the relationship between martensitic transformation and tensile deformation. The martensite phase transformed due to plastic deformation was detected by the SQUID magnetometer and X-ray diffraction. Coercive force was measured by vibrating sample magnetometry (VSM).

3. Results and Discussion

Results of tensile test are shown in fig.2. It shows work hardening in each sample. Sample 2 with coarse grains show striking work hardening at near 400MPa. This may be due to an larger extent of martensitic transformation in sample 2. Yield stress is about 270 MPa for Sample 1 and about 220 MPa for sample 2.

Fig. 3 shows the relationship between magnetization and magnetic field measured by SQUID magnetometer in sample 1. Magnetization increases with the increase of tensile stress. At rupture stress the saturation magnetization reached a maximum of about 55emu/g.

Fig.4 shows the relationship between magnetization and magnetic field for sample 2. In case of sample 2 with coarse grain size very high value of magnetization (110 emu/g) was obtained at rupture stress. As compared to sample 1 the value of magnetization is about twice in sample 2.

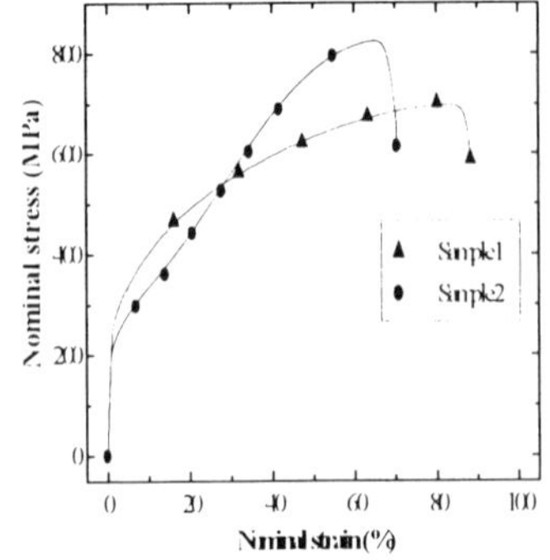

Fig.2 Stress-strain curves of samples

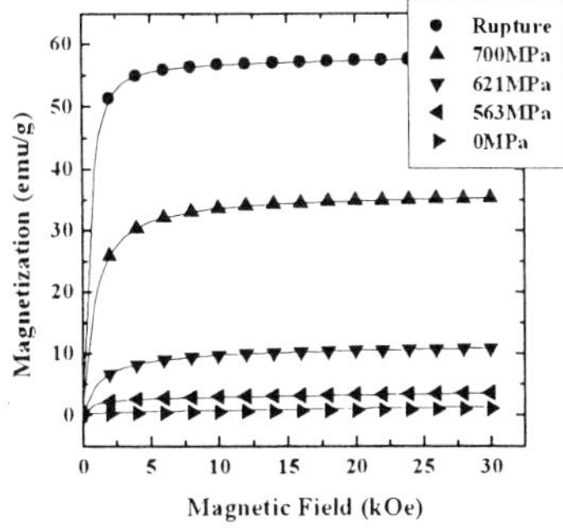

Fig.3 Magnetic curves of sample 1

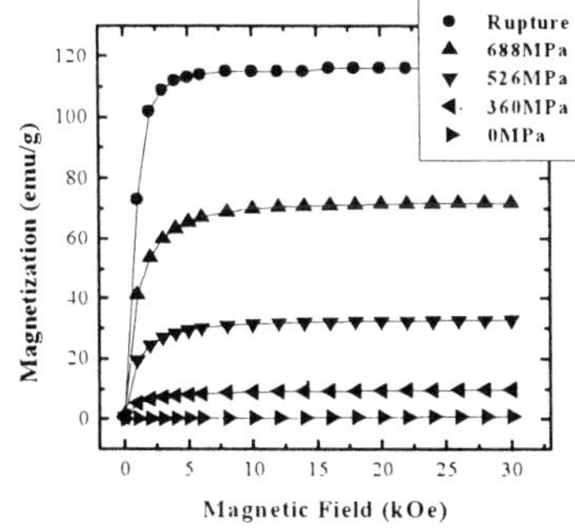

Fig. 4 Magnetic curves of sample 2

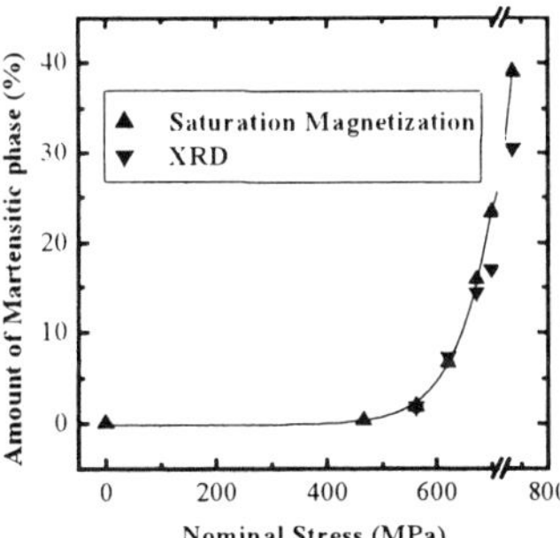

Fig.5 Amount of martensite phase versus
nominal stress of sample 1

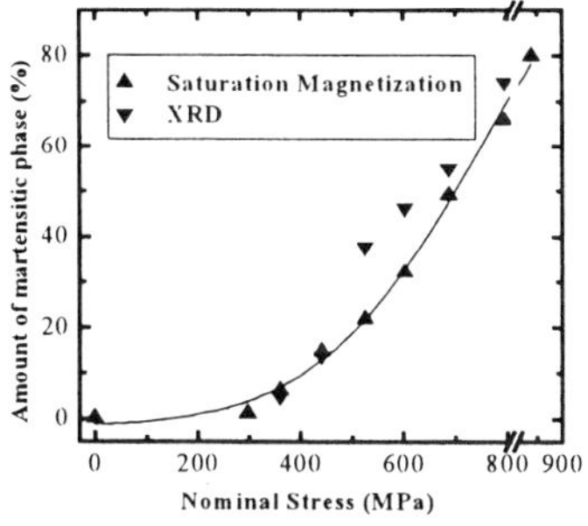

Fig.6 Amount of martensite phase versus
nominal stress of sample 2

Figs. 5 and 6 shows the amount of martensite phase (%) versus tensile deformation for sample 1 and 2. The amount of martensite phase (%) was measured from the saturation magnetization and by X-ray diffraction methods, and was represented by ▲ and ▼ triangles, respectively. At rupture stress about 80 % martensitic phase was detected in sample 2 whereas in sample 1 the martensitic phase was about 40 %.

Magnetic techniques are more sensitive to the amount of ferromagnetic martensitic phase even at very low tensile stress where the X-ray diffraction method fail to detect the amount of martensite. For X-ray diffraction pattern, martensite phase is not detected in low strain

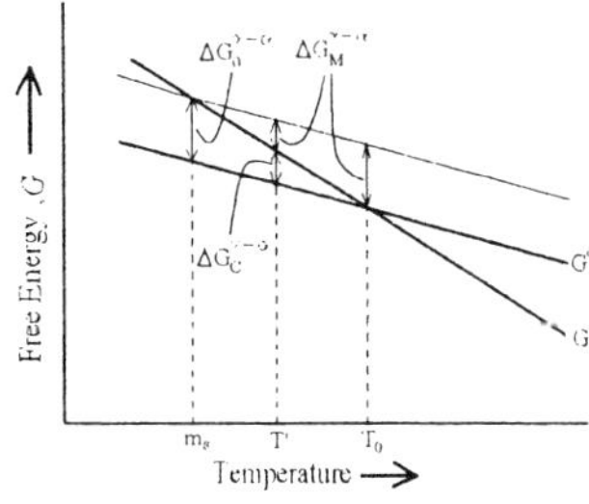

Fig. 7 Temperature dependence of the free
energies of α and γ

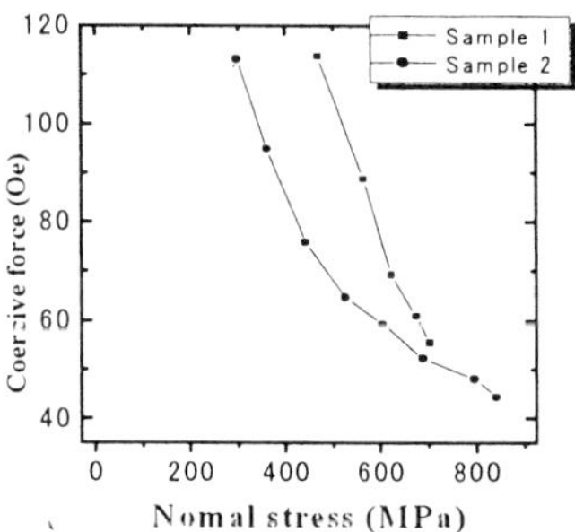

Fig. 8 Nominal stress versus
coercive force of sample 1

area. Martensite phase less than 3 % cannot be measured by X-ray method, because in low strain area the amount of martensite phase is small. Amount of martensite phase is about 40% at rupture stress and increases rapidly after 600MPa in sample 1.

In sample 2 martensitic phase increases rapidly after tensile stress of a 350 MPa. As compared with sample 1 martensitic transformation in sample 2 occurred in the early stage of plastic deformation.

Generally saturation magnetization of magnetic material is represented by equation (1) [2-3].

$$M_S = \frac{M_B \cdot \mu_B \cdot b}{\rho \cdot a^3} \quad , \qquad (1)$$

where, M_s is the saturation magnetization, M_B is the saturation magnetic moment of atom, μ_B is Bohr magneton, ρ is the density of the magnetic material, a is the lattice constant, and b is the number of atom in a unit lattice of the magnetic material.

For example, the calculated saturation magnetization of pure iron is about 222 emu/g by using saturation magnetic moment of Fe atom $2.22 \mu_B$. In the case of stainless steel considering the effect of Ni, Cr etc., saturation magnetization is about 175~154 emu/g [2]. In this study, saturation magnetization of 100% martensite phase is 154 emu/g.

The martensitic transformation depends on the relative magnitude of free energies of α-(martensitic) and γ-(austinite) phases; these free energies are a function of the chemical compositions. Fig.7 shows temperature dependence of the free energies of α- and γ-phases at homogenous chemical composition. At temperature below T_0 driving force does not depend on diffusion from parent phase. Martensitic transformation is usually treated as diffusion-less phase transformation. The driving force for transformation from α to γ phase grow under temperature T_0, the lower the temperature is the larger the driving force will be.

m_s is the martensite start temperature; the martensitic transformation occurs spontaneously below m_s. The difference between two free energies $\Delta G_0^{\gamma-\alpha}$ at m_s is equivalent to the energy for the martensitic transformation overcoming the interface energy between the two phases at m_s. Between m_s and T_0 the mechanical driving force $\Delta G_M^{\gamma-\alpha}$, is necessary for martensitic transformation by plastic deformation, which is equivalent to the difference $\Delta G_0^{\gamma-\alpha}$ and the chemical driving force $\Delta G_C^{\gamma-\alpha}$. The mechanical driving force is caused of $\Delta G_M^{\gamma-\alpha}$ at T'. In tensile deformation the driving force necessary for the initiation of martensitic transformation is attained above the normal m_s temperature. Therefore, a change in m_s can be accounted for simply by taking into consideration the contribution of additional mechanical free energy of the reaction.

In a wider sense grain boundary is a site of lattice fault, and once nucleation of martensite occurs it may transform easily there. Atom at grain boundary is stable comparatively

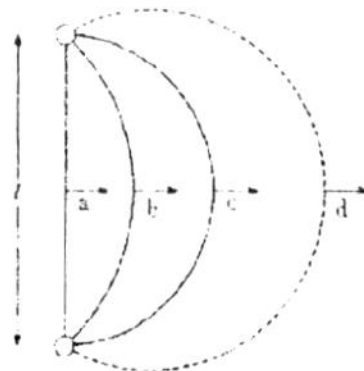

Fig.9 Incontinuously movement domain wall are placed under restraint both ends

Fig.10 Domain wall are placed under restraint both ends by martensite phase

because neighboring atomic restriction is small and lattice fault at near grain boundary disappeared in grain boundary, therefore, it is hard to for nucleation to occur. Accordingly, if grain size is small, parent phase is stable and amount of martensitic transformation is low.

Fig. 8 shows the relationship between coercive force measured by VSM and the deformation stress. The direction of magnetic field is parallel to the tensile stress direction. In each sample, coercive force decreases with the increase of tensile deformation.

Coercive force is given by the following equation:

$$(H_0)_{max} = \frac{\pi \lambda \sigma_0}{2I_s \cos\theta} \quad , \quad (2)$$

where λ is magnetostriction constant, σ_0 is internal stress, and I_s is spontaneous magnetization of magnetic domain.

Domain wall are placed under restraint at both ends. In case of $H=0$ domain wall is straight like **a** in fig.9, but as H increases, there is a pressure P extending on the domain wall which is given by:

$$P = 2I_s H \cos\theta \quad \cdot \quad (3)$$

The increase in pressure make domain wall curve like **b** in fig. 9. At this time, the radius of curvature is given by:

$$\frac{\gamma}{r} = 2I_s H \cos\theta \quad , \quad (4)$$

where γ is the energy of domain wall at unit volume, r is the radius of curvature.

Accordingly, coercive force at **c** is given by:

$$H_0 = \frac{\gamma}{I_s l \cos\theta} \quad (5)$$

Discontinuous movements of domain wall are placed under restraint at both ends, martensite phase and austenite phase exist at constant crystallographic direction. Accordingly coercive force decreases as tensile deformation increases.

4. Conclusions

Magnetic methods noticed sensitively a very small quantity of martensite phase, which can't be noticed by X-ray method. Accordingly, the use of magnetic methods are effective for NDE of martensitic transformation induced by plastic deformation.

In work-induced martensitic transformation, grain boundaries obstruct the growth of martensite phase. Coarse grain decreases grain boundaries and resulted in martensite phase. Accordingly, grain boundary makes m_s temperature lower.

Finally, the initial condition of SUS 304 stainless steel is important in NDE using magnetic methods.

References

[1] S.Takahashi, J.Echigoya and Z.Motoki " Magnetization curves of plastically deformed Fe metals and alloys " JOURNAL OF APPLIED PHYSICS, Vol.87, No.2, pp 805, January 2000.
[2] S.S.Hecker, M.G.Stout, K.P.Staudhammer and J.Smith: Metall. Trans. A. vol.13A, p619. (1982)
[3] L.E Murr, K.P.Staudhammer and S.S. Hecker: Metall. Trans. A. vol.13, p627. (1982)

Electromagnetic Nondestructive Evaluation (VI)
F. Kojima et al. (Eds.)
IOS Press, 2002

The Potential of NDE by Magnetic
Methods in SUS 304 Stainless Steel

K. Mumtaz, S. Takahashi, J. Echigoya and T. Ueda

Faculty of Engineering, Iwate University, 020-8551 Morioka, Japan

Abstract. The present work demonstrates that magnetic non-destructive evaluation techniques can be useful for detecting the presence and extent of ferromagnetic α' martensitic phase induced in SUS 304 stainless steel specimens due to plastic deformation after annealing at various temperatures and heat treatment sensitization.

The amount of α' martensite phase transformation is found to be dependent on annealing temperature and the increase in martensitic phase is observed at annealing temperature between 1003 and 1103 K. A remarkable increase in α' martensitic phase transformation is observed at annealing temperature of 1553 K. A good correlation between ferromagnetic martensitic phase and magnetic field has been obtained. Saturation magnetization is increased and coercive is decreased depending on the volume of ferromagnetic phase transformation.

Chromium carbide is precipitated along the grain boundaries during sensitization at 950 to 1023 K and at the same time, a zone depleted of chromium and carbon is formed in the region adjacent to chromium carbide. The ferromagnetic martensitic phase is also induced in the specimen by sensitization at temperature between 950 and 1023 K and high value of saturation magnetization is obtained at sensitization temperature of 1023 K. The magnetic transition indicates that enough volume of ferromagnetic martensitic phase is formed in the specimen at 1023 K. The ferromagnetic phase percentage decreases to a low value above 1023 K.

1. Introduction

SUS 304 stainless steel is commonly used in chemical and nuclear power plant components because of its excellent tensile strength and resistance to corrosion. However, the SUS 304 stainless steel undergoes structural degradation in service due to irradiation, metal fatigue and stress corrosion. Degradation causes transformation of paramagnetic austenite to ferromagnetic martensite in SUS 304 stainless steel.

The importance of studying NDE techniques is to determine the secular degradation i.e., the crack initiation before it appears in SUS 304 stainless steel. The NDE techniques useful for crack detection are magnetic measurements of martensitic transformation due to internal changes in the material and forecasting [1-2]. By monitoring the amount and nature of phase changes occurring in stainless steel structure, one can determine when to take preventive measures and perhaps, intervene at early stages of the process. Thus the risk of cracking can be minimized while at the same time increasing reliability of stainless steel components.

The main objective of this research is to get the quantitative relationship between material damage and martensitic transformation induced by plastic deformation at various annealing temperatures, and heat treatment sensitization by high precision magnetic measurement which is a powerful non destructive evaluation technique.

2. Experimental Procedure

The material used for the investigation was cold rolled plate of SUS 304 stainless steel. The chemical composition of the SUS 304 stainless steel is shown in Table 1. All the specimens used in these experiments are first solution annealed in vacuum at a temperature of 1323 K for 1 hour and then water quenched.

Table 1 Chemical composition of SUS304 stainless steel

	C	Cr	Ni	Mn	P	S	Si	Fe
Wt%	0.05	18.36	8.32	0.79	0.031	0.003	0.57	Bal.

Before tensile deformation, specimens are annealed at different temperature (850 to 1253 K) for 1 hour. Tensile deformation tests are performed at a strain rate of 0.5 mm/min at room temperature. For heat treatment sensitization, specimens are annealed at temperature between 773 and 1173 K for 2 hours with a heating rate of 323 K/h. The saturation magnetization and amount of martensitic phase are obtained from magnetization curves by SQUID magnetometer and the coercive force is obtained from magnetic hysteresis loops with VSM magnetometer. X-ray diffraction is also carried out to determine the crystallographic features and amount of austenitic to martensitic phase transformation. Micro structural characterizations of the specimens have been done using optical microscopy.

3. Results

3.1 Effect of annealing temperatures on plastic deformation

The stress-strain curves for SUS 304 stainless steel at room temperature are shown in Fig. 1. A most remarkable fact in Fig. 1 is the variation of tensile stress, which decreases with increasing annealing temperatures between 300 and 1453 K. At room temperature and 853 K with less than 3 % strain a rapid increase in flow stress occurs and the tensile stress increases quite rapidly. Analogous stress-strain curves are obtained at annealing temperatures between 1003 K and 1453 K which are characterized by an easy tensile deformation up to fracture. It can be seen that, in the annealing temperatures between 1003 K and 1453 K, the strain is high (up to 55 %). The curves are parabolic type and are smooth up to 55 % plastic strain. The temperature at which the maximum ductility occurs is between 1203 K and 1453 K. At these annealing temperatures, the strain-hardening increase continuously with the strain and the martensite is constantly formed little by little during deformation up to fracture. When the total amount of martensite formed is small, the strain percentage is relatively large and the total elongation increased. The elongation does not simply depend on the total amount of martensite induced during tensile deformation but also on the conditions of martensite formation. It should be noted that the dislocation density and twins are not sensitive to crack initiation up to 55 % plastic strain without showing micro cracks above annealing temperatures of 1103 to 1443 K. In the case of specimen annealed at 1553 K with large grain size two stages of deformation can be seen in Fig. 1 an early stage followed by rapid hardening stage. High amount of martensite transformation might cause both easy deformations in the early stage and rapid hardening in the later stage of deformation. The martensite is formed plentifully in the specimen annealed at this temperature with plastic strain is about 35 %. The decline in percentage strain at 1553 K and the increase in tensile stress is found to be essentially dependent on the martensitic transformation. The total elongation to failure is also decreased. Crack initiation may be induced prematurely at this annealing temperature, due to the presence of large amount of martensite.

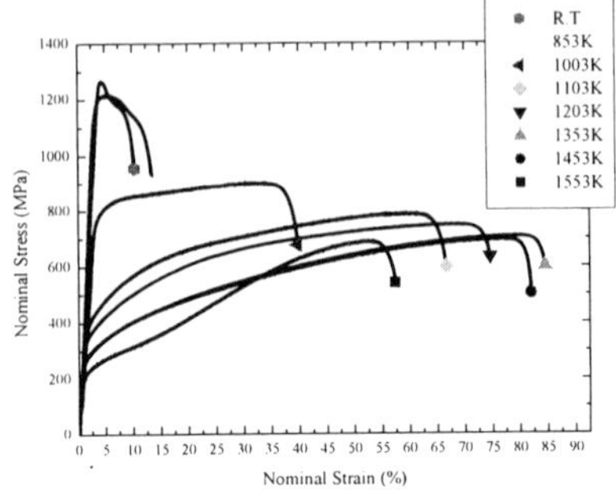

Fig.1 The stress-strain curves of SUS 304 stainless steel deformed at room temperature after annealing at various temperature

3.2 Effect of annealing temperatures on saturation magnetization

Fig.2 shows the maximum saturation magnetization as a function of annealing temperature in the maximum plastic strain condition. In all the specimens after tensile deformation the coexistence of ferromagnetic and paramagnetic states are observed reasonably depending on the amount of martensite phase. The value of saturation magnetization is 40 emu/g below 853 K annealing temperature, however at annealing temperature of 1553 K, the value of saturation magnetization becomes 130 emu/g which is the maximum value. The value of saturation magnetization is 85 emu/g at annealing temperatures between 1353 K and 1453 K. Before plastic deformation, the magnetism is paramagnetic at annealing temperature below 923 K; however at annealing temperatures between 950 and 1023 K and at 1553 K the slight increased in ferromagnetism is observed. The reason for this high value of saturation magnetization at annealing temperature of 1553 K by plastic deformation seems to be due to increase in the grain size which resulted in the transformation of large amount of ferromagnetic α' martensite. On the other hand, at annealing temperatures between 1003 and 1103 K in homogeneity chemical composition due to sensitization resulted in the formation of chromium carbide and transformation of α' martensite preferentially in the depleted zone along the grain boundaries, may be the reason of increase in saturation magnetization. At annealing temperatures above 1100 and below 1453 K where desensitization prevented the depletion of chromium and carbon and therefore transformation of martensite is not occurred to that extent even by plastic deformation and resulted in low value of saturation magnetization. The saturation magnetization increased significantly and approaches to value of 130 emu/g is much higher than that of the specimens plastically deformed at between 300 and 853 K. The α' martensite is a parasitic ferromagnetism and has a saturation magnetization of 4 emu/g in non deformed specimen and 62 emu/g in plastically deformed specimen at 300 K. This leads to conclusion that the magnetization in the plastically deformed specimens annealed at various temperature originates from α' martensite transformation.

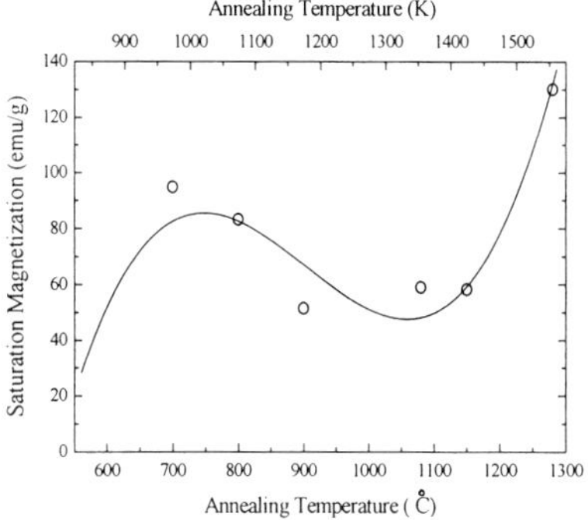

Fig. 2 The saturation magnetization versus annealing temperature
in SUS304 stainless steel

3.3 Effect of annealing temperatures on amount of martensitic transformation

The amount of martensite percentage is plotted against annealing temperature and is shown in Fig. 3. SUS 304 stainless steel annealed at different temperatures are found to undergo α' martensitic transformation on tensile deformation. The amount of martensitic transformation primarily depends on the annealing temperature. By magnetic measurement and X-ray methods 5-20 % α' martensitic phase is observed at different annealing temperatures in the non-deformed specimens. X-ray analyses also indicated no evidence of the presence of ε martensite.

The magnetic measurement and X-ray analysis showed that with the increase in annealing temperatures amount of α' martensite percentage increases as compared to the specimen that is tensile deformed at 300 K. Figure 3 shows that at annealing temperature between 1003 and 1103 K about 70 % martensitic phase is induced in the specimen, which is quite high as compared to 45 % martensitic phase transformation at annealing temperatures between 1203 and 1453 K. At annealing temperature of 1553 K the amount of α' martensite phase reached a maximum value about 100 %.

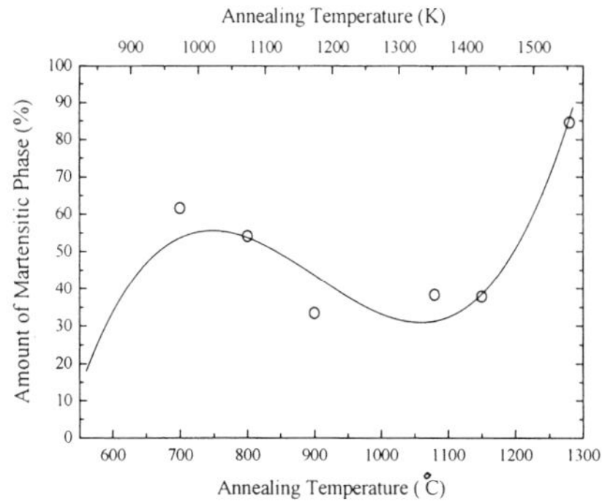

Fig.3 The amount of martensite phase versus annealing temperature in
SUS304 stainless steel

3.4 Effect of annealing temperatures on the coercive force (Hc)

The magnetic measurement is performed in three directions: parallel to the tensile axis (X), perpendicular to the surface of specimens (Y) and perpendicular to X and Y, (Z) as shown in Fig.4.

Figure 4 shows the coercive force (Hc) in the maximum plastic strain condition plotted against annealing temperatures. Coercive force depends on annealing temperature and it ranges from 20 to 190 Oe. High value of coercive force is obtained at annealing temperatures between 300 and 853 K. It depends on the magnetization direction and relatively scattered results are obtained. The coercive force shows the smallest value of 20 Oe at annealing temperature of 1553 K, where the saturation magnetization takes the maximum value. The same value of coercive force is obtained in all three directions of magnetization at annealing temperature of 1553 K as shown in Fig. 4. Coercive force depends not only on the temperature and tensile axes but also on the grain size.

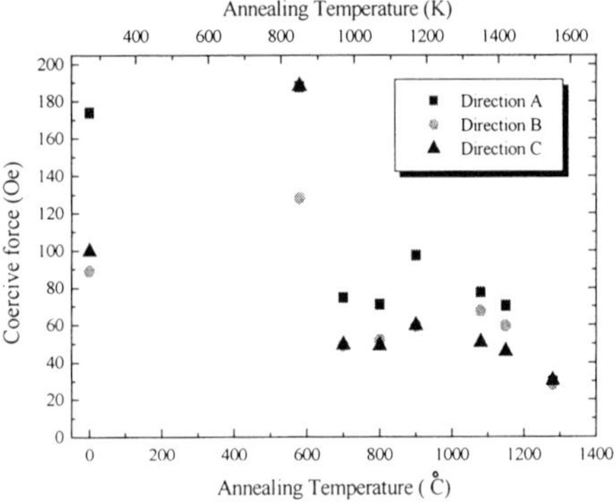

Fig. 4. The coercive force versus the annealing temperature in SUS 304 stainless steel

3.5 Effect of heat treatment sensitization on chromium carbide formation

The chemical heterogeneity is induced in the welding. At these sensitization temperatures, the diffusion of carbon and chromium results in the formation of chromium carbide $Cr_{23}C_6$. The formation of chromium carbide at the grain boundaries is shown schematically in Fig.5(a). This diffuse segregated constituent seems to be not good for structural use because of a_high susceptibility to stress corrosion cracking (SCC). The segregation of chromium carbide, which leads to the depletion of Cr in austenitic solid solution at the grain boundary, reduces the corrosion resistance. The region of stainless steel in which the local chromium composition falls below about 12 % have a diminished ability to form a passive film and hence corrode preferentially Fig. 5 (b).

Figure 6(a)-(c) demonstrates representative microstructure of specimens sensitized at 773, 973 and 1023 K for 2 hours. The chromium carbide is seen at the grain boundaries and with increasing sensitization temperature the size of chromium carbide precipitate increases dramatically. No changes in grain size are revealed by microscopic examination after sensitization at these temperatures. Martensite nucleation is said to occur by anomalous lattice vibration in the neighborhood of certain imperfection. We may expect that around the chromium carbide precipitate the defects resulted in the formation of martensite phase.

In the specimen sensitized at 773 K Fig. 6 (a), very little precipitation of chromium carbide. Chromium carbide formation not occurs in every grain and the grain boundaries are not completely covered with chromium carbide thus resulted in little chromium depletion. Therefore, it does not transform much martensite. In the specimen sensitized at 973 K annealing temperature the amount of chromium carbide formation varies greatly from grain to grain and continuous chromium carbide is clearly delineated, Fig.6 (b). Sensitization at 1023 K underwent profuse precipitation of chromium carbide along the grain boundaries as shown in Fig.6 (c). The chromium carbide essentially cover the grain boundaries. Streaked are also observed in the grains. During sensitization chromium carbide ($Cr_{23}C_6$) form along the boundaries and depleted the adjacent regions of both chromium and carbon. Martensite can then easily form along the grain boundaries of the fully sensitized specimen. The chromium carbide and martensite coexist along the grain boundaries of the sensitized SUS 304 stainless steel. Annealing above 1023 K prevented the depletion of chromium and carbon, and therefore the martensite does not occurred along the grain boundaries and the specimen is in desensitized state.

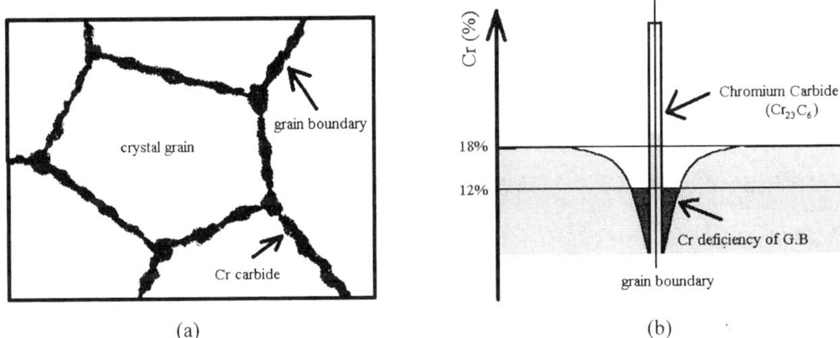

Fig.5 Schematic diagrams (a) showing $Cr_{23}C_6$ precipitates along grain boundaries and (b) the chemical heterogeneity process by HTS near the grain boundary in SUS304 stainless steel

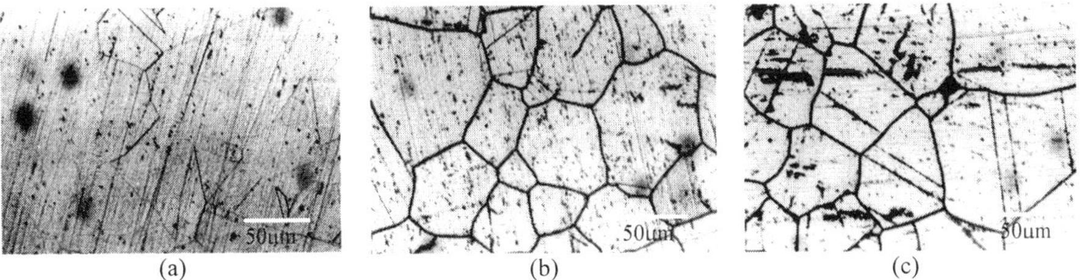

Fig.6 Micrographs of specimens sensitized at different annealing temperatures.
(a) 773K   (b) 973K and (c) 1023K

3.6 Effect of heat treatment sensitization on saturation magnetization

The magnetic transition due to heat treatment sensitization is measured by the analysis of magnetization curves. The saturation magnetization, which is in proportion to the volume of ferromagnetic phase is obtained from extrapolating the linear parts of magnetization curves back to the zero applied field. The value of saturation magnetization is 0.09 emu/g before annealing which shows that 6×10^{-2} % ferromagnetic state in volume is included in the specimen.

Figure 7 shows the saturation magnetization dependence of the annealing temperature without plastic deformation. The saturation magnetization does not change by the annealing below the temperature of 923 K. The saturation magnetization increases by the annealing above 923 K and takes the maximum value of 0.23 emu/g at 1023 K. The magnetic transition indicates that 0.1% phase changes to martensitic by the annealing at 1023 K.

4. Discussion

The above results have shown that magnetic methods offered the possibility of detecting secular degradation i.e., martensitic phase transformation in the SUS 304 stainless steel by an applied magnetic field. The martensite is the phase with superior magnetic properties. The ferromagnetic susceptibility is larger, about tens of thousands times in comparison with paramagnetic state. The magnetic susceptibility is very sensitive in the phase transition. Result shows specifically how the diffusionless transformation nucleates from austenite, given the free energy difference between the austenite and martensite phases at the M_S temperature. It is evident from the result that α' martensitic phase occurred only regardless of whether the martensite transformation is induced by plastic deformation after annealing at various temperatures or heat treatment sensitization.

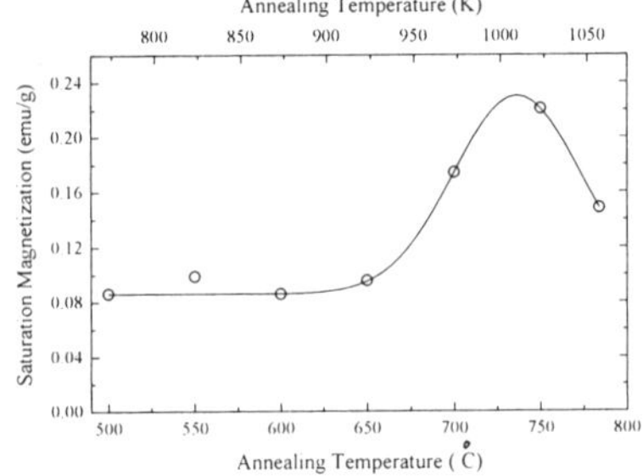

Fig.7 The saturation magnetization versus sensitization temperature in SUS304 stainless steel.

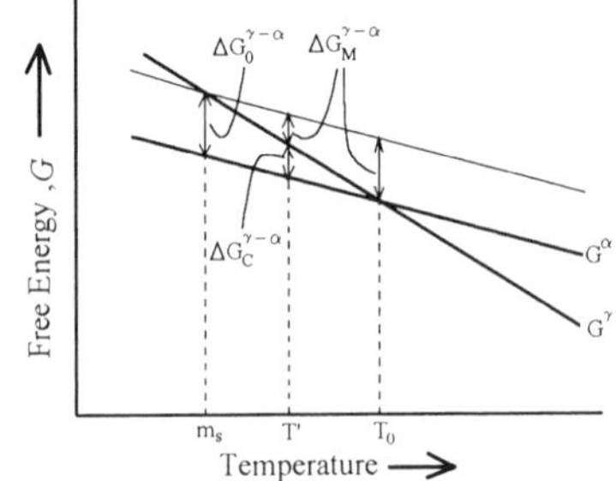

Fig.8 Temperature dependence of the free energies of α' and γ phases

All known attempts for calculating the transformation of austenite into martensite had started from the Gibbs free energies of austenite face centered cubic (γ) and martensite body centered cubic (α') phases. These free energies are a function of chemical composition. Figure 8 shows temperature dependence of the free energies of α' and γ phases. Considering the two-phase α'- γ field, there is a critical temperature T_0 where the free energies of austenite and equally composed martensite are identical. Above T_0, α' martensite is of higher energy than γ austenite. Beneath T_0, the energy will be released when γ austenite transforms into α' martensite. At martensite start temperature $M_S < T_0$ an energy difference is available as a driving force for the transformation into α' martensite.

The martensitic transformation does not occur by the external factor such as plastic deformation at above the critical temperature T_0. The martensitic transformation occurs spontaneously below M_S. The difference between two free energies $\Delta G_0{}^{\gamma-\alpha'}$ at M_S is equivalent to the energy for the martensitic transformation overcoming the interface energy between the two phases at M_S. Between M_S and T_0 the mechanical energy, $\Delta G_M{}^{\gamma-\alpha'}$, is necessary for martensitic transformation by plastic deformation, which is equivalent to the difference $\Delta G_0{}^{\gamma-\alpha'}$ and the chemical potential $\Delta G_c{}^{\gamma-\alpha'}$. The mechanical driving force is caused of $\Delta G_M{}^{\gamma-\alpha'}$ at T'. In tensile deformation the driving force necessary for the initiation of the martensite transformation is attained above the normal M_S temperature. Therefore a change in M_S can be accounted for by simply taking into consideration the contribution of additional mechanical free energy of the reaction.

We have investigated the martensitic transformation due to plastic deformation after annealing at various temperatures. By plastic deformation, the martensitic transformation occurs completely only in the specimen annealed at 1553 K. It is confirmed by the microscopic observation that the grain size becomes large (100 μm) by annealing at 1553 K. The present study shows that the larger the grain size is, the more stable the martensitic state becomes.

Results show that the value of saturation magnetization and coercive force depends on the grain size. The general aspect of grain boundary is different from the present result that large grain boundaries make the martensite phase stable. The grain boundary may exert two influences: 1) dislocation pile up at the grain boundary and 2) internal stresses. With increasing grain size the dislocations at the grain boundary accumulates, which in turn increase the amount of martensite transformation. As the grain size is increased, new martensite is nucleated near the grain boundary. The presence of dislocation pile ups have the role of increasing dislocation resistance, hence hardening the matrix, at the same time generating considerable internal stress. It is thought that the generation of more defects and the increase of local pile up stresses leads to the observed increase in the martensitic transformation after plastic deformation at 1553 K annealing.

These dislocations pile up may be acted as pinning sites to domain wall moment and domain wall energy increases at these pinning site. In response to large grain size due to high annealing temperature, the domains within individual grains must reorient to minimize their energy, therefore the magneto crystalline anisotropy energy also decreases. As a result, low value of coercive force Hc (20 Oe) is obtained at 1553 K.

The reason why large grain size is so much more susceptible to martensitic transformation may be the changes in slip behavior, dislocations, lattice defects and tensile properties within this phase have been suggested as possibilities.

Coercive force decreases with increase of saturation magnetization in Fig. 4. Coercive force gives us information of the shape of martensitic phase. Below 853 K annealing temperature, the martensitic phase would have the uni-axial form such as the needle shape or the bi-axial form such as plane shape and they distribute in seclusion. This martensitic phase has a strong magnetic anisotropy and coercive force is large, when its volume is small. As the martensitic transformation advances, the uni-axial and bi-axial phases unite with each other and make a big magnetic cluster. The shape magnetic anisotropy decreases as the saturation magnetization increases and disappears completely at annealing temperature of 1553 K.

The heat treatment sensitization phenomenon is important in the mechanism of weld sensitization of SUS 304 stainless. The chemical heterogeneity is induced in the specimens due to sensitization. It seems that diffusion of chromium and carbon and formation of chromium carbide along the grain boundaries apparently enhanced thermally activated martensite nucleation, by changing the driving force resulting in more martensitic transformation at 1023 K. Spatially in homogenous concentration profiles naturally appear in the sensitization treatment where growing cavities act as sources or sinks for the diffusion fields describing moving vacancies and interstitial atoms. The extended lattice defects, for example, grain boundaries, affect the motion of point defects and defect clusters in a radically different way. In the vicinity of grain boundaries the distribution of growing cavities becomes highly in homogenous. The growth of cavities is maximum along the grain boundaries and at a certain distance from the boundary the rate of growth is decreasing, and it is suppressed in the interior region of the grain. Following the formation of chromium carbide precipitates, therefore, it is likely that existing lattice imperfections on the atomic scale like vacancies and vacancy clusters due to depletion of chromium and carbon play a role in subsequent martensite transformation.

The important effect of heat treatment sensitization is that of martensite transformation which increases the saturation magnetization. Transformation of martensite preferentially occurred in the depleted zone along the grain boundaries, because of loss of chromium and carbon which stabilize the austenite in the zone. Due to rapid grain boundary diffusion, the chromium depletion may be reasonably uniform at 1023 K sensitization temperature. According to Eichermann [3] increase in chromium and nickel content will result in the decrease in M_S temperature. Therefore, the formation of chromium carbide around the grain boundaries due to sensitization resulted in the increase in M_S temperature, which increased the amount of martensitic transformation. Furthermore, the formation of chromium carbide precipitate lower the chromium content near the grain boundaries which reduces the stacking fault energy, and resulted in martensitic transformation.

It is evident from Fig.6 (c) that the sensitizing heat treatment enhances the chromium carbide formation at 1023 K. But the desensitization at temperature above 1023 K prevented the depletion of chromium and carbon, and therefore the martensitic transformation does not occurred.

It is concluded that non destructive techniques by magnetic methods shows potential to obtain the information of secular degradation through the martensitic transformation in SUS 304 stainless steel.

References

1. S. Takahashi, J. Echigoya and Z. Motoki, J. of applied Physics, 2000, Vol. 87 No. 2, p. 805
2 S. Takahashi, J. Echigoya, T. Ueda, X. Li and H. Hatafuku, J. of Materials Processing Technology, 2001, Vol. 108, p. 213
3. G. H. Eichermann and F. C. Hall, Tran. Am. S. M, 1953, Vol. 45, p.77

New Sensors

Development of a Microwave Focusing Sensor for Nondestructive Evaluation of Materials

Yang JU[1], Kenji SARUTA[1], Masumi SAKA[1] and Hiroyuki ABÉ[2]
[1]*Department of Mechanical Engineering, Tohoku University, Aoba 01, Aramaki, Aoba-ku, Sendai 980-8579, Japan*
[2]*Tohoku University, 2-1-1 Katahira, Aoba-ku, Sendai 980-8577, Japan*

Abstract. A microwave focusing sensor utilizing an ellipsoidal reflector was developed for nondestructive evaluation of materials. By using the focusing sensor, the standoff distance between the sensor and sample was enlarged effectively, while a relative higher spatial resolution was obtained. By using the focusing sensor, fatigue cracks were detected successfully.

1. Introduction

Microwave nondestructive evaluation (NDE) has become more and more an important technique with the emergences and applications of new and advanced materials. The primary advantage of microwave NDE is that the inspection result is not based on the density but the intrinsic properties of the materials, thus, materials can be inspected sensitively. For example, in the microwave region, the variation of permittivity for dielectric materials is significantly larger than the contrast of density. That is why microwave inspection is more sensitive than the other techniques for testing dielectric materials. Moreover, microwave NDE does not need any coupling medium because microwave can propagate well in air [1]. In the recent years, microwave NDE has been used for detecting cracks [2], CT imaging [3], moisture determination [4], fiber orientation testing [5], thickness measurement of thin films [6], and so on.

Microwave signal is an electromagnetic wave having a frequency between 300 MHz and 300 GHz (wave length between 1 m and 1 mm). It has a divergent property which usually decreases the sensitivity and the spatial resolution of the measurement. Therefore, in some cases, to obtain higher sensitivity and spatial resolution, the standoff distance between the sensor and sample has to be sufficiently small. For example, in the case of microwave NDE of the delamination in IC packages, the standoff distance was required to be 0.5 mm [7]. However, in some practical environments, such as the detection of fatigue cracks in the piping of power plants, a larger standoff distance is extremely needed.

In the present paper, an ellipsoidal focusing sensor was developed. A larger standoff distance was obtained with a higher sensitivity and spatial resolution. By using the focusing sensor, closed fatigue cracks were detected successfully.

2. Focusing Sensor

The developed focusing sensor consists of a reflector and a feed as shown in Fig. 1. The geometry of the reflector is shown in Fig. 2. The use of the reflector is based on the concepts

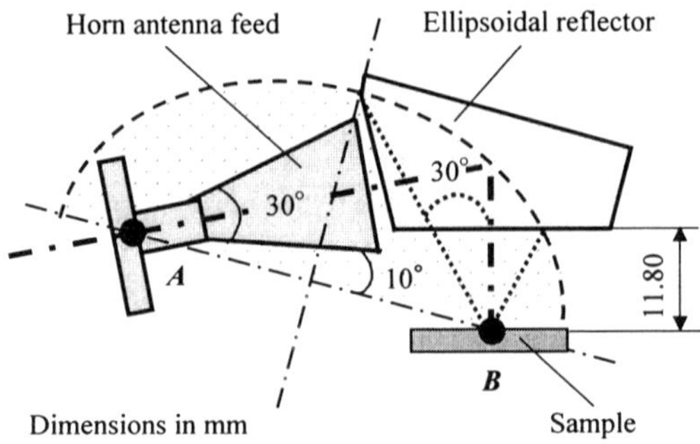

Fig. 1 Configuration of the ellipsoidal focusing sensor

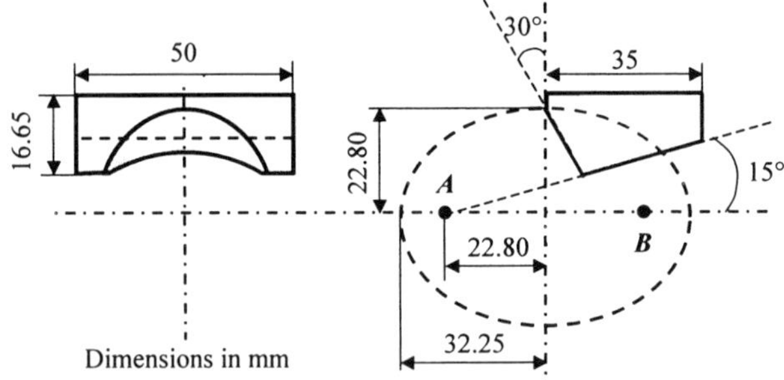

Fig. 2 Descriptive geometry of the ellipsoidal reflector

of ray optics [8], according to which the reflector takes the divergent rays from a point source and converts them into a required beam. The reflector has an ellipsoidal reflecting surface which is made of metal. A horn antenna is used as the feed, and is placed at one focus, A, of the ellipsoid. The incident microwave irradiated from the horn antenna is reflected by the ellipsoidal reflector and is focused at another focus, B, of the ellipsoid. The standoff distance of the focusing sensor is 11.8 mm. Since the sample is placed far from the sensor aperture, the higher order modes excited due to the reflector located in the near field of the feed can be ignored. The higher order modes are evanescent and decay rapidly from the aperture.

3. Experimental Procedure

The configuration of the microwave measurement system is shown in Fig. 3. It consists of a network analyzer, *x-y-z* stage, waveguide, sensor and computer. The network analyzer was used to generate a continuous wave signal fed to the sensor and to measure the phase or the amplitude of the reflection coefficient. A frequency in W-band (75 ~ 110 GHz) was used in the experiment. The computer was used to control the stage and to create a one-dimensional graph using the result measured by the network analyzer. To compare the focusing sensor with other sensors, an open-ended waveguide was also used. The geometry of the waveguide sensor is shown in Fig. 4.

To evaluate the spatial resolution of the focusing sensor, an aluminum strip sample was used. The descriptive geometry of the sample is shown in Fig. 5. The width of aluminum strips, α, was enlarged gradually with a step of 0.5 mm until the sensor was able to

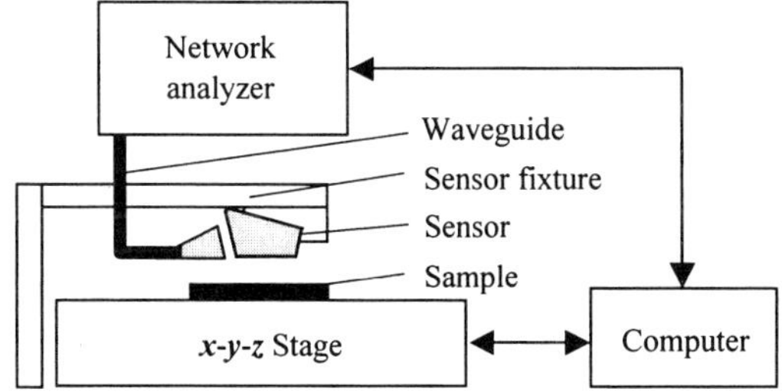

Fig. 3 Configuration of the microwave measurement system

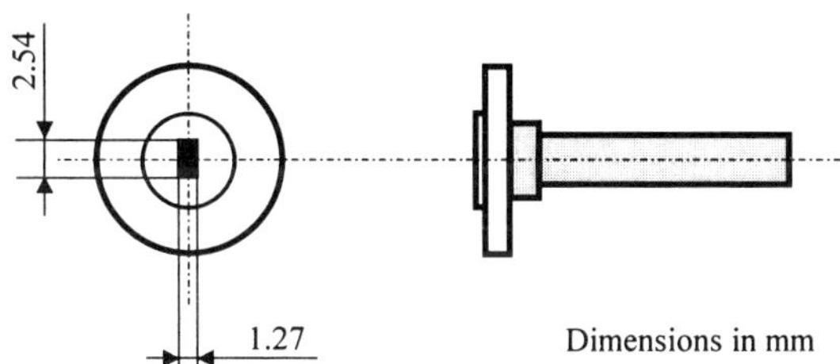

Fig. 4 Descriptive geometry of the open-ended waveguide sensor

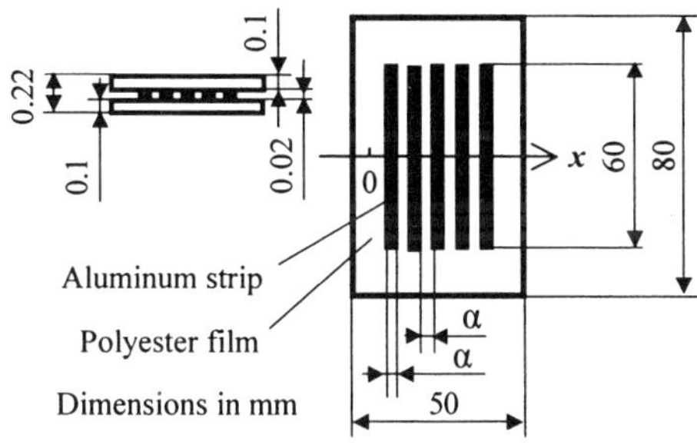

Fig. 5 Descriptive geometry of the aluminum strip sample

distinguish the five strips. Here, α is used as a measure of the spatial resolution of the sensor. The sample was measured by the focusing sensor and the open-ended waveguide sensor, respectively, with a standoff distance of 11.80 mm, where the scanning pitch in the *x*-direction was 0.25 mm.

In addition, to verify the capability of the focusing sensor to detect small defects, fatigue cracks were inspected. The descriptive geometry of the fatigue cracked specimen is shown in Fig. 6. Two small fatigue cracks were introduced in different specimens, *S1* and *S2*, respectively. Specimens were prepared as plates having the initial dimensions of 330×35×25 mm, from austenitic stainless steel, AISI 304. To introduce a fatigue crack, an initial notch was situated in the *L-S* orientation of the plate, where the direction of crack growth is considered perpendicular to the longitudinal rolling direction, *L*, and parallel to the short transverse direction, *S*, of the material (ASTM code for crack plane orientation).

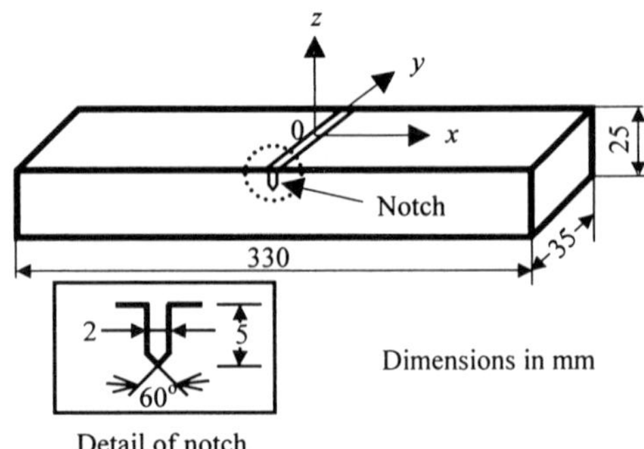

Fig. 6 Descriptive geometry of the fatigue cracked specimen

Table 1 Conditions used for introducing fatigue cracks

Specimen	Maximum stress intensity factor K_{Imax} ($MPa \cdot m^{1/2}$)	Frequency (Hz)	Stress ratio R	Crack depth d (mm)
S1	22	6	0.1	2.0
S2	22	6	0.1	2.8

The fatigue crack in each specimen was grown from the tip of the initial notch by cyclically loading the plate in four points bending of tension and tension on the dynamic testing machine. During the process of fatigue, crack growth was monitored from both sides of the specimen and the maximum value of the stress intensity factor, K_{Imax}, was determined with the average value of the crack depths measured on both sides of the specimen. The stress ratio, R, defined as a ratio of the minimum to the maximum value of the stress intensity factor, was maintained during the crack growth. After the desired depth of the crack was reached, the plate was machined and polished to remove the initial notch, leaving a true fatigue crack in the remaining material. The conditions used for introducing fatigue cracks in the specimens are listed in Table 1. Since fatigue cracks introduced in the same material under the same conditions as the present experiment had been evaluated to be closed by ultrasonic technique [8], the fatigue cracks introduced in *S1* and *S2* are considered to be closed. The depth of the fatigue cracks, d, indicated in Table 1, is the average value of the crack depths measured on the both sides of the specimen. In the experiment, the scanning origin was located at the center of the crack as shown in Fig. 6, and the scanning pitch was 0.1 mm. The standoff distance was 11.80 mm.

4. Results and Discussion

Figure 7 shows the phase measured by using the focusing sensor and the waveguide sensor for the aluminum strip sample. When α is 1.5 mm, the result obtained by using the focusing sensor shows five sharp peaks significantly corresponding to the aluminum strips as indicated by the arrows. However, in the case of the waveguide sensor, the five strips cannot be

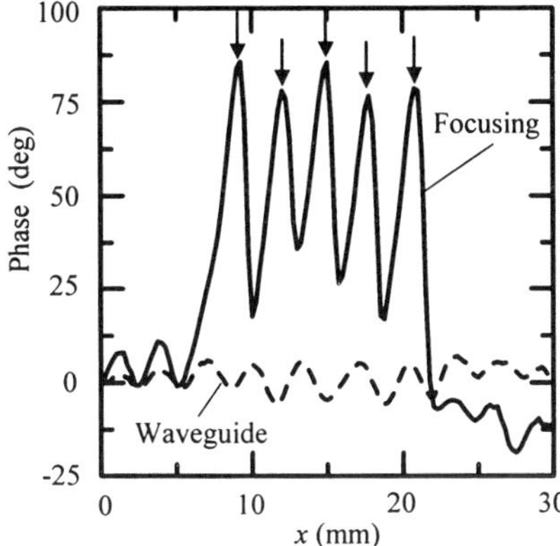

Fig. 7 Measurement results of the phase for the aluminum strip sample obtained by using the focusing sensor and the open-ended waveguide sensor

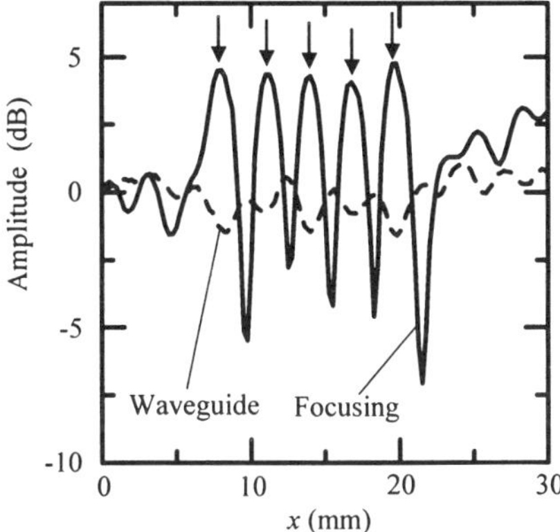

Fig. 8 Measurement results of the amplitude for the aluminum strip sample obtained by using the focusing sensor and the open-ended waveguide sensor

separated. Therefore, the spatial resolution of the focusing sensor regarding to the phase measurement was evaluated to be 1.5 mm. Figure 8 shows the amplitude measured by thefocusing sensor and the waveguide sensor for the aluminum strip sample. Same as the results of the phase measurement, when α is 1.5 mm, the result shows five sharp peaks significantly in the case of the focusing sensor, however the five strips cannot be separated in the case of the waveguide sensor. Therefore, the spatial resolution of the focusing sensor for the amplitude measurement was also evaluated to be 1.5 mm.

In the measurements of the phase and the amplitude, operating frequency was 109.8 GHz. At this frequency, a resonance occurred due to the reflection of microwaves from the aluminum strips and the sensor aperture. By using this resonant frequency, the sensitivity and the spatial resolution become higher.

Figure 9 shows the measurement results for the fatigue cracks obtained by using the focusing sensor. An operating frequency was 103.4 GHz. Around $x=0$ mm, where the fatigue crack is located, a peak can be observed. It indicates that the focusing sensor can detect fatigue

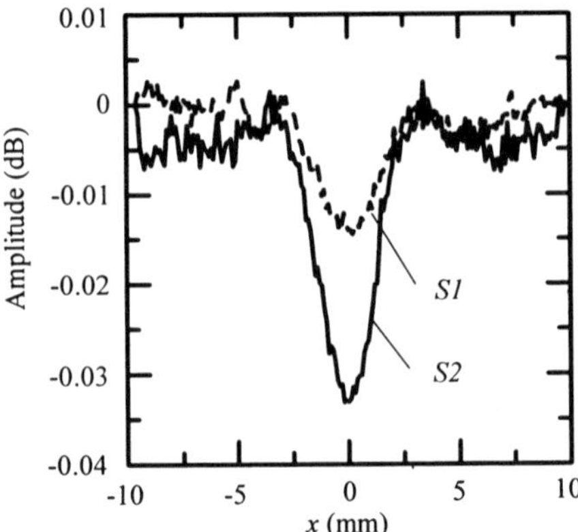

Fig. 9 Measurement results for the fatigue cracked specimens obtained by using the focusing sensor

cracks on the metal surface effectively. It should be noted that so far the usual microwave techniques were not able to detect such fatigue cracks. In addition, the magnitude of the peak as shown in Fig. 9 enlarges corresponding to the increase of the depth of the fatigue crack. It indicates a possibility to evaluate the depth of the fatigue cracks. In this case, since only the variation of microwave responses are used, the calibration of the network analyzer is not needed.

5. Conclusions

To increase the standoff distance between the sensor and sample, a focusing sensor utilizing an ellipsoidal reflector was developed. The spatial resolution of the focusing sensor was evaluated to be 1.5 mm when the standoff distance was 11.8 mm. By using the focusing sensor, fatigue cracks on the metal surface were detected successfully. The experimental results show a possibility to evaluate the size of fatigue cracks. The present technique provides a good prospect for the further applications of microwaves in NDE of materials.

Acknowledgements

The authors would like to thank Mr. M. Mikami of Tohoku University for preparing the specimen. This work was partly supported by Japan Society for the Promotion of Science under Grant-in-Aid for Scientific Research (B)(2) 13555023 and 13555192, and the Ministry of Education, Culture, Sports, Science, and Technology under Grant-in-Aid for COE Research 11CE2003.

References

[1]   D. M. Pozar, Microwave Engineering, John Wiley & Sons, Inc., 1998.

[2] N. Qaddoumi, E. Ranu, J. D. McColskey, R. Mirshahi and R. Zoughi, Microwave Detection of Stress-Induced Fatigue Cracks in Steel and Potential for Crack Opening Determination, *Reserch in Nondestructive Evaluation* **12**(2) (2000) 87-103.

[3] T. Nakajima, H. Sawada and I. Yamaura, Microwave CT Imaging for a Human Forearm at 3 GHz, *IEICE Transactions on Communications* **E78**-B(6) (1995) 874-876.

[4] Z. Ma and S. Okamura, Analysis and Elimination of the Reflection Influence on Microwave Attenuation Measurement for Moisture Determination, *IEICE Transactions on Electronics* **E80**-C(10) (1997) 1324-1329.

[5] K. Urabe and S. Yomoda, A Nondestructive Testing Method of Fiber Orientation by Microwave, *Advanced Composite Matrials* **1**(3) (1991) 193-208.

[6] H. C. Han and E. S. Mansueto, Thickness Measurement for Thin Films and Coatings Using Millimeter Waves, *Research in Nondestructive Evaluation* **9**(2) (1997) 97-118.

[7] Y. Ju, M. Saka and H. Abé, NDI of Delamination in IC Packages Using Milimeter-Waves, *IEEE Transactions on Instrumentation and Measurement* **50**(4) (2001) 1019-1023.

[8] S. Silver, Microwave Antenna Theory and Design, Peter Peregrinus Ltd., 1984.

[9] S. R. Ahmed and M. Saka, Quantitative Nondestructive Testing of Small, Tight Cracks Using Ultrasonic Angle Beam Technique, *Materials Evaluation* **58**(4) (2000) 564-574.

Development of the electromagnetic micro sensor and its application to the inspection of cast duplex stainless steel

Yasuo KUROZUMI

Institute of Nuclear Safety Systems, inc., 64 Sata, Mihama-cho, Fukui 919-1205, Japan
Takashi WAKABAYASHI, Masayoshi HIGASHI, Yusuke OHTSUKA,
Masahiro NISHIKAWA
Course of Electromagnetic Electronics, Information Systems and Energy Engineering,
Graduate School of Engineering, Osaka University
2-1 Yamada-oka , Suita, Osaka 565-0871, Japan

Abstract. The diagnosis of material degradation and the detection of defects in cast duplex stainless steel have been studied using ultrasonic techniques. This paper describes an electromagnetic micro sensor consisting of a coil and ferrite core as a transducer for detecting ultrasonic waves. When this sensor is used in combination with the ultrasonic transmitter, it can detect the electromotive force caused by the oscillation at the surface of the material. Results demonstrating the ability of the transducers are presented.

1. Introduction

Due to the good corrosion resistance, high strength and good weld ability of cast duplex stainless steel, it is widely used in chemical plants, nuclear power plants and other applications. Cast duplex stainless steel has a double-phase structure, comprised of ferrite phase and austenite phase. The ferrite content of cast duplex stainless steel ranges from 8 to 30%.

When cast duplex stainless steel is exposed to high temperatures of about 300℃ for an extended period of time of 30,000 hours or longer, the material gets affected by thermal aging in the ferrite phase. Once thermal aging occurs, in metallurgical terms, spinodal decomposition progresses in the ferrite phase, finally causing change in chromium concentration. At the same time, magnetic permeability of the ferrite phase also changes. As for mechanical changes, fracture toughness decreases, and hardness increases.(1-4) These mechanical degradations must influence the transmission characteristic of ultrasonic waves that pass through the material.

In this research, an ultrasonic tool for assessing thermal aging in cast duplex stainless steel has been studied. As the first step, an electromagnetic micro sensor consisting of coil and ferrite core was developed.

When this sensor is used in combination with the ultrasonic transmitter, it can detect the electromotive force due to the oscillation at the surface of the material with the magnetic bias field. Several kinds of electromagnetic acoustic transducer(EMAT) or piezo-electric ultrasonic transducer were used to load the oscillation on the surface of the material. Since the sensor has dimensions that are smaller than 1mm, it can detect the oscillation in a very small part on the material. Further, it is possible to evaluate the degradations of the mechanical properties of cast duplex stainless steel due to thermal aging by using the electromagnetic micro sensor.

2. Electromagnetic micro-sensor (EMS) design and experimental set-up

In this study, an electromagnetic micro-sensor (EMS) has been developed as an ultrasonic detection sensor. Figure 1 shows the design and dimensions of the sensor. The sensor consists of a ferrite core and coil, and is extremely small in size: 2 mm in width, 0.3 mm in thickness and 3 mm in height. There is a gap of 0.1 mm (maximum) at the bottom where the ferrite core is set on the material. Figure 2 shows the measurement system set up on the test piece. A permanent magnet is placed near the sensor to apply a magnetic bias field to the sensor. An EMAT or a piezo-electric transducer is used as a transmitter, and shear horizontal-waves(SH-waves) were generated to pass through the material surface.

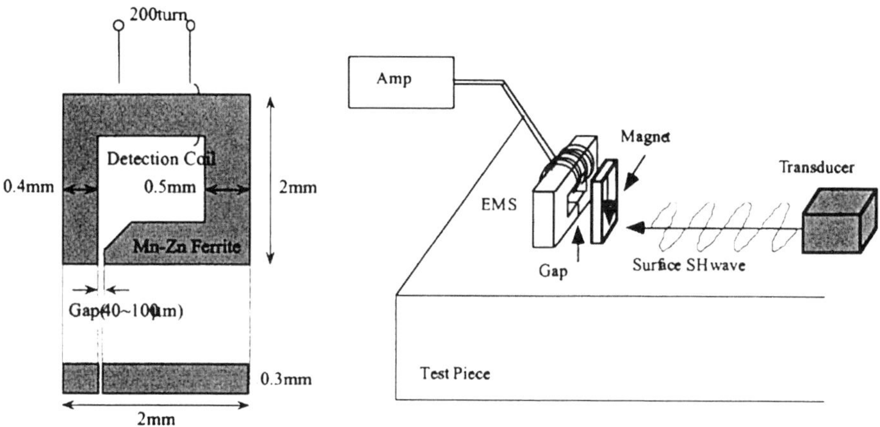

Figure.1 Design of EMS   Figure.2 Entire measurement system

Figure 3 shows the principle of detection using the electromagnetic micro-sensor. When SH-waves propagate to the material surface under the EMS, an electromotive force is induced due to the interaction between the magnetic bias field and the surface shear oscillation resulting in the induction of the eddy current in the material. Induced magnetic field caused by the eddy current is detected by the pick up coil. In reality, detection is believed to be far more complicated. The resolution is found to be nearly equal to the dimensions of the bottom part including the gap.

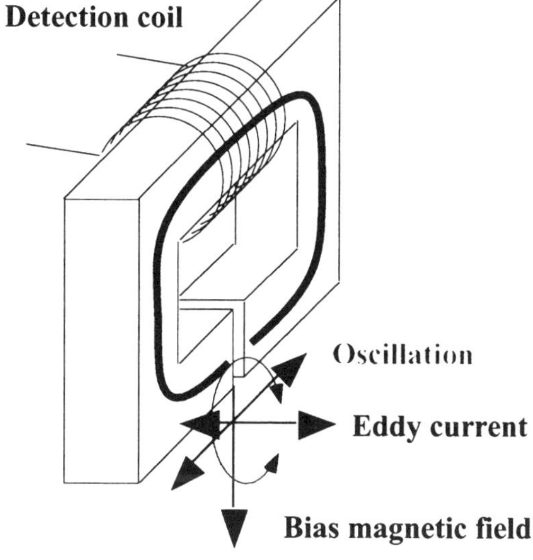

Figure.3 The principle of the detection of EMS

3. Experimental
3.1 Ultrasonic detection signals by EMS

Figure 4 shows the ultrasonic detection signals measured with a forged stainless steel test piece. The distance between transmitter and receiver has been varied from 20 mm to 40 mm. Ultrasonic waves were detected with sufficient sensitivity. Acoustic velocity can be calculated based on relation between the flight time of the first wave and the distance. In this case, acoustic velocity is approximately 3,100 m/sec.

Figure 5 shows the ultrasonic detection signals measured with a cast stainless steel test piece. The distance between the transmission and reception probes had been varied from 11 mm to 20 mm. Ultrasonic waves were detected with sufficient sensitivity even in the case of cast stainless steel. Acoustic velocity was measured by the same method and was approximately 2,900 m/sec.

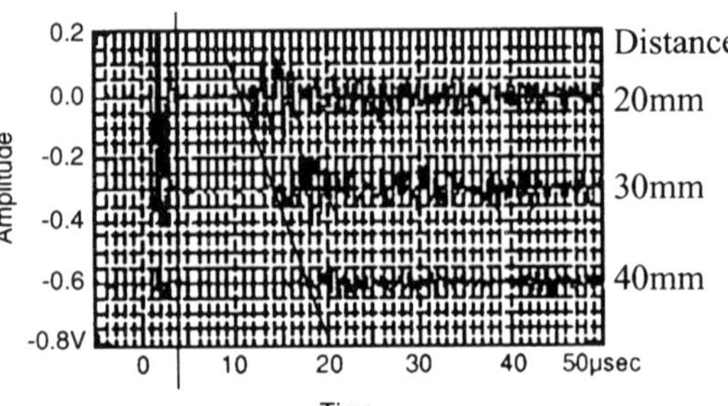

Figure.4   Ultrasonic detection signal (Forged Steel)

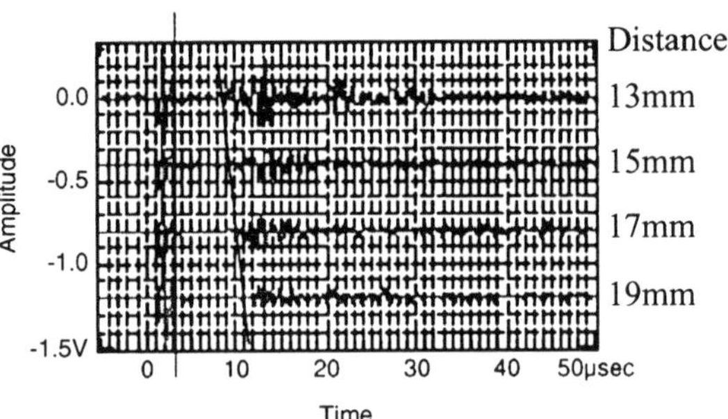

Figure 5 Typical ultrasonic detection signal (Cast Steel)

These data indicate that the EMS has sufficient sensitivity as an ultrasonic wave detection sensor.

3.2 Spatial resolution test of EMS

To verify the spatial resolution of the EMS, a test was carried out using a hole. As shown in Figure 6, a hole with a diameter of 3 mm was bored in a forged stainless steel test piece with a thickness of 25 mm. SH wave propagated toward this hole, and the EMS was scanned in the direction of traversing the straight line passing through the center of the hole. A piezo-electric transducer was used as the transmitter. The frequency was 1 MHz.

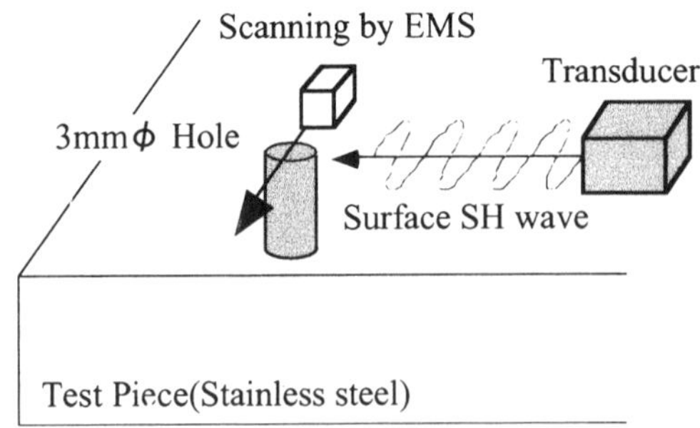

Figure.6 Measurement of the drilled hole

Figure 7 shows the result of spatial resolution test. On this graph, the amplitudes of detection signals measured at 0.5 mm intervals are plotted. The maximum of the detection signal amplitude decreased suddenly, as soon as the EMS was scanned at the edge of the hole, and dropped to the minimum at the center of the hole.

The resolution becomes 0.5mm if we employ a definition that the change of the signal amplitude should be three times as much as the fluctuations caused by the surface irregularities to identify the hole.

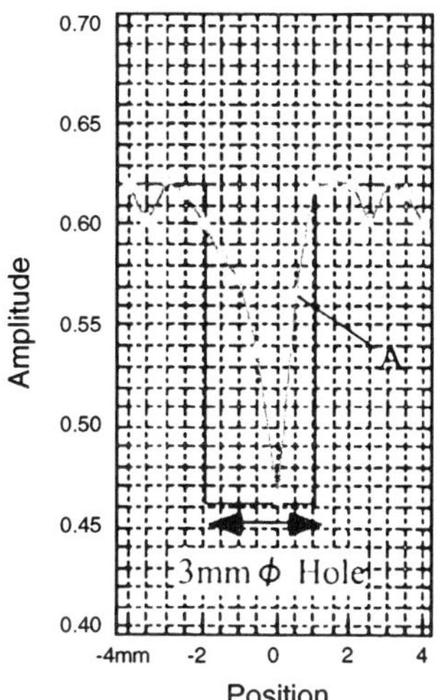

Figure.7 Measurement result of the drilled hole

3.3 C-scan image of wave amplitude in forged stainless steel

Using this sensor, the distribution of intensity of ultrasonic waves that are transmitted through the material was measured. Figure 8 shows the results of measurement in front of a piezo-electric transducer placed on the surface of a 25-mm thick forged stainless steel test piece. The specimen was scanned at 2-mm pitch in the vertical direction and 1-mm pitch in the horizontal direction. The figure indicates that ultrasonic waves propagated while slightly expanding in the vertical direction, instead of propagating in the horizontal direction, i.e. propagation by the edge effect of the rectangular transducer.

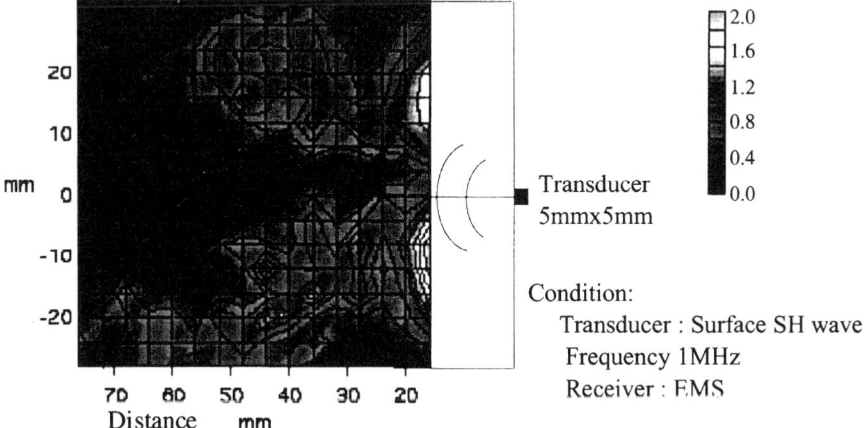

Figure 8. C-scan image of signal amplitude in forged steel

3.4 C-scan image of wave amplitude in cast steel

Figure 9 shows the results of measurement conducted under the same conditions on the surface of a 25-mm thick cast duplex stainless steel test piece. The graph on the left shows data of a non-aged test piece, and the graph on the right shows data of a test piece aged for 2,000 hours at 400℃. The amplitude of the detection signal was larger than that of forged stainless steel. The transmission of ultrasonic waves is more complex. The attenuation is considerably larger than that of forged stainless steel. The amplitude disparity depending on location is also great. This shows that ultrasonic waves are scattered at grain boundaries of cast stainless steel. The figure on the right shows how waves are divided and propagate in the vertical direction. In these two figures, ultrasonic wave transmission appears different. In order to determine whether this difference is due to thermal aging or simply due to location, it is necessary to study the problem in greater detail.

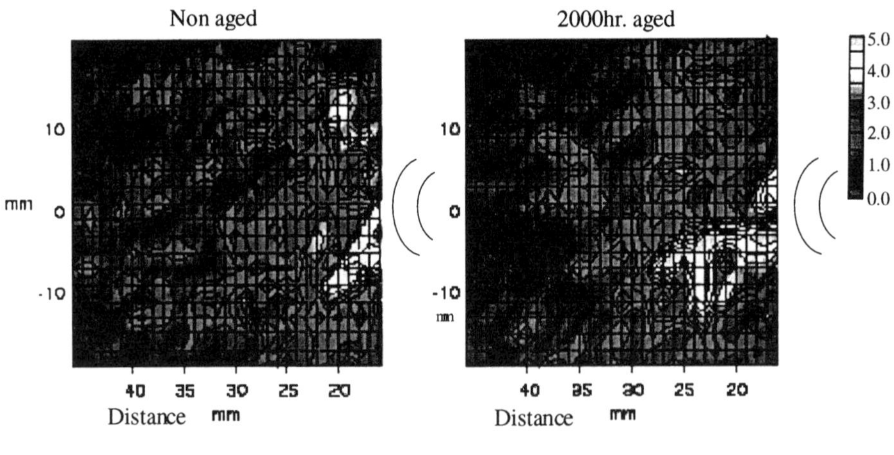

Condition:
Transducer : Surface SH wave (Frequency 1MHz)
Receiver : EMS

Figure 9. C-scan image of signal amplitude in cast steel

4. Conclusion

The fundamental characteristic of the electromagnetic micro-sensor were studied and C-scan images of ultrasonic wave propagation by non-aged and aged cast stainless steel test pieces were obtained by the sensor. Conclusions made at this point include the following:

- The electromagnetic micro-sensor has sufficient sensitivity as an ultrasonic wave detection sensor.
- The resolution of the electromagnetic micro-sensor is 0.5 mm (maximum).
- The distribution of intensity of ultrasonic waves on the surface of cast stainless steel has been obtained.

Future activities will be focused on:
- Development of electromagnetic micro-sensors with a smaller resolution.
- Collection and analysis of distribution data from aged and non-aged cast duplex stainless steel test pieces.

This paper describes the application of the electromagnetic micro-sensor to cast duplex stainless steel. This sensor is expected to be applicable to many other cases.

References

1. O. K. Corpra, H. M. Chung, "Long-Term Aging of Cast Stainless Steels: Mechanisms and Resulting Properties", NUREG/CP-0091, Vol.2, pp407-428, 1987
2. T. Hardin, W. Pavinich, W. Server, "Implication of Thermal Ageing of Cast Stainless Steel and Weldments", Proc of Int Nucl Plant Aging Sympo, pp353-362, 1988
3. O K Corpra, H M Chung, "Kinetics and mechanism of thermal aging embrittlement of Duplex Stainless Steels", DOE Report, CONF-870839-6, 32p, 1987
4. O K Corpra, H M Chung, "Long Term Embrittlement of Cast Duplex Stainless Steel in LWR Systems : Semiannual Report October 1986-March 1987", NUREG/CR-4744, Vol.2, ANL-87-45, 1987

Signal Detection

Electromagnetic Nondestructive Evaluation (VI)
F. Kojima et al. (Eds.)
IOS Press, 2002

285

Generalized Vector Sampled Pattern Matching Method
–Theory and Applications

Hisashi ENDO, Seiji HAYANO, and Yoshifuru SAITO
Graduate School of Engineering, HOSEI University
3-7-2 Kajino, Koganei, 184-8584 Tokyo, Japan

Kenzo MIYA
International Institute of Universality,
1-4-6 Nezu, Bunkyo, 113-0031 Tokyo, Japan

Abstract. A powerful iterative solver for system of equations has been developed. This solver, Generalized Vector Sampled Pattern Matching (GVSPM), enables us to obtain the solution of any ill-posed linear system equations, e.g., having rectangular or singular matrix. The key idea is that the objective function is the angle obtained by an inner product between the input vector and solution system of equations. This paper reviews the Generalized Sampled Pattern Matching (GSPM) method, which is the original of proposed method, and then introduces the GVSPM method along with solution algorithm. The theoretical background and applications to inverse source problems on magnetic fields are described. From the applications, the GSPM and GVSPM methods tend to lead the discontinuous and continuous solutions, respectively. However, the GVSPM has the advantages of stability and convergence.

1. Introduction

When we address the non-destructive evaluation (NDE) and electromagnetic compatibility (EMC) problems, it is essential to reduce into solving for inverse problems. Because of the nature of their formulations, the number of equations and that of solutions are not always the same. Therefore, grappling with the inverse problems results in solving for ill-posed linear system of equations. In order to solve them, some approaches may be considered. For instance, least square approach is applicable when the number of equations is greater than that of solutions. Inversely, when the number of solutions is greater than that of equations, some constraint conditions should be imposed [1]. In addition, the solution by the neural network (NN) has dependence on the number of training as well as structure of the networks.

The early 1990's, Saito proposed a novel solution strategy called Sampled Pattern Matching (SPM) method to apply the inverse problems which result in solving for the ill-posed linear system of equations. The principle is that the patterns of column vectors constituting the system matrix are investigated. This idea has worked out the electric as well as magnetic source searching problems [2][3], shape design of magnetic cores [4] and so on. Afterward, this inverse problem solver was generalized to solve for various types of linear system of equations. The generalized solver, Generalized Sampled Pattern Matching (GSPM), has been applied to optimize the dose of electron beams [5][6] and so on. Though the GSPM is possible to solve for any types of linear equations, the convergence process as well as mathematical background was unclear at the time. Recently, mathematical background has been clarified, and the convergence process has been proved [7]. This leads to improvement of the GSPM.

This paper reviews the GSPM method, and then introduces the improved method, Generalized Vector Sampled Pattern Matching (GVSPM). The key idea of the GVSPM is that the objective function is the angle obtained by means of inner product between the input vector and solution system of equations. The convergence process is described in analytical way. The applications and comparisons of the GSPM and GVSPM are demonstrated along with the inverse source problems on magnetic fields. As a result, the applications reveal that the GSPM and GVSPM methods tend to lead the discontinuous and continuous solutions, respectively. Moreover, the solution by means of GVSPM converges with smaller number of iterations for computation.

2. Generalized Sampled Pattern Matching (GSPM) Method

2.1 Basic Equation

Solving the inverse problems results in handling the ill-posed linear system of equations. The basic equation we have to solve is as follows:

$$\mathbf{Y} = C\mathbf{X}, \tag{1}$$

where $\mathbf{Y}$ and $\mathbf{X}$ denote the n-th order input- and m-th order solution/output- vectors, respectively; C is an n by m rectangular matrix. Even if $m=n$ case, the matrix C is not always positive definite due to the formulation of inverse problems. (1) can be rewritten by

$$\mathbf{Y} = \sum_{i=1}^{m} x_i \mathbf{C}_i, \quad \mathbf{X} = \begin{bmatrix} x_1 & x_2 & . & x_m \end{bmatrix}^T, C = \begin{bmatrix} \mathbf{C}_1 & \mathbf{C}_2 & . & \mathbf{C}_m \end{bmatrix}. \tag{2}$$

(2) means that the input vector $\mathbf{Y}$ is represented by means of linear combination of column vectors $\mathbf{C}_i$, $i=1, 2,\ldots, m$, in the system matrix C. The principle of the SPM method is to search for the patterns representing by the column vectors $\mathbf{C}_i$s in order to satisfy the input vector $\mathbf{Y}$ in (1).

2.2 GSPM Method

In the GSPM, the elements in solution vector $\mathbf{X}$ are approximated by quantized discrete values as step width. At first, matching rates between the vector $\mathbf{Y}$ and column vector $\mathbf{C}_i$, $i=1, 2,\ldots, m$, are calculated using the Cauchy-Schwarz formula given as (3) to select the most dominant pattern.

$$\gamma_i = \frac{\mathbf{Y} \bullet \mathbf{C}_i}{|\mathbf{Y}\|\mathbf{C}_i|}, \quad i = 1,2,\ldots,m, \tag{3}$$

where γ is called pattern matching index. If a column vector $\mathbf{C}_i$ is the most dominant, then correspoding element of the solution vector $\mathbf{X}$ is increased by the quantized discrete value. From the next calculation steps, the searching patterns are overlapped by the selected column vector, namely,

$$\gamma_i^{(k)} = \frac{\mathbf{Y} \bullet \begin{bmatrix} \mathbf{C}_i + \mathbf{C}_k \end{bmatrix}}{|\mathbf{Y}\|\mathbf{C}_i + \mathbf{C}_k|}, \quad i = 1,2,\ldots,m, \tag{4}$$

where the superscript (k) refers to the k-th iteration. Moreover, $\mathbf{C}_k$ is the selected pattern in terms of k-1-th iteration. (4) means that the amplitude of the solution is approximated by the concentrating rate of similar solution pattern of $\mathbf{C}_k$. The calculation continues until the pattern matching index $\gamma_i^{(k)}$ takes 1. This means the obtained pattern $(\mathbf{C}_i+\mathbf{C}_k)$ gives the best pattern matching to the input vector $\mathbf{Y}$. Finally, increment the elements constituting the output vector $\mathbf{X}$ in each of iterations yields the GSPM solution. In this method, calculating (4) carries out to each of column vectors, therefore, it is essentially required enormous time and number of iterations for computation.

3. Generalized Vector Sampled Pattern Matching (GVSPM) Method

3.1 Key Idea

Normalzing (2) by the input vector amplitude $|\mathbf{Y}|$ gives the following relationship:

$$\frac{\mathbf{Y}}{|\mathbf{Y}|} = \sum_{i=1}^{m} x_i \frac{|\mathbf{C}_i|}{|\mathbf{Y}|} \frac{\mathbf{C}_i}{|\mathbf{C}_i|} \quad \text{or} \quad \mathbf{Y}' = C'\mathbf{X}', \tag{5}$$

where the prime (') denotes the normalized quantities. (5) means that the normalized input vector $\mathbf{Y}$' is obtained as a linear combination of the weighted solutions $x_i|\mathbf{C}_i|/|\mathbf{Y}|$, i=1, 2,…, m, with the normalized column vectors $\mathbf{C}_i/|\mathbf{C}_i|$, i= 1, 2,…, m. It should be noted that the solution vector $\mathbf{X}$ could be obtained when an inner product between $\mathbf{Y}$' and C'$\mathbf{X}$' becomes 1. This is the key idea of the GVSPM method.

3.2 Objective Function

Define a function f derived from an angle between the input vector $\mathbf{Y}$ and $C\mathbf{X}^{(k)}$ given in terms of the k-th iterative solution $\mathbf{X}^{(k)}$, as given by

$$f\left(\mathbf{X}^{(k)}\right) = \frac{\mathbf{Y}}{|\mathbf{Y}|} \bullet \frac{C\mathbf{X}^{(k)}}{\left|C\mathbf{X}^{(k)}\right|} = \mathbf{Y}' \bullet \frac{C'\mathbf{X}'^{(k)}}{\left|C'\mathbf{X}'^{(k)}\right|} . \tag{6}$$

Then the solution $\mathbf{X}^{(k)}$ is obtained when the function $f(\mathbf{X}^{(k)})$ converges to

$$f\left(\mathbf{X}^{(k)}\right) \to 1 . \tag{7}$$

This is the objective function of the GVSPM solution. The objective function of the GSPM is the angle between the input vector and pattern incrementally composed of column vectors while those of the GVSPM is the angle between the normalized input vector and output system of equations. Thereby, evaluation of (7) needs only once an iteration.

3.3 Iteration Algorithm

Let $\mathbf{X}'^{(0)}$ be an initial solution vector given by

$$\mathbf{X}'^{(0)} = C'^{T} \mathbf{Y}' , \tag{8}$$

then the first deviation vector $\Delta\mathbf{Y}'^{(1)}$ is obtained as

$$\Delta\mathbf{Y}'^{(1)} = \mathbf{Y}' - \frac{C'\mathbf{X}'^{(0)}}{\left|C'\mathbf{X}'^{(0)}\right|} . \tag{9}$$

When the deviation $\Delta\mathbf{Y}'$ becomes zero vector, the objective function (7) is automatically satisfied. Modification by the deviation vector $\Delta\mathbf{Y}'^{(k-1)}$ gives the k-th iterative solution vector $\mathbf{X}'^{(k)}$, namely,

$$\mathbf{X}'^{(k)} = \mathbf{X}'^{(k-1)} + C'^T \Delta\mathbf{Y}'^{(k-1)}$$

$$= C'^T \mathbf{Y}' + \left(I_m - \frac{C'^T C'}{\left| C' \mathbf{X}'^{(k-1)} \right|} \right) \mathbf{X}'^{(k-1)}, \tag{10}$$

where I_m denotes a m by m unit matrix.

3.4 Convergence Condition

The convergence condition of the GVSPM iterative strategy is that the modulus of all characteristic values of state transition matrix in (10) must be less than 1 [8]. The state transition matrix S is given by

$$S = I_m - \frac{C'^T C'}{\left| C' \mathbf{X}'^{(k-1)} \right|} = I_m - \frac{C'^T C'}{\left| \mathbf{Y}'^{(k-1)} \right|}. \tag{11}$$

Since the vector $\mathbf{Y}'^{(k-1)}$ is normalized, (11) can be rewritten by

$$S = I_m - C'^T C'. \tag{12}$$

This means the convergence condition is independent of initial solution in (8). Furthermore, the solution is always evaluated by means of (6) so that the solution only depends on the system matrix C.

Let λ be characteristic value of the state transition matrix S. Then the determinant of symmetrical matrix is obtained:

$$\left| \lambda I_m - S \right| = \begin{vmatrix} \lambda & \varepsilon_{12} & . & \varepsilon_{1m} \\ \varepsilon_{12} & \lambda & . & \varepsilon_{2m} \\ . & . & . & . \\ \varepsilon_{1m} & \varepsilon_{2m} & . & \lambda \end{vmatrix} = 0. \tag{13}$$

It is obvious that the moduli of off-diagonal elements in (13) take less than 1 because of the normalized column vectors of matrix C', namely,

$$\left| \varepsilon_{ij} \right| < 1, \qquad i=1,2,\ldots,m, j=1,2,\ldots,m. \tag{14}$$

Suppose the modulus characteristic value $|\lambda|$ takes more than 1. Then the column vectors in (13) become linear independence because of (14). In such a case, the determinant in (13) is not zero so that the condition $|\lambda| < 1$ should be satisfied. Therefore, it is proved that the GVSPM is always carried out on stable iteration.

4. Applications

4.1 Visualization of 2-Dimensional Current Distribution

To visualize the current vector distribution from locally measured magnetic field is one of the typical inverse problems and fundamental of NDE. Let us consider a unit loop current model illustrated in Fig. 1[9]. The magnetic field H caused by a loop coil shown in Fig. 1(b) can be analytically obtained as follows:

$$H = \frac{I}{2\pi}\left[\frac{1}{\sqrt{(a+r)^2 + z^2}}\right]\left[\frac{a^2 - r^2 - z^2}{(a-r)^2 + z^2} E(\kappa) + K(\kappa)\right]$$

$$\kappa^2 = \frac{4ra}{(r+a)^2 + z^2}, \tag{15}$$

where $K(\kappa)$ and $E(\kappa)$ are the first- and second- kind elliptic integrals, respectively. The system of equations is based on (15) in this example. In this case, the vectors **Y** and **X** in (1) are composed of the measured magnetic field H and current I expressed by (15), respectively.

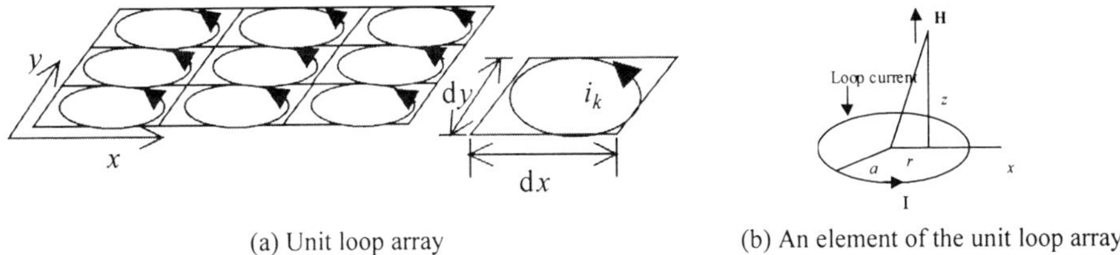

<table>
<tr><td align="center">(a) Unit loop array</td><td align="center">(b) An element of the unit loop array</td></tr>
</table>

Fig. 1 Unit loop current model

Fig. 2 shows the measured magnetic field distribution generated by 3 excited coils. At first, each component of the 3-dimensional magnetic field is measured at a surface over the target, as shown in Figs. 2(b)-(d). Number of measured points is 16 by 16. Therefore, the input vector **Y** in (1) becomes 256th order, namely $n=256$. Second, the system of equations obtained by (15) is solved in terms of the z component of magnetic field by the GSPM and GVSPM methods. The estimation points are set to 10 by10, 16 by 16 and 32 by 32 in this example. In these cases, numbers of elements in output vectors are 100, 256 and 1024. Finally, 2-dimensional current vector distribution can be obtained as in Figs. 3 and 4.

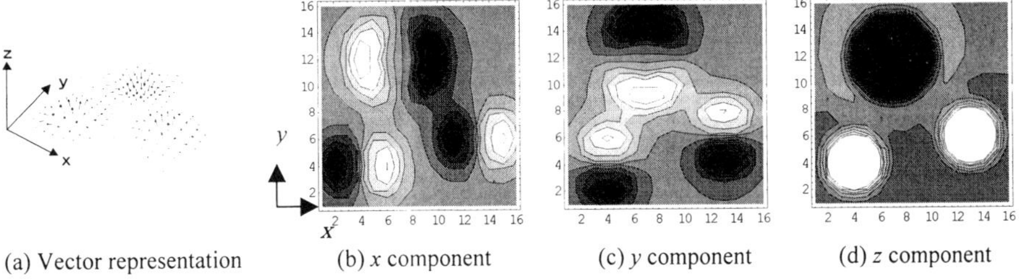

<table>
<tr><td align="center">(a) Vector representation</td><td align="center">(b) x component</td><td align="center">(c) y component</td><td align="center">(d) z component</td></tr>
</table>

Fig. 2 Measured magnetic field of 3 excited coils

The GSPM and GVSPM realize Figs. 3 and 4, respectively. In cases of $m<n$ and $m=n$, each of the solutions yields fairly good approximation. On the other hand, in case of $m>n$, the GVSPM gives better result than the GSPM in respect of current flowing directions. Since the GSPM method carries out element-to-element evaluation described as (4), then it needs a large number of iterations. The results of GSPM and GVSPM have spent 200000 and 100 iterations on computation, respectively. The GVSPM realizes faster computation as well as good approximation in this case.

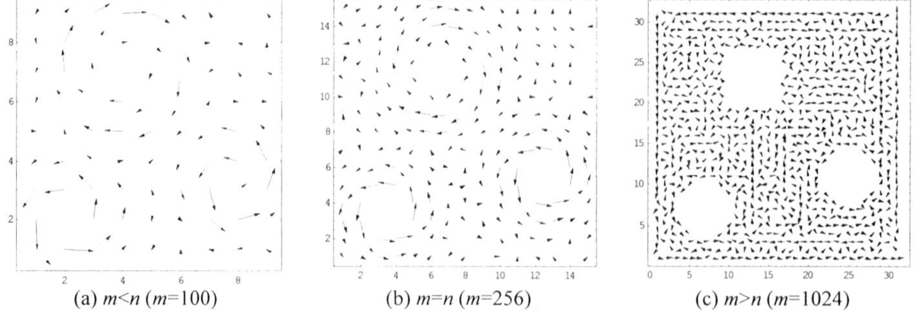

(a) *m<n* (*m*=100)   (b) *m=n* (*m*=256)   (c) *m>n* (*m*=1024)

Fig. 3 2-dimensional current vector estimation by means of GSPM (*n*=256)

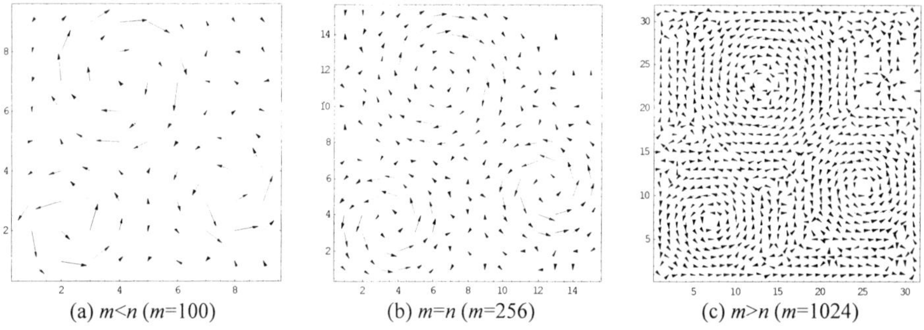

(a) *m<n* (*m*=100)   (b) *m=n* (*m*=256)   (c) *m>n* (*m*=1024)

Fig. 4 2-dimensional current vector estimation by means of GVSPM (*n*=256)

4.2 Current Position Searching by Deconvolution

Next, searching the current flowing position is carried out using measured magnetic field. Information coming from the current can only be observed by specific sensor such as solenoidal coils. In most cases, the current distribution is observed by sensor signal including some physical space characteristics such as Green's relationship.

Let us consider the one dimensional magnetic field distribution scanned by a solenoidal sensor coil. Then, the output voltage would be obtained like Fig. 5, and it takes maximum voltage at the position 100 cm. Using this physical space characteristic of the sensor coil, the exact current flowing positions are investigated. In this case, the vectors $\mathbf{Y}$ and $\mathbf{X}$ in (1) are corresponding to the measured magnetic field obtained by the sensor and the exact current distribution to be searched, respectively. The system matrix C, shown in Fig. 6, is composed of the physical space characteristic. This scheme is called deconvolution.

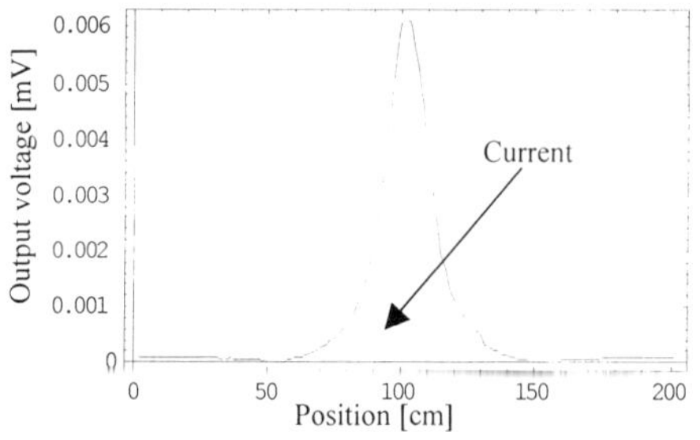

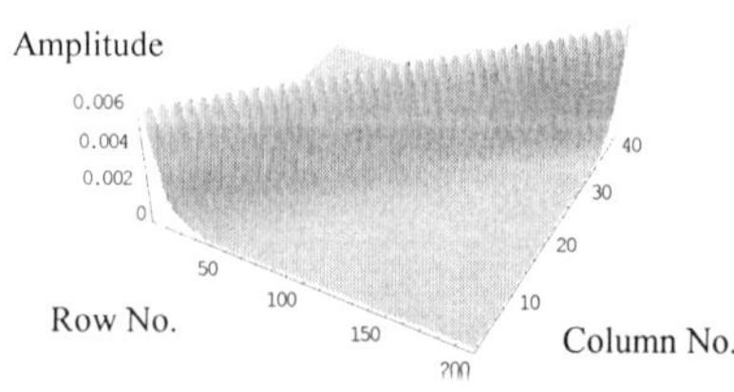

Fig. 5 Physical space characteristic of a sensor coil   Fig. 6 Elements of system matrix

Fig. 7 shows the measured magnetic field caused by two current lines. The measurement is carried out by a solenoidal coil over the current line surface. Number of measured points is 41 and number of measured points of physical space characteristic is 201. Therefore, **Y** and **X** respectively become the vectors with 41 and 201 orders, namely, $m=201$, $n=41$. By means of the GSPM and GVSPM, Fig. 8 shows the deconvoluted results. The result of GSPM in Fig. 8(a) reveals the positions of the exact currents has been estimated with much higher accuracy than that of GVSPM in Fig. 8(b). This nature is caused by iteration algorithm. Because the GSPM method carries out element-to-element evaluation, the solution having discontinuous characteristics is easy to be obtained. On the other hand, since the GVSPM evaluates with vector type iteration described above, then it tends to be continuous solutions. The results of GSPM and GVSPM have spent 200000 and 100 iterations on computation, respectively.

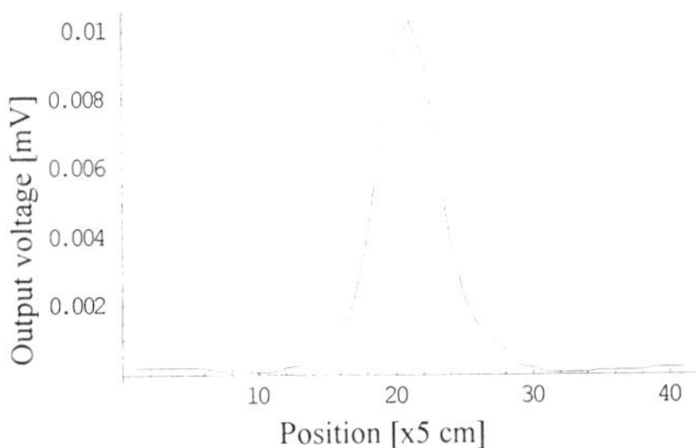

Fig. 7 Measured magnetic field distribution by the sensor coil

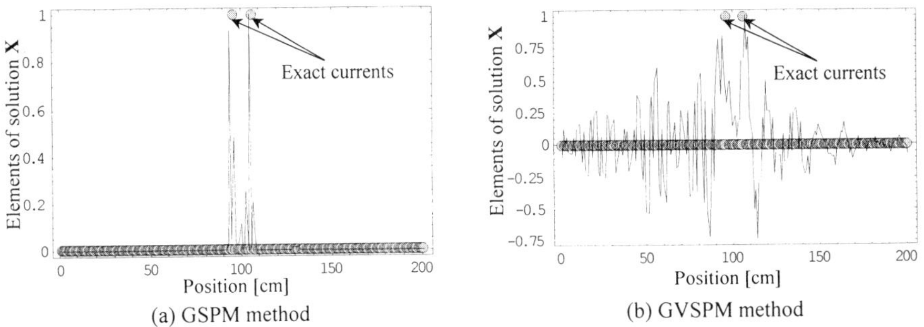

(a) GSPM method

(b) GVSPM method

Fig. 8 Current flowing position searching by means of deconvolution

5. Conclusions

This paper has introduced the GVSPM as one of the powerful iterative solvers for linear system of equations. The original version of the GSPM requires an enormous computation time while the GVSPM algorithm makes it possible to reduce the computation time and the reliablity of the solution is clearly verified. The distinguished feature is that the objective function is the angle between the right and left hand terms in linear system of equations. This makes it possible to solve any types of simultaneous equations, i.e., having row- as well as column- wide type rectangular system matrices. The convergence process has been described analytically. As the applications of GSPM and GVSPM, the current searching problems from locally measured magnetic fields have demonstrated the usefulness of our solution strategy. Solving for the ill-posed system of equations is inevitable for the inverse problems. However, the GSPM and GVSPM make it possible to select the physically existing solution although this solver never uses the matrix inversion. Particularly, the GSPM and GVSPM tend to be discontinuous and continuous solution, respectively.

References

[1] G. Anger, Inverse Problems in Differential Equations, Plenum Press, New York, 1990.
[2] Y. Saito, E. Itagaki, and S. Hayano, A Formulation of the Inverse Problems in Magnetostatic Fields and Its Application to a Source Position Searching of the Human Eye Fields, J. Appl. Phys., Vol. 67, 1990, pp. 5830-5832.
[3] H. Saotome, T. Doi, S. Hayano, and Y. Saito, Crack Identification in Metallic Materials, IEEE Trans. Magn., Vol. 29, No. 2, 1993, pp. 1861-1864.
[4] H. Saotome, J. L. Coulomb, Y. Saito, and J. C. Sabonnadiere, Magnetic Core Shape Design by the Sampled Pattern Matching Method, IEEE Trans. Magn., Vol. 31, 1995, pp.1976-1979.
[5] Y. Saito and K. Yoda, Method of Generating Energy Distribution, Japanese Patent Application 408315279, 1996.
[6] K. Yoda, Y. Saito, and H. Sakamoto, Dose Optimization of Proton and Heavy Ion Therapy Using Generalized Sampled Pattern Matching, Phys. Med. Biol., IOP Publishing, Vol. 42, 1997, pp 2411-2420.
[7] D. Sekijima, S. Miyahara, S. Hayano, and Y. Saito, A Study on the Quasi-3D Current Estimation" Trans. IEE of Japan, Vol. 120-A, No. 10, 2000, pp. 907-912 (in Japanese).
[8] G. Strang, Linear Algebra and its Applications, Academic Press, New York, 1976.
[9] H.Takahashi, S.Hayano, and Y.Saito, Visualization of the currents on the printed circuit boards, IEEE Vis99 Late Breaking Hot Topics Proceedings, 1999, p37-40.

Electromagnetic Nondestructive Evaluation (VI)
F. Kojima et al. (Eds.)
IOS Press, 2002

293

Crack Detection by a Small Loop Antenna

Yasumoto Sato, Kenichi Yagi, Tetsuo Shoji
Fracture Research Institute, Graduate School of Engineering,
Tohoku University, Aobayama 01, Aobaku, Sendai/980-8579, JAPAN

Abstract. We have developed a new technique using micro antennas in order to
realise an advanced non-destructive inspection technique that incorporates a high
level of sensitivity and high accuracy for the detection and characterisation of
surface cracks in metals. The principle of the new technique is presented. A new
design of probe combined with an induction wire in which an excitation signal
flows was developed to enable simple, high-speed measurement. Experiments to
detect and determine the size of fatigue cracks were performed on a 304 stainless
steel specimen.

1. Introduction

The detection and sizing of defects is one of the key technologies for mechanical plant life
management based on fracture mechanics. Improved performance of the available
non-destructive inspection techniques is always required for better detection sensitivity,
better quantitative evaluation and better field applicability. In this work we report on a new
non-destructive inspection (NDI) and evaluation (NDE) technique, based on high frequency
electromagnetic field measurement. In [1], the characterisation of artificial surface defects in
metals was performed by using antennas. The crack depth was evaluated based on the
analysis of data for electromagnetic waves reflected from the metal. However, the angle of
incidence equals the angle of reflection for such waves. Taking into account the fact that the
walls of real cracks are neither flat nor perpendicular to the surface, techniques utilising
reflection are not really very suitable for the characterisation of fatigue cracks. Therefore, a
new technique for the detection and characterisation of surface cracks in metals using a Small
Loop Antenna was developed [2, 3]. When an alternating field is applied to a metallic
material, either directly or indirectly, the existence of defects on the surface of the material
causes changes in the field around them. As a result, the changes in the electromagnetic field
generated by this change in the alternating field can be detected on or near the surface of the
material by using a small antenna. The advantage of this system is that it is possible for the
probe to measure defects without direct contact with the specimen. Additionally, such a
system can detect cracks in ferromagnetic material as well as paramagnetic material [4]. On
the other hand, the measurement time is relatively long in comparison with other non-contact
NDI techniques, because this system requires a conductive wire to be attached to the
specimen surface in order to induce the alternating field.

In this paper we describe the use of a Small Loop Antenna combined with an
induction wire to create a novel non-contact NDI technique. The ability of this sensor to
detect real defects is demonstrated by measurements on an actual metallic specimen with a
fatigue crack of 29mm in length.

2. Principles of the Small Loop Antenna technique and measurement system

A schematic illustration of the principles of the Small Loop Antenna technique is shown in Fig. 1. When an alternating signal is applied in the conductive wire (induction wire), an induced signal is generated in the opposite direction beneath the surface of the metal. The changes in the induced signal flow around a crack lead to a change in the magnetic field in the vicinity of a crack. This magnetic field has its largest magnitude inside the crack, but part of it penetrates outside the crack and can therefore be detected by the Small Loop Antenna.

The measurement system consists of a network analyser, an X-Y-Z sensor-positioning table and a personal computer (Fig. 2). The network analyser measures the amplitude ratio and phase difference of the transmitted wave compared to the incident wave [5]. Its operating frequency range is between 30kHz and 6GHz. In this work, the frequency of the transmission wave was chosen to be 1GHz, based on our preliminary experiment [6]. A computer controlled X-Y-Z positioning table controls the motion of the antennas.

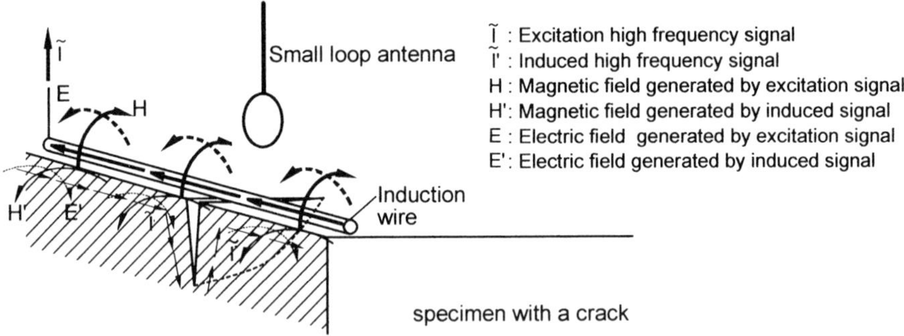

Fig. 1 Principle of electromagnetic field measurement by a Small Loop Antenna

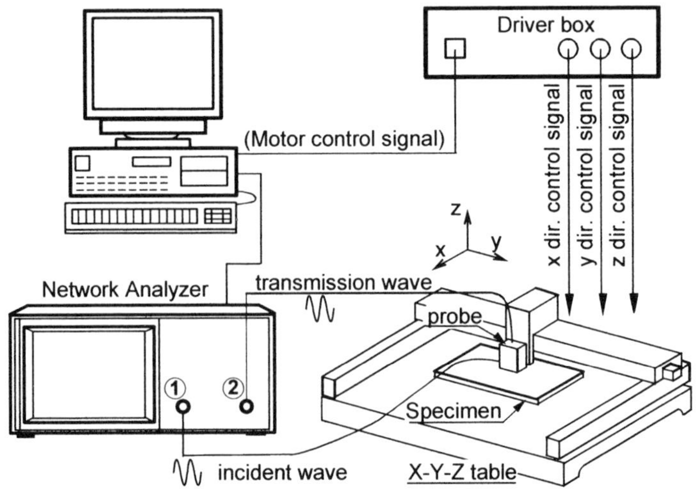

Fig. 2 Measurement system for the Small Loop Antenna technique

3. Development of a new sensor

3.1. Loop Antenna

A loop antenna was used to measure changes in the electromagnetic field in the vicinity of the cracks. The loop antenna contains a coaxial line with a Cu inner conductor and a Cu outer conductor, as well as a loop that is also closed by a solid Cu conductor. The solid Cu conductor is soldered at both ends to the inner and outer conductors of the coaxial line. A schematic illustration of the loop antenna is shown in Fig. 3 (a).

This antenna enables us to measure the potential drop U *(t)* generated at the output of the loop antenna, which is expressed as follows [7, 8]:

$$U\ (t)=j\omega NAB \tag{1}$$

where j is the imaginary unit, ω is the angular frequency, N is the number of coils, A is the loop area and B is the value of the component of the vector of the magnetic flux density within the loop, in the direction perpendicular to the loop. Accordingly, the amplitude of U is proportional to B within the loop antenna [9].

3.2. Determination of the design parameters of the new sensor

In [2], [3] and [4], it is necessary to fix an induction wire onto the specimen surface in order to make measurements, which prevents the implementation of simple, high-speed measurements for a quick inspection process. Accordingly, a new design of antenna sensor combined with an induction wire has been developed. A schematic illustration of this new sensor is shown in Fig. 3 (b).

The main design parameters are the length of the induction wire (L), the distance between the antenna and the induction wire (δ), and the loop diameter of the antenna (D).

The main criterion for improving the sensitivity of the new sensor is to increase the frequency of the applied signal, because a higher frequency yields larger measured values. However, it is difficult to supply a high frequency signal to the conductive wire without any devices being connected to it. Accordingly, the resonance phenomenon of the incident and reflected waves was considered to supply a high frequency alternating signal and to from the standing wave within the induction wire. If a standing wave is formed within the induction wire, a high-power alternating signal exists in the induction wire. A 1GHz-alternating signal can form a standing wave in an induction wire 125mm in length.

The measured signals are strongly affected by the distance between the antenna and the induction wire, δ, because an electromagnetic field is formed by both the supplied and the induced signals. Furthermore, the electromagnetic field generated by the supplied signal only has a negative effect on the measured signal, and only the electromagnetic field generated by the induced signal has to be measured because information about the target material can only be included on this signal. Six different values of δ (3.5, 4, 4.5, 5, 5.5, and 6mm) were tested, and the maximum signal changes measured above the defect were compared. We found that a δ value of 5mm was the optimum distance to obtain sufficient signal intensity and to avoid the effects of the supplied signal.

The antenna diameter was set at 7mm, and diameters of 5mm and 10mm were also tested. The results of these showed that the smaller diameter antenna gave very small measurement values, while the larger diameter decreased the sensitivity.

A Cu conductive wire with a rectangular cross section of 2mm x 0.035mm was fixed at the bottom of the sensor as an induction wire, and the tip of the loop antenna was set at a distance of 0.2mm from the surface of the specimen.

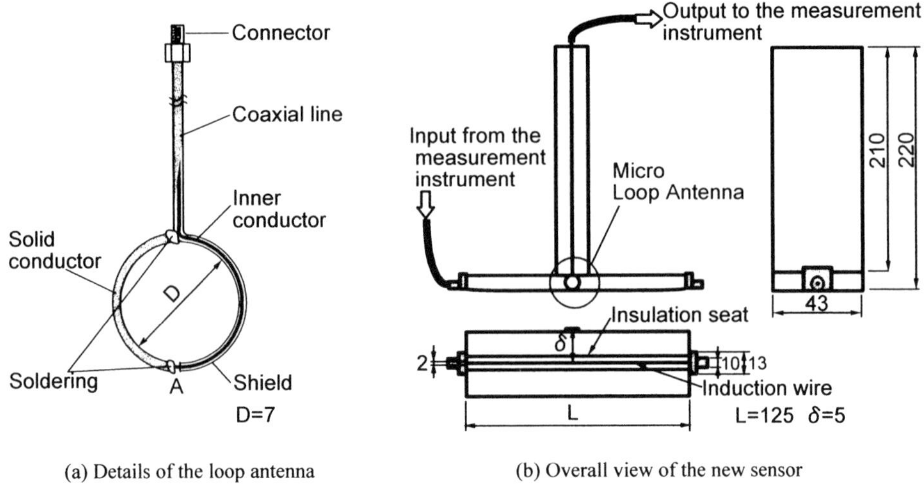

(a) Details of the loop antenna   (b) Overall view of the new sensor

Fig. 3 New sensor combined with induction wire (units: mm)

4. Measurement procedure

Specimens containing two types of defects were prepared in order to evaluate the applicability of the Small Loop Antenna technique. One of them had artificial defects and the other contained a fatigue crack.

The design of the specimens with artificial defects, which were made of a paramagnetic material (Type-316 stainless steel), is shown in Fig. 4 (a). The specimen contained three artificial defects (Electric Discharge Machining slits, known as EDM slits) of different depths (called TP-1). These defects all had the same length (10mm) and width (0.5mm), but had depths of 1mm, 3mm and 5mm respectively.

A fatigue crack was produced by using 4 point bending fatigue crack propagation test on a plate of paramagnetic material (Type-304 stainless steel), which contained an EDM slit as a crack initiator, (starting ΔK was 17.8 MPa$\sqrt{m}$ at the bottom of the initial crack). In order to make a specimen that contained only a fatigue crack the EDM slit was sliced off after finishing the fatigue test. The shape and dimensions of the specimen with the fatigue crack (called TP-2) are shown in Fig. 4 (b). According to surface observations and measurement by an optical microscope, the length of the fatigue crack was about 29mm.

The origin of the X-Y-Z co-ordinate system was set at the center of the specimen. The X-axis was defined as being directed along the length of the defects, with the Y-axis perpendicular to the length of the defects, as shown in Fig. 4 (a). The Z-axis was directed perpendicular to the surface of the specimen. During the measurements, the loop antenna was oriented parallel to the Y-axis.

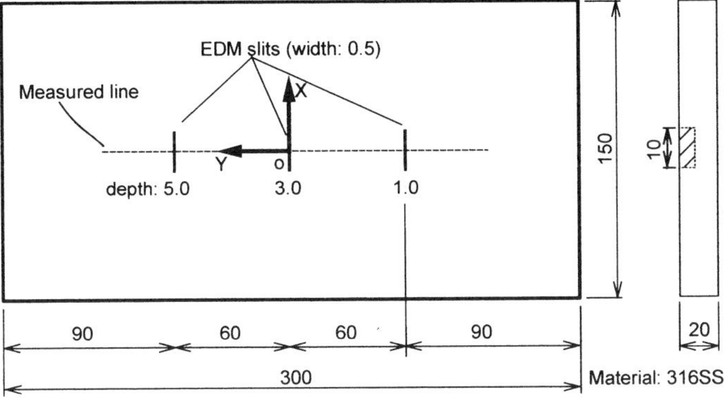

(a) Specimen containing artificial defect (EDM slit)

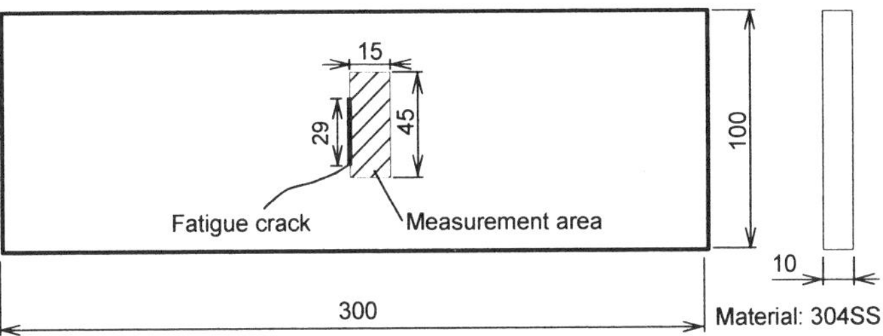

(b) Specimen containing fatigue crack

Fig. 4 Geometry and dimensions of specimens (units: mm)

5. Results and discussion

5.1. Detection of artificial defects by the new sensor

The first set of the measurements with the new probe was performed on specimen TP-1. Fig. 5 (a) shows the distribution of the ratio of the amplitude, R, across the artificial defect. R is plotted against the position of the centre of the induction wire. R represents the ratio of the amplitude of the potential drop $|U(t)|$ of the transmitted wave to that of the incident wave (Fig. 2). The measurement line is parallel to the Y-axis. Measurement was performed when the induction wire was positioned along the line of the centre of the crack (X=0) and the Small Loop Antenna was aligned along the crack edge (X=-5). It can be seen from Fig. 5 that as the sensor approaches the slit, R drops slightly and then rises up again. When the sensor is above a defect, R reaches a maximum value.

A schematic illustration of the electrical signals and magnetic fields in the vicinity of a crack is shown in Fig. 6. A change in the induced signal flow around a defect leads to a change in the magnetic field in the vicinity of the defect. This magnetic field has its largest magnitude within the defect, but part of it penetrates outside the crack and can be detected by the Loop Antenna. Consequently, the maximum value for R can be measured above the defect.

The detection sensitivity, ΔdB, is defined as the maximum change of R above the crack. Fig. 7 shows the relationship between the detection sensitivity and the slit depth obtained from Fig. 5. The defects can be detected quantitatively.

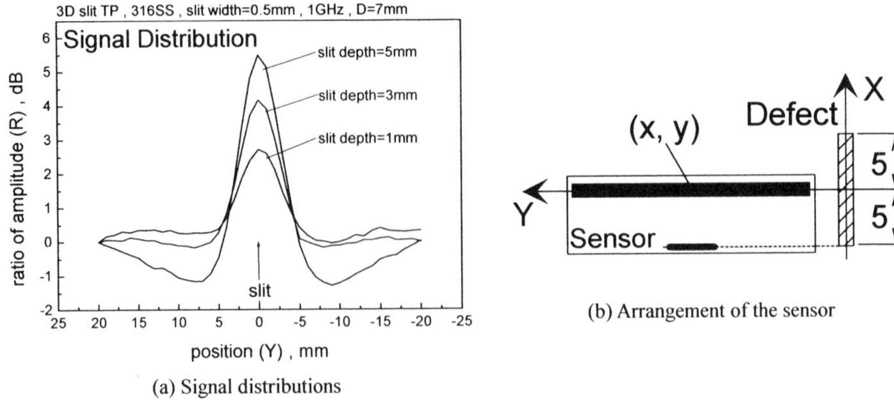

(a) Signal distributions

(b) Arrangement of the sensor

Fig. 5 Ratio of amplitude distribution measured by the new sensor
around the artificial defect with a depth of 3mm

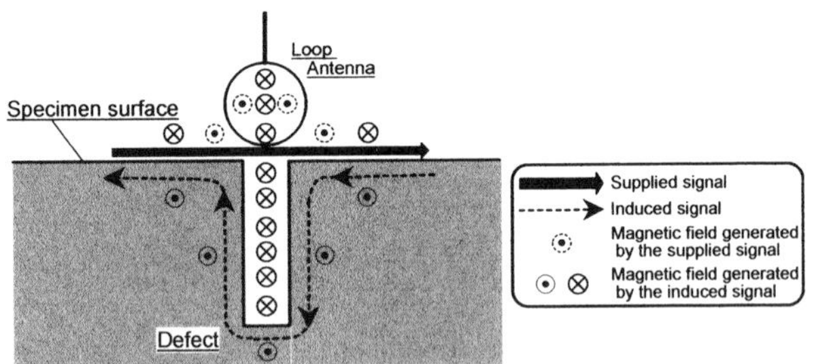

Fig. 6 Magnetic field measured by the Small Loop Antenna

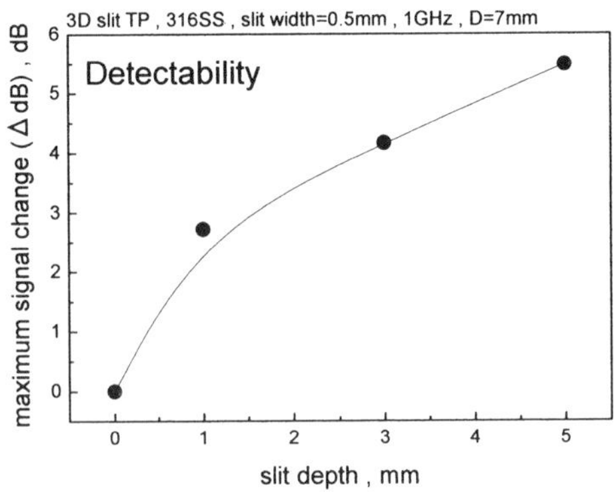

Fig. 7 Relationship between crack detection sensitivity and crack depth

5.2. *Detection of fatigue cracks by the new sensor*

Fig. 8. is a map showing the ratio of amplitudes measured on the specimen with a fatigue crack by the new sensor. The fatigue crack is located along the line Y=0, from -14.5 to 14.5 in the X direction. The fatigue crack is clearly detected by the new sensor, although it is difficult to determine the crack length from the signal distribution. This is due to the asymmetry of the sensor. Accordingly, in order to improve the signal quality, further development of the sensor is suggested with special attention to the sensor symmetry.

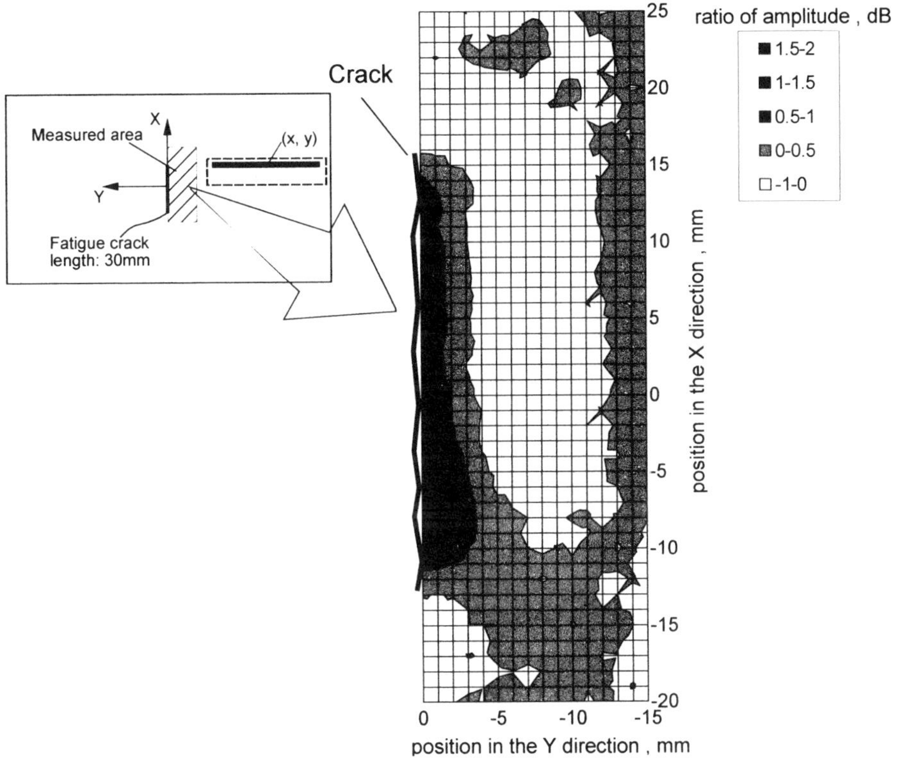

Fig. 8 Ratio of amplitude maps measured by the new sensor on the specimen with fatigue crack

6. Conclusions

The development of our Small Antenna technique has been summarised. A new type of sensor, a Small Loop Antenna probe with an induction wire has been developed for non-contact NDI. The detection performance of this sensor for real defects was demonstrated by the measurement of a specimen with fatigue crack 29mm in length.

In order to improve the signal quality, further development of the sensor is suggested, with special attention being given to the sensor symmetry.

Acknowledgements

Partial support for this work by the Grant-in-Aid for COE (Centre of Excellence) Research (No.11CE2003) from the Japanese Ministry of Education, Culture, Sports, Science and Technology is gratefully acknowledged.

References

[1] M. Saka, Materials Evaluation by Using Microwave -Feasibility Proceedings of the Annual Meeting of the Japanese Society of Mechanical Engineering, Tokyo, JSME, 2000, pp.11-12.

[2] N. Sato, T Shoji, Y. Sato, M. Sato, Development of a novel electromagnetic technique for detection of cracks by use of micro antenna, Transactions of the Japan Society of Mechanical Engineers, 63-632, 1999, pp.925-931

[3] Y. Sato, T. Shoji, Nondestructive Inspection Based on Electromagnetic Technique -Potential Drop and Micro Antenna techniques-, Proceedings of The Sixth International Conference on Material Issues in Design, Manufacturing and Operation of Nuclear Power Plants Equipment, 1, 2000, pp.271-281

[4] K. Yagi, Y. Sato, T. Shoji, Inspection of Metal Surfaces Containing Cracks by Small Antennas, Review of Progress in QNDE20, in press

[5] Vector Network Analyser seminar text, Hewlett Packard Co, 1997

[6] K. Yagi, Y. Sato, T. Shoji, Crack Detection by Means of Electromagnetic Field Measurement with Micro Antennas, Proceedings of APCFS & ATEM01, 2001, pp.280-285

[7] R. C. Johnson, H. Jasik, Antenna engineering handbook, second edition, McGraw-Hill, 1984, pp.2-7

[8] L. V. Blake, ANTENNAS, Artech House, 1984, p.332

[9] Nondestructive Testing of Materials, R. Collins et al., IOS Press Amsterdam, 1995

Author Index